Basic Electronics

Electronic Devices and Circuits

How They Work and How They Are Used

By
Gene McWhorter
Alvis J. Evans

This book was developed and published by:
Master Publishing, Inc.
Niles, Illinois

Edited by:
Gerald Luecke
Charles Battle

Printing by:
Arby Graphic Service, Inc.
Niles, Illinois

Photograph Credit:
All photographs that do not have a source identification are either courtesy of Radio Shack, the authors, or Master Publishing, Inc.

Copyright © 1994, 2000
Master Publishing, Inc.
6019 W. Howard St.
Niles, Illinois 60714
voice: (847) 763-0916
fax: (847) 763-0918
e-mail: MasterPubl@aol.com
All Rights Reserved

Visit Master Publishing
on the World Wide Web at:
www.MasterPublishing.com

In Memory of Alvis J. Evans:
In early 1997, Al Evans passed away. He had survived for 25 years thanks to a kidney transplant—the kidney had been donated by his sister, Fern. Mr. Evans was an excellent educator, teacher and author—and an outstanding individual. Master Publishing hereby dedicates this and all future printings of *Basic Electronics* to his memory.

REGARDING THESE BOOK MATERIALS
Reproduction, publication, or duplication of this book, or any part thereof, in any manner, mechanically, electronically, photographically is prohibited without the express written permission of the publisher. For permission and other rights under this copyright, write Master Publishing, Inc.

The Author, Publisher and Seller assume no liability with respect to the use of the information contained herein.

15 14 13 12 11 10

Table of Contents

		Page
Preface		iv
Chapter 1.	Learning About Electronics	1
Chapter 2.	What is DC Electricity?	17
Chapter 3.	What is AC Electricity?	37
Chapter 4.	Diodes and Transistors—How They Work	59
Chapter 5.	How Amplifiers and Oscillators Work	87
Chapter 6.	Radio Transmitters and Receivers	103
Chapter 7.	What Digital Circuits Do	121
Chapter 8.	Digital Logic Circuits—How They Work	135
Chapter 9.	How Electronic Memories Operate	153
Chapter 10.	Computers and Other Digital Systems	177
Chapter 11.	How Photoelectric Devices Work	193
	Appendix	209
	Answers to Chapter Questions and Problems	209
	Glossary	214
	Index	218

Preface

Basic Electronics has been written to explain basic concepts and fundamentals about electronic devices and electronic circuits. The science and technology of electronics have had the greatest era of change in the last 30 years, and have penetrated nearly every phase of our lives. Development of new electronic devices and applying these devices in circuits have led to this change.

Basic Electronics tells you how electronic devices work and how they are used in circuits to perform useful functions. In a straightforward, easy-to-read language, using detailed illustrations, practical worked-out application examples solidify the understanding of basic concepts and fundamentals as they are discussed. If you have a curiosity about electricity and electronics, or a desire to understand how engineers design electronic circuits, *Basic Electronics'* simple explanations should meet your needs.

Basic Electronics begins by explaining the meaning of electronics and current; DC electricity and how current relates to the voltage applied to a circuit; AC electricity and how it differs from DC, and how AC circuit values must be calculated using vector arithmetic. Following a discussion of how diodes and transistors work, *Basic Electronics* explains the application of such devices in amplifiers and oscillators, radio transmitters and radio receivers.

The discussion then shifts to the application of electronics in digital circuits—logic circuits, both serial and parallel transmission, combinational and sequential, and electronic memories. A chapter covers how digital circuits can be combined into larger circuits and applied in systems.

Basic Electronics concludes with an explanation of devices that are sensitive to or emit light.

Multiple-choice quizzes and a set of questions and problems at the end of each chapter reinforce the learning. Answers to quizzes are on the quiz page; worked-out answers to questions and problems are given in the Appendix.

Studying *Basic Electronics,* understanding the worked-out examples, and working the quizzes and problem sets should lead to proper understanding of available electronic devices, circuit applications, and the functions circuits perform. That was our goal; we hope we have succeeded.

G.McW
A.J.E.
M.P.I.

CHAPTER 1
Learning About Electronics

As we begin to study electronics, let's define what we are talking about. We say that a flashlight or a vacuum cleaner is "electric." But a radio or a computer is more often called "electronic." What's the difference?

Electron Devices: Electron Tubes and Semiconductors

Anything that works with electricity is electric, including both a vacuum cleaner and a radio, but not all electric things are electronic. *"Electronic" refers to electron tubes and semiconductor devices.* As shown in *Figure 1-1,* electron tubes and semiconductor devices are known as *electron devices.* They are named as such because they depend on the flow of electrons for their operation.

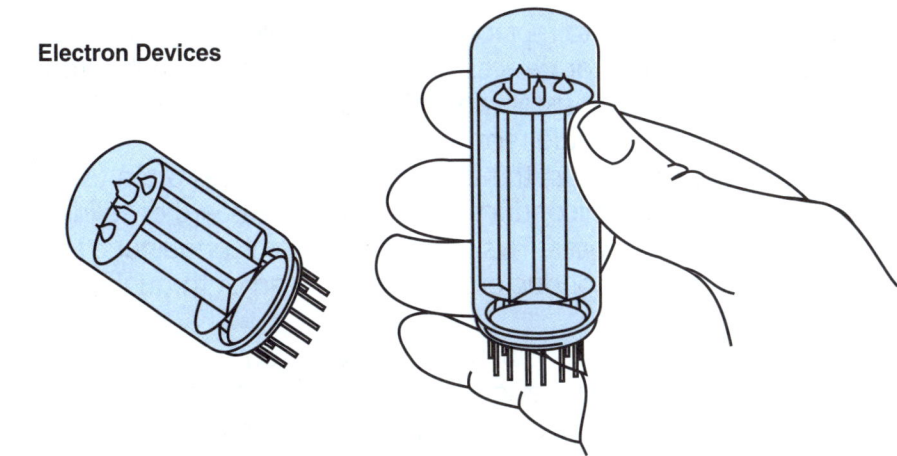

a. Electron Tubes (Vacuum and Gas-Discharge Electron Devices)

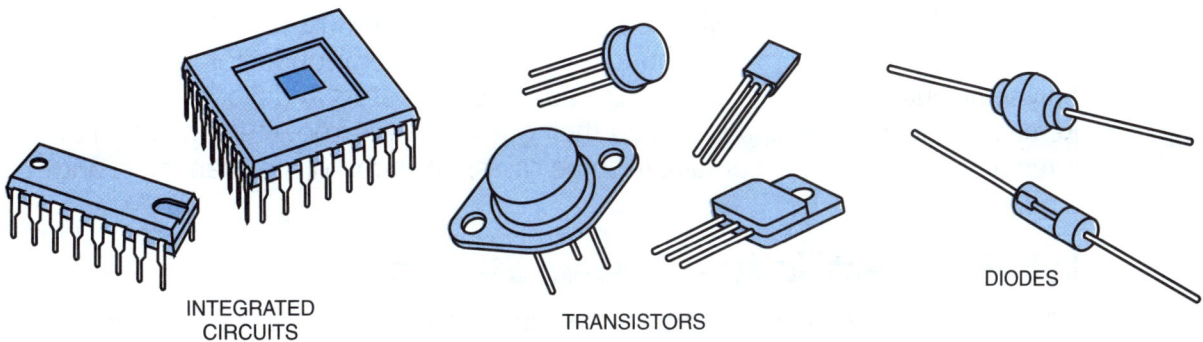

b. Semiconductor Devices (Solid-State Electron Devices)

Figure 1-1. Electronic circuits use electron devices, which are electron tubes and semiconductor devices.

Basic Electronics

Electron Tubes

Electron tubes, commonly called vacuum tubes, are devices in which electrons flow through a vacuum or a low-pressure gas. Before 1965, most kinds of electronic devices, such as radios, televisions, stereos, and public address systems, used electron tubes instead of semiconductor devices. Most vacuum tubes are glass cylinders or bulbs that have physical parts inside connected to conductor pins that come out the bottom of the tube. The tubes plug into sockets to connect to other circuit components and a source of power. A few kinds of electron tubes are still used today. Some examples are television picture tubes, high-power transmitting tubes for radio signals, and fluorescent lamp tubes.

Semiconductor Devices

Semiconductor devices, sometimes just called semiconductors, are solid-state electron devices in which electrons flow in and through a semiconductor material. There is no vacuum or gaseous chamber. The types of semiconductor devices which you are most likely to see, as shown in *Figure 1-1,* are transistors, diodes, and integrated circuits. We assume you have heard of integrated circuits. In an integrated circuit, there are many semiconductor devices built at the same time into the same piece of semiconductor material.

The semiconductor material which is most commonly used is silicon. Silicon is an element whose atoms resemble carbon atoms. It is found abundantly in the earth's surface. Semiconductor materials conduct electric current in special ways, different from the way electrons pass through metals. We will learn how they work in Chapter 4, after we learn more about direct current (dc) and alternating current (ac) electricity.

The transistor was invented about 1948. During the 1950s, circuits using semiconductor devices began to replace circuits using electron tubes. Electronic equipment using semiconductor devices was often said to be "transistorized." For instance, people spoke of "transistor radios." To some people, in fact, a transistor was a small battery-powered radio receiver.

During the 1960s and 1970s, the term "solid state" was often used in advertising and marketing semiconductor products. One heard of solid-state electronics and solid-state devices. This is because in semiconductors, electrons flow through solid crystals of semiconductor materials instead of through a vacuum or a gas.

By about 1975, hardly any equipment using electron tubes was being manufactured. The exceptions, as noted previously, include a few specialized purposes—television picture tubes, high-power and high-frequency radio signal tubes, and gas-discharge lamps. Nowadays, everyone takes for granted that electronic circuits are based on semiconductor devices. It is becoming unnecessary to speak of semiconductor or transistorized or solid-state electronics. Instead, we just say electronics.

This book, then, is mostly about circuits that use semiconductor devices, and the only semiconductor devices we will consider are transistors and diodes. Remembering, of course, if we know their operation, we know the operation of integrated circuits which contain, in some cases, up to millions of transistors and diodes.

What Is Electricity – Atomic Structure

What is electricity? To provide a basis for explaining its nature, let's begin with the fundamental atomic structure of matter. Matter is commonly thought of as anything that has mass and occupies space. It can take several forms or states—the three common forms being solid, liquid and gas. A fourth state of matter is plasma. In its common form, matter has mass or weight. We know weight as the force of the gravitational pull of the earth on an object. Anything that has mass and is near another mass, such as the earth, is pulled toward the other mass by a gravitational force.

Structure of Matter

Matter consists of extremely tiny particles grouped together to form atoms. There are 92 different *naturally occurring atoms* called *elements*. They are placed in a periodic table in sequence by their atomic number and atomic weight. There are also some 14 man-made elements which do not occur in nature, making about 106 elements known to date. *Elements cannot be changed by chemical means.* They can be changed only by an atomic or nuclear reaction. However, they can be combined to make the countless number of compounds which we experience every day.

Structure of an Atom

An atom may be pictorially or symbolically represented by a model that resembles our solar system, which has the sun at its center and the planets rotating around it. This atomic model, shown in *Figure 1-2a,* was proposed by the Danish physicist, Niels Bohr, in 1913. Quantum mechanics has shown that this model is not exactly correct, but it is still useful in visualizing the atom.

The center of an atom is called the nucleus and is composed principally of particles called protons and neutrons. They make up by far most of the mass of an atom. Orbiting around every nucleus are tiny particles called electrons. These electrons are many times smaller *in mass* than either the proton or neutron. Other subatomic particles are studied in atomic physics, but these three will be sufficient for our purpose.

All of the elements shown in a periodic table are constructed of these three subatomic particles with the sole exception of hydrogen. As shown in *Figure 1-2c,* the nucleus of a hydrogen atom is a single proton, outside of which there is a single electron. The proton and the neutron have a mass about 1840 times that of the electron.

Atomic Number and Atomic Weight of Elements

The elements are identified by their atomic number and atomic weight. Normally, an atom has an equal number of protons in the nucleus and electrons around the nucleus. This *number of protons or electrons* is called the *atomic number*. The *atomic weight* of an element is the *total number* of particles—*protons and neutrons*—in the nucleus, rounded to the nearest integer.

One of the first elements to be investigated by scientists was oxygen. Upon examination, the nucleus of oxygen was found to contain eight protons and eight neutrons, so it was assigned an atomic weight of 16 and an atomic number of 8.

Orbits and Shells

A more simplified and easier to understand atomic structure is shown in *Figure 1-2b*. We will be discussing such diagrams in much more detail in Chapter 4 when we learn about semiconductor diodes and transistors, but for now, we will just deal with enough so that we understand the basics of electricity. As we see in *Figure 1-2b*, the electrons which surround the nucleus of any atom are arranged in orbits or rings. These orbits are referred to as shells. The shells, and their contained subshells, represent certain energy levels away from the nucleus that are only permitted at particular levels. The total number of electrons contained in each shell or subshell cannot exceed a certain maximum. The first shell can contain 2 electrons, the second shell can contain 8 (2, 6), the third can contain 18 (2, 6, 10), the fourth can contain 32 (2, 6, 10, 14), etc.

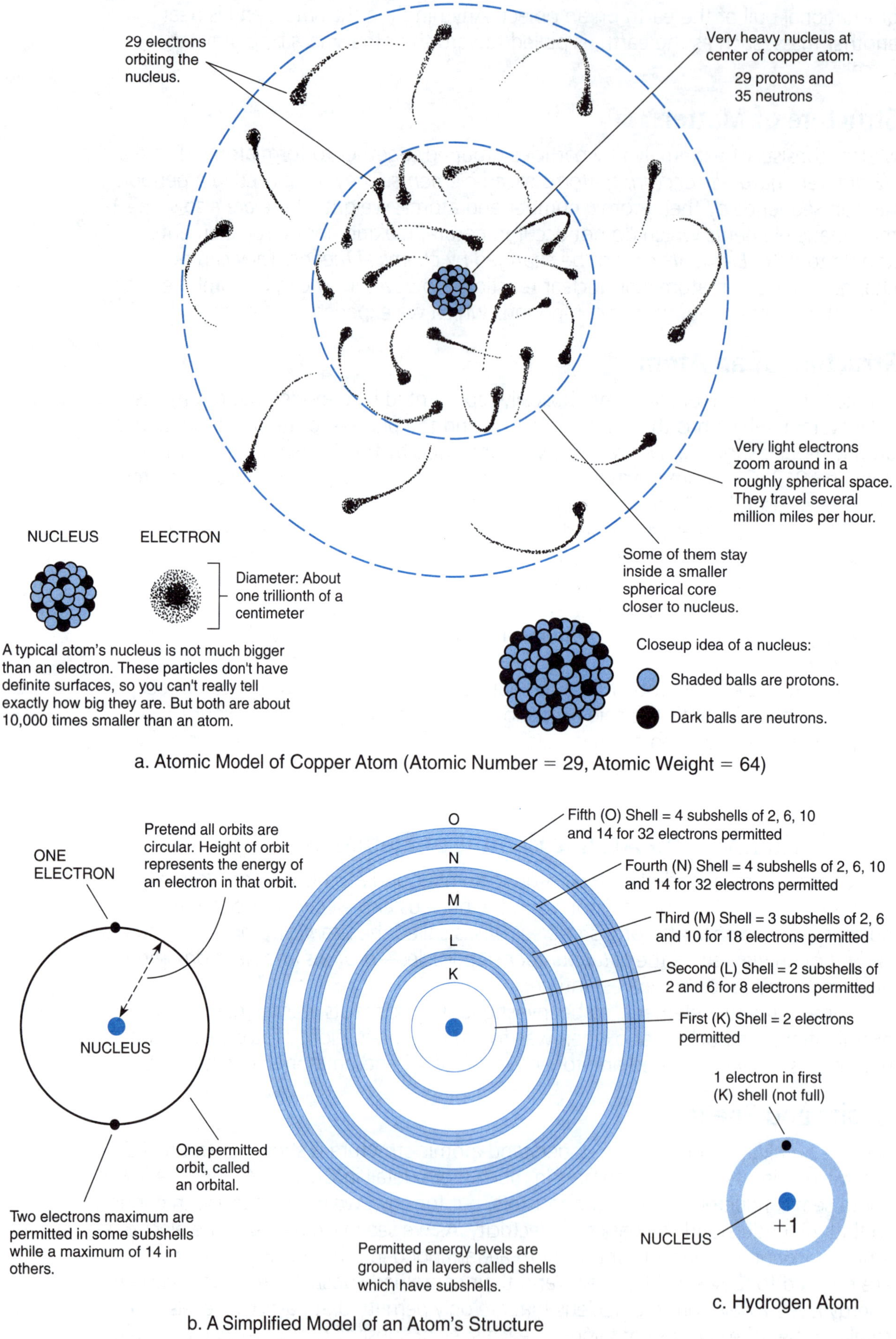

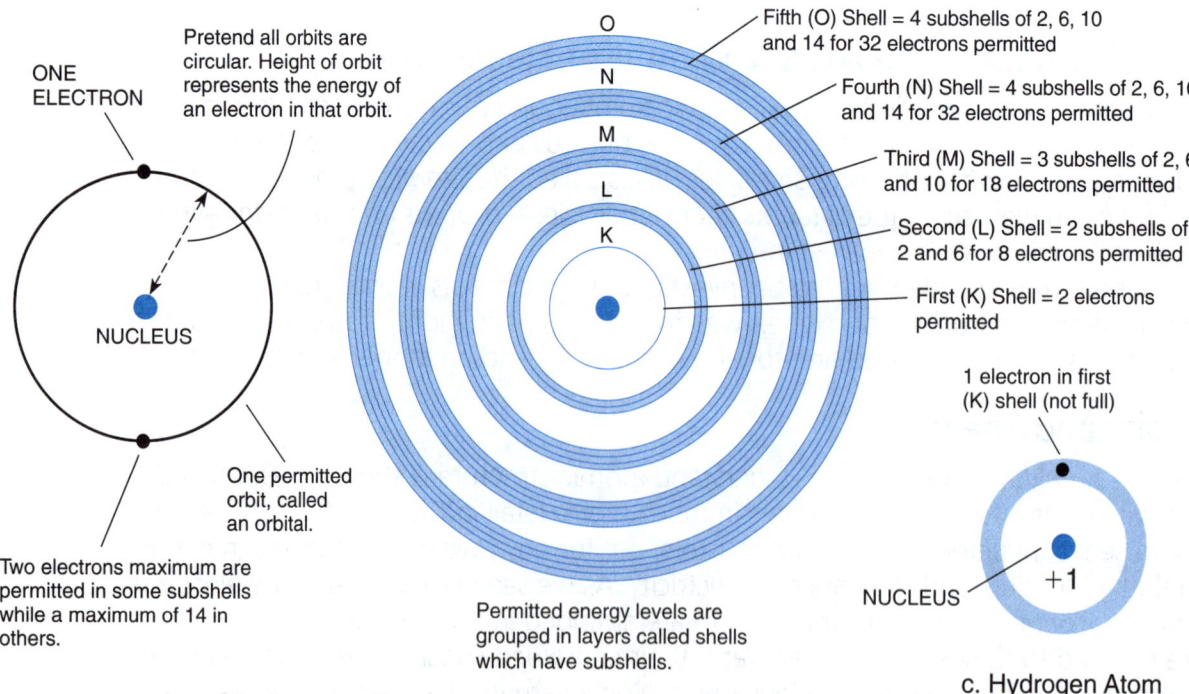

Figure 1-2. A Bohr Atomic Model has electrons orbiting around nucleus, resembling our solar system.

These major shells are usually named K (Helium), L (Neon), M (Argon), and N (Krypton), etc. The name in parentheses represents the element formed when the outer subshell is completely filled to maximum with electrons. The last or outermost shell for any atom is called the *valence shell*. An aluminum atom is shown in *Figure 1-3*. The electrons in the outermost shell are referred to as *valence electrons*.

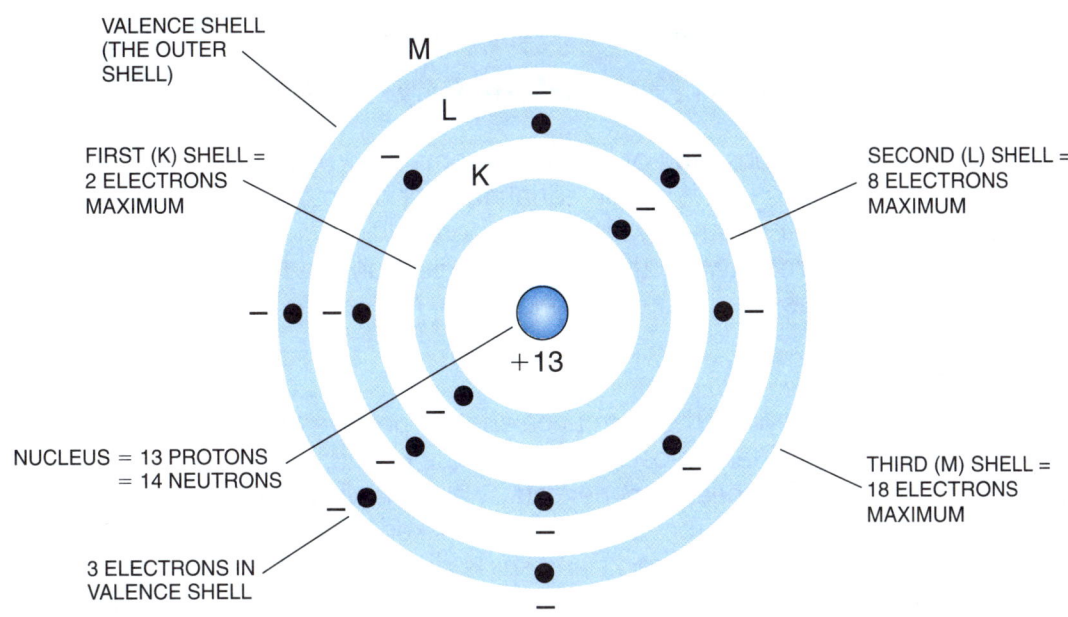

Figure 1-3. A Bohr Atomic Model of an aluminum atom has 13 orbiting electrons with three electrons in its valence shell.

Example 1. Valence Electrons in a Germanium Atom

How many valence electrons does a germanium atom have? (It has an atomic number of 32 and an atomic weight of 72.)

Since it has an atomic number of 32, it has 32 total electrons. Placing these electrons according to the rules above, we find that the first shell has two electrons, the second shell has eight, the third shell has 18, and the fourth shell has the remaining four in two subshells. These outer electrons constitute a germanium atom's valence electrons.

Incidentally, in some elements, electrons may appear in the next major shell subshell before the present major shell is filled to a maximum. It's complicated so we will leave a more complete study of subshells to atomic and solid-state physics.

Electric Charge (q = It)

Protons and electrons exert forces on one another, over and above the forces of gravitational attraction between them. It has been determined that besides mass, electrons and protons carry an electric charge, and these additional forces are attributed to the electric charge that they carry—in other words, *it is an electrical force*. However, there is a difference in the forces. Between masses, the gravitational force is always one of attraction; between electrical forces, both attraction and repulsion exists. Protons and electrons attract one another. Protons exert forces of repulsion on other protons, and electrons exert forces of repulsion on other electrons.

Thus, there appears to be two kinds of electric charge. Protons have arbitrarily been designated as carrying a positive (+) charge and electrons a negative (−) charge. The neutron, as its name implies, is neutral in charge. So, simply stated, *electricity is based on electric charge*. The directional quality of electricity based on the type of charge is called *polarity*. This, then, leads us to the first law of electrostatics, a basic law of nature, and one of the most vital laws in electricity which states:

> ## LAW OF ELECTROSTATICS
> *Unlike charges attract each other; like charges repel each other.*

An atom or a body consisting of billion of atoms is *normally electrically neutral* or uncharged. This means that it has an equal number of positive and negative charges, thus it is neither electrically attracted nor repelled by other objects. If by some means the balance between the number of electrons and protons is upset, that is, if a body has an excess or deficiency of electrons, *it is said to be charged*.

Charged Atoms

There are many ways in which the normal balance of positive and negative charges can be altered. The oldest, and most common, is the phenomenon of charging by friction. Friction charge has been experienced by almost everyone at one time or another by walking across a carpet and getting shocked when touching a doorknob. Or opening the dryer and finding socks and shirts clinging together and hearing crackling sounds when the clothes are separated. If the room is dark, small sparks, like miniature lightning flashes, can be seen as the clothes are pulled apart. In fact, another example of friction charges is lightning. As clouds move past each other, particles of moisture rub electrons off of some and become charged; that is, some have a deficiency of electrons. The result can be the buildup of a tremendous force. When the force is great enough, the electrons can jump from the negatively-charged cloud to a positively-charged cloud or to the earth, which we see as lightning. All of these are examples of a buildup of static electricity—the gathering of static electric charge of either positive or negative polarity on separate objects. The charge remains (is static) until discharged, when the charge moves and forms a current.

> ### Example 2. Static Electricity
> A way to generate static electricity is to comb your hair with a plastic or hard rubber comb. This is particularly effective for a person with long hair. It is even more effective when the air is warm and very dry.
>
> Place some small bits of paper or aluminum foil on a table or similar non-metallic surface. Comb through your hair several times, then bring the comb close to the bits of paper or foil. The pieces will jump to the comb and stick to it.
>
> Why does this happen? As you comb your hair, electrons are pulled from the atoms in the hair and stick to the comb. The comb becomes charged negatively and the hair is left with a positive charge. As you bring the comb near the paper or foil, the force from the electric charge (electrostatic charge) causes the positive charges in the paper or foil to be separated so that the positive charge is attracted by the negative charge of the comb, and the paper or foil is picked up.

Until about fifty years ago static electric charge was no more than an annoyance or a curiosity for laboratory study. More recently, a number of important applications for static electricity have been invented, one of which is the cathode-ray tube (picture tube) that is commonly used in television sets and computer monitors.

Fundamental Charge

All electrons have precisely the same negative charge and all protons have precisely the same equal and opposite positive charge. No charges have ever been observed of smaller magnitude than those of a proton or electron. The quantity of this small charge is called the elementary or fundamental charge. Note that the charges of a proton or an electron are equal in size or magnitude, while opposite in sign or direction. Recall the directional quality is called polarity. Any object, even a normal atom, containing equal numbers of protons and electrons exhibits no net charge.

So, a charged body is one having an excess or deficiency of electrons compared to its number of protons. The magnitude of the body's net charge might be described with a number that states the difference. However, in practice, scientists and engineers need a unit much larger than the charge of an individual electron or proton. In honor of Charles Coulomb, a large unit of measure of electrical charge, the coulomb, was adopted for practical use. It is equivalent to 6.28 billion billion elementary charges (positive or negative) or, written in scientific notation:

One coulomb = 6.28×10^{18} electrons or protons

The lowercase letter q or uppercase letter Q will be used to represent the excess positive or negative charge on a body.

Atomic Electrostatic Forces

The name coulomb was chosen because the first quantitative investigation of the law of force between charged bodies was done about 1785 in France by Charles Augustin Coulomb (1736-1806). Though the concept of charge was not clear at that time, he showed that the force of attraction or repulsion between two charged bodies followed an inverse square law. This was similar to Newton's law of gravitational force between masses. Coulomb's law may be stated as follows:

> **COULOMB'S LAW OF ELECTROSTATIC FORCES**
> *The force (F) of attraction or repulsion exerted between two charged bodies is directly proportional to the product of their charges (Q) and inversely proportional to the square of the distance (s) between them.*
>
> $$F_{Q_1Q_2} = k \frac{Q_1 Q_2}{s^2}$$

The magnitude of the proportionality constant (k) depends on the units of force (F), charge (Q), distance or space (s) and the medium or material between the charges. When the unit of force (F) is in **newtons** and the distance (s) is in **meters**, k is approximately 9×10^9 **newton meter^2/coulomb^2** when the medium is a vacuum.

Note that the algebraic signs of the charges have not been included in the calculations. In this case, the direction of the force is one of attraction along the line joining the charges. If both charges had been positive, the force would have the same magnitude, but the charges would have repulsed each other.

Electrostatic Fields

The behavior of electrically-charged bodies has been observed for centuries. The ancient Greeks observed that an invisible force was produced when certain materials, particularly amber, were rubbed with a piece of fur. This force caused bits of wood and paper to be attracted and held to the non-metallic objects. The Greek term

Example 3. Force Between Two Charges

What is the magnitude of force between two charges Q1 and Q2 when they are 5.0 cm apart in a vacuum?

Given: $Q_1 = -50$ coulomb
$Q_2 = +25$ coulomb

Since s is in a unit of meters, use $k = 9 \times 10^9$ newton meter^2/coulomb^2

$$F_{Q1Q2} = k\left(\frac{Q_2Q_2}{s^2}\right) = (9 \times 10^9) \times \left(\frac{50 \times 25}{(0.05)^2}\right) = (45 \times 10^{14})$$

$$= 45 \times 10^{14} \text{ newtons (attraction in this case)}$$

"elektron" actually means amber and is the root word for several of our related words, including electric, electron, and electronic. The action and reaction of charged bodies is the subject of static electricity; that is, the reaction between stationary charges.

The attraction or repulsion of electrically-charged bodies is due to an invisible force called an electrostatic field, which surrounds the charged body. We can picture this electrostatic field and the force it creates with lines as shown in *Figure 1-4*. These lines are called electrostatic lines of force, and they graphically represent the force that exists around a charged body. This method of representing the force with imaginary lines was first devised by Michael Faraday (1791-1867). The conventional method of drawing the force lines is for the arrowheads to point away from the positive charge and toward the negative charge.

Earlier we talked about how an object or body can be charged with friction. Now let's talk about how bodies can be charged using other forms of energy. If the valence shell of an atom contains one or two electrons, those electrons can be broken free from the atom with additional energy. When a valence shell lacks one or two electrons, electrons can easily drop into the holes in the band when broken free with additional energy. This added energy may be in the form of heat, light, and magnetic energy. When the valence electron acquires enough energy, it can break free of its atom and become a free electron. When this happens, the atom is no longer electrically neutral. It has more protons in its nucleus than it has electrons in orbit. This unbalanced atom is called a *positive ion*. In contrast, if an atom gains an additional electron instead of losing one, it becomes a *negative ion*. An ion is, therefore, a charged atom, and it can have either a negative or a positive net charge. Different elements have varying ability to give up or acquire electrons. Some atoms give up electrons readily while others are reluctant to give up their electrons. This brings us to our next topic.

Conductors, Semiconductors and Insulators

In our study of the atomic structure of matter, we have seen that all matter is basically electrical in nature. Each atom has its own definite characteristics. Some atoms have complete valence shells and other atoms have valence shells that are incomplete. We saw that a free electron can be generated when additional energy is applied to a material. Adding additional energy causes the freed electrons to move from one atom to another, the result is a flow of negative electric charges through the material. This *flow of charges* is an *electric current*. Some substances conduct an electric current readily with little energy applied. They are known as conductors. Other substances require a large amount of energy to conduct an electric current only very slightly. These are known as insulators.

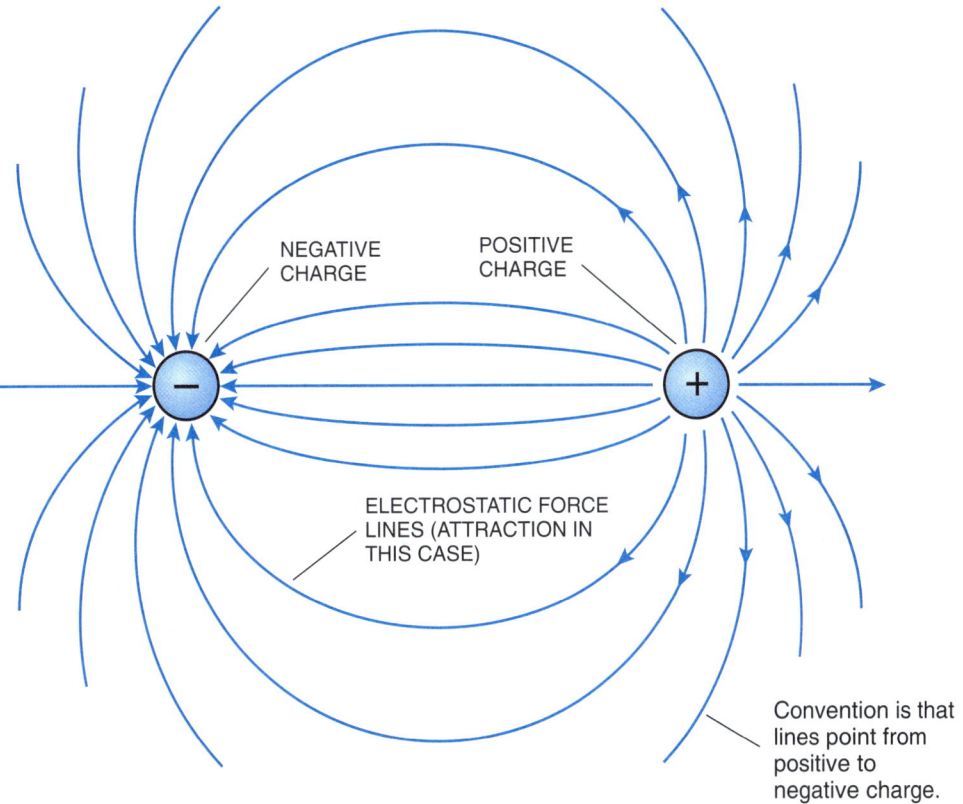

Figure 1-4. Force between charged particles can be represented by imaginary electrostatic force lines from the positive charge to the negative charge.

As a general rule, a conductor is a material whose atomic structure is such that the valence shell contains *less than half* of the total allowable number of electrons while an insulator is a material whose valence shell contains *more than half* of the allowable total number. We may say that a good conductor of electricity is a material that has valence electrons which may be easily freed. A good conductor has high conductivity or conductance; an insulator has low conductivity or conductance.

Elements that have valence electrons which can be easily removed are good electrical conductors. Among materials that are good conductors are copper, silver and gold, but cost and other physical properties considered, copper is by far the most popular conductor in use. In between the conductors and insulators are the semiconductors—carbon, silicon, germanium, etc. These elements are common in manufacturing resistors and solid-state devices, such as diodes, transistors and integrated circuits.

Superconductivity

The most exciting scientific breakthrough since computer chips and lasers is superconductivity, where electrons move through a material with no opposition at all. This phenomenon has been a matter of scientific knowledge since 1911; however, until recently it was possible only at a temperature of -452°F. Now, however, new materials have made superconductivity possible at temperatures higher than that. Physicists hope that they will be able to create materials that are superconductive at room temperature.

Energy and Electromotive Force (Voltage)

Energy is defined as the ability or *capacity to do work*. Work is accomplished, in a physics sense, only when a force acting on a body (mass) moves that body a definite distance. Stated mathematically:

$W = Fs$ where: W = work accomplished
F = force applied
s = distance moved in space

Units of Measure

	Metric System	English System
W (Work)	**joules**	**foot-pounds**
F (Force)	**newtons**	**pounds**
s (Distance)	**meters**	**feet**

Example 4. Determining Work Done

What is the amount of work done (or energy used) to move a mass with a force of 3 newtons a distance of 2 meters?

$W = F \times s = 3 \text{ newtons} \times 2 \text{ meters} = 6 \text{ joules}$

Types of Energy

There are two distinct classifications of energy: potential energy and kinetic energy. *Potential energy* is the energy stored in a body, or the energy a body possesses because of its position relative to another position, called a reference level. As an example, consider a coiled spring with a steel block attached as seen in *Figure 1-5*. The position of the block with the spring neither stretched nor compressed represents the zero reference, S_0. Moving the weight from the reference level S_0 in either direction transforms the work done (moving the block) into potential (stored) energy. The work done with a force, F, moving the block from S_0 to S_1 (ΔS) stores potential energy in the spring attached to the block. This potential energy may then be transformed into the second classification of energy, which is kinetic energy. *Kinetic energy* is energy due to motion. Whenever a physical object is in motion, it has kinetic energy and is capable of doing work by coming to rest. A common example is a hammer that comes to rest on the head of a nail.

Power

Throughout our discussion of work and energy, it may be noted that the time required to perform a task is not a factor in determining the amount of work done. Time, however, is very important when considering power. *The time rate of doing work is power.* Said another way, a certain amount of work, done in a given amount of time, determines the power used. That is, it takes a definite amount of power when a given quantity of work is performed in a certain time interval. Expressed mathematically:

$P = \dfrac{dw}{dt}$ where: dw is the quantity or amount of change in work or energy in **joules**
dt is the time interval over which the work is done in **seconds**
P is the power in **watts** or **joules per second**.
(the d means "a change in")

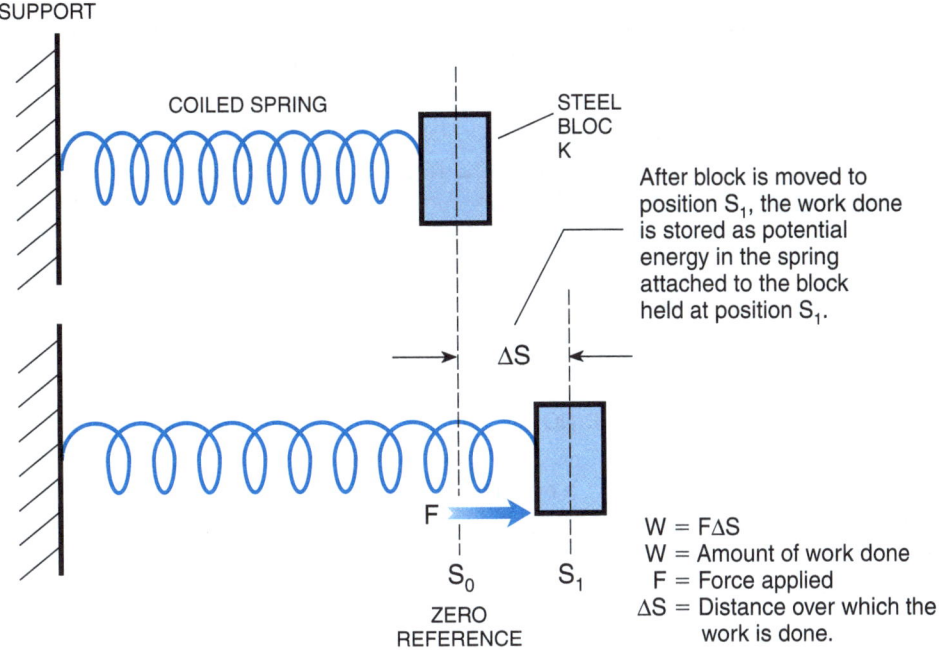

Figure 1-5. When moving a steel block connected to a coil spring, work is done by the force used to move the block from S_0 to S_1. The potential energy is stored in the spring attached to the block.

When you expend dw joules of energy to do work, and the work is done in dt seconds, the watts or joules/second of power used is dw/dt. The watt (joules per second) is a common unit used to measure the amount of electrical power used. For example, your electric power bill is based on the number of watts used.

The watt may be converted into mechanical units by the relationship:

1 horsepower = 746 watts

The transformation of energy from one form (mechanical, chemical, electrical, heat, light, atomic or mass) to another obeys the conservation of energy law which states:

CONSERVATION OF ENERGY LAW
Energy can be changed from one form to another but it cannot be created or destroyed.

Example 5. Determining Power Used

How much power is used to move a mass with a force of 6 newtons a distance of 4 meters in 2 seconds?

dw = 6 newtons × 4 meters = 24 joules
dt = 2 seconds
P = dw / dt = 24 joules / 2 seconds = 12 joules/second
P = 12 watts

Compare the gain in the potential energy of an electron as it is moved from a reference level toward a positive or a negative charge as shown in *Figure 1-6* with the steel block and spring in the example described above in *Figure 1-5*.

Voltage or Potential Difference

The *electric intensity* at a point in an electric field is the ratio of the force on a test charge at a point to the magnitude of the charge, or it is the *force per unit charge*. Similarly, the *potential* at a point in an electric field is defined as the ratio of potential energy of a test charge to the magnitude of the charge, or as the *potential energy per unit charge*. If a coulomb of charge is moved and a joule of energy is used, then the potential energy has changed by one volt. A potential energy, or potential, for short, of *one joule per coulomb is called one volt*. The unit of volt is named in honor of Alessandro Volta (1745-1827), an Italian scientist and the inventor of the "voltaic pile," the first battery cell.

Also, *the potential difference*, V_{ab}, between points a and b in *Figure 1-6* equals the work done per unit charge against electrical forces when the charge is moved from point a to point b. So, the potential difference between points a and b is one volt if one joule of work per coulomb of charge is done against electrical forces when a charge is moved from point a to point b.

The concept of potential difference is an extremely important one, both in electrostatics and in electronic circuits. It has the unit of volts (V), and has a polarity associated with it. It is commonly referred to as voltage, voltage drop, voltage difference, electromotive force (electron moving force), or EMF. For example, the potential difference between the terminals of a common automobile storage battery is about 12 volts. The terminal with the higher potential is designated by a "+" sign; that with the lower potential by a "−" sign. The potential difference between two points is measured by a voltmeter.

Electric Current

In electrostatic problems such as those considered earlier in this chapter, we were chiefly concerned with the forces between charges, and the final, steady-state distribution of these charges brought about by these forces. Now, we are ready to discuss the motion of charge in a conductor when an electric field is maintained within the conductor. This motion of charges constitutes an electric current.

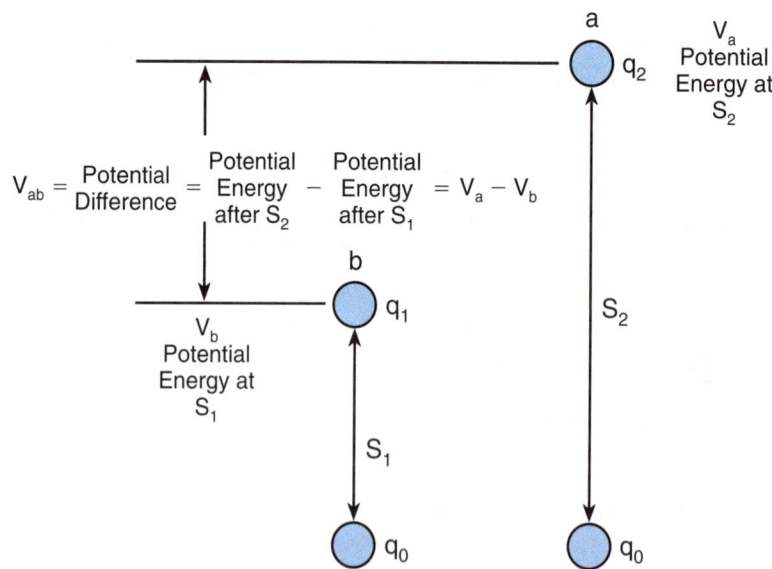

Figure 1-6. Potential energy stored at a point is equal to the work done to move the charge to that point. Potential difference is the difference in potential energy between two points.

Random Drift of Electrons

We have learned that all materials are composed of atoms, each of which is capable of being ionized. If some form of energy, such as heat, is applied to a material, some electrons acquire sufficient energy to move to a higher energy level. As a result, some electrons are freed from their parent atoms and the atoms become ions. Other forms of energy, particularly light, a magnetic or an electric field, can cause ionization to occur. The number of free electrons resulting from ionization is dependent upon the amount of energy applied to the material as well as the atomic structure of the material. Suppose heat is applied to a conductor made up of atoms that have loosely bound electrons in their valence shell. As the energy of the valence electrons increases due to the added energy and they break away from their atoms, they freely drift about the material in a haphazard manner. Their path is not predictable, and so such movement is termed *random drift*.

Directed Drift or Current

An electron, specifically a free electron, will be affected by an electrostatic field in exactly the same manner as any negatively charged body. It is repelled by a negative charge and attracted by a positive charge. If a conductor has a difference in potential impressed across it as shown in *Figure 1-7,* a direction is imparted to the random drift. This *directed drift of free electrons* due to the potential difference constitutes an *electric current*.

Prior to the use of atomic theory to explain the composition of matter, scientists mistakenly defined electric current as the motion of positive charges in a conductor from a point of positive polarity to a point of negative polarity. In many electrical engineering and physics textbooks, current is still considered to be conventionally from positive to negative. Such a current is termed *conventional current* and many concepts and explanations are based on this theory. We since have learned that current is based on the flow of electrons, or a negative charge. Therefore, electron current is in the opposite direction from conventional current. Whichever direction that charge is assumed to move or flow, it constitutes an electric current.

NOTE

The term "current flow" is used erroneously in many texts. Current does not flow; current is already defined as "a flow of charge." Though "current flow" is well established in the electrical and electronics field, it will not be used in this book.

Intensity of Current

The magnitude or intensity of current is measured in *amperes*. The unit symbol for amperes is A. The symbol for current in equations and on diagrams is either the lowercase i or the uppercase I. Amperes is a measure of the rate at which charge is transported across a section of a wire or other conductor. *One ampere is a coulomb of charge moving past a point in one second*. Stated mathematically:

$$I = \frac{\Delta q}{\Delta t}$$

where: I is the average current in **amperes**
Δq is the change in charge measured in **coulombs**
Δt is the time interval in **seconds**

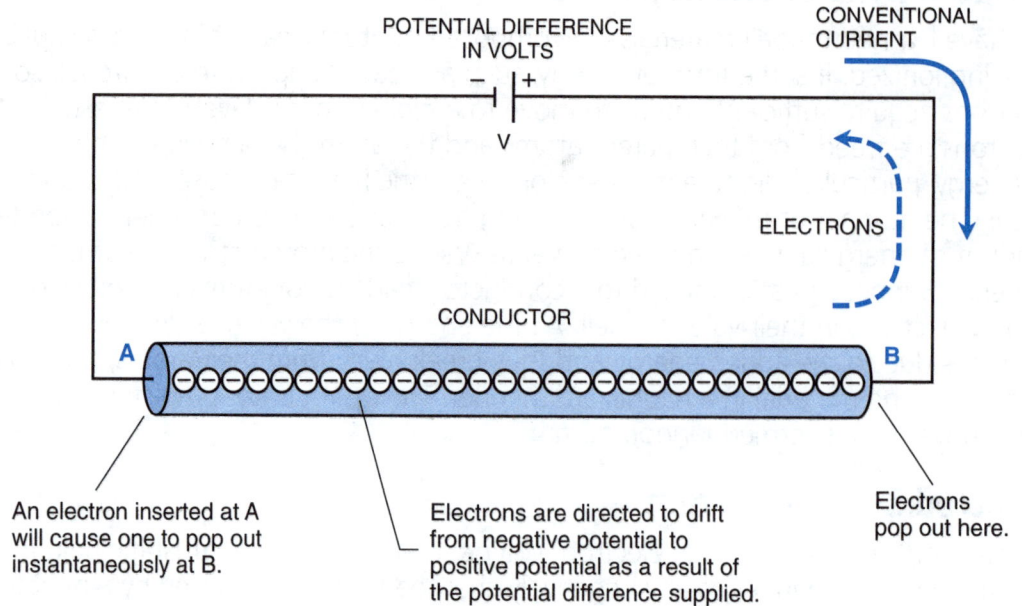

Figure 1-7. Potential difference (in volts) causes a current of electrons (in amperes) to drift through the conductor from negative to positive potential.

Example 6. Determine Amount of Current

What is the current in a wire that is carrying an average of 12 coulombs in 4 seconds?

I = Δq / Δt = 12 coulombs / 4 seconds = 3 coulombs / second or 3 amperes

An electric current seems to move instantaneously through a wire. Let's explain it this way. The concept of electric current is analogous to a pipe filled with red marbles, where the pipe is the conductor (such as wire) and the marbles are free electrons in the wire. When a white marble is inserted in one end of the pipe, a red marble is pushed out the opposite end. Thus, the effect of inserting a marble at one end of the pipe appears almost instantaneously at the opposite end, though the white marble would move relatively slowly through the pipe. The concept is shown in *Figure 1-7* using electrons. Even though the effect of the moving charge seems to be instantaneous, the movement of a single electron is much slower. Just as the accumulation of marbles depends on the rate of movement of the marbles and the amount of time, the amount of electricity (coulombs of charge) depends on the amount of current and the length of time.

Example 7. Determining Amount of Charge

How much charge will pass a point in a circuit that is passing 4 amperes for 5 seconds?

q = I × t = 4 amperes × 5 seconds = 20 coulombs

Summary

Knowing the basic principles, we will now put these concepts of voltage and current to work as we consider dc and ac electricity in the next two chapters.

Quiz for Chapter 1

1. The semiconductor material which is most commonly used is:
   a. silver
   b. copper
   c. silicon
   d. tin

2. The center of an atom is called the:
   a. neutron
   b. nucleus
   c. electron
   d. valence

3. The subatomic particle with the least mass is the:
   a. electron
   b. proton
   c. neutron
   d. nucleus

4. The total number of electrons contained in the third shell cannot exceed:
   a. 2
   b. 8
   c. 18
   d. 28

5. The electrons in the outermost shell are referred to as _____ electrons.
   a. major
   b. secondary
   c. primary
   d. valence

6. The directional quality of electricity is called:
   a. polarity
   b. voltage
   c. power
   d. newton

7. Static electricity is:
   a. conventional current
   b. stationary charge
   c. a flow of electrons
   d. a flow of protons

8. A charged or unbalanced atom is called:
   a. an ion
   b. inert
   c. a neutron
   d. a valence

9. Substances that conduct an electric current only very slightly are known as:
   a. insulators
   b. conductors
   c. metals
   d. solids

10. The ability or capacity to do work is the definition of:
    a. force
    b. power
    c. current
    d. energy

11. The magnitude or intensity of current is measured in:
    a. volts
    b. watts
    c. amperes
    d. ohms

12. A potential of one joule per coulomb is called one:
    a. volt
    b. watt
    c. ampere
    d. ohm

Answers: 1 c, 2 b, 3 a, 4 c, 5 d, 6 a, 7 b, 8 a, 9 a, 10 d, 11 c, 12 a

Questions and Problems for Chapter 1

1. How many valence electrons does a sodium atom have (atomic number 11)? What about a chlorine atom (atomic number 17)?

2. What is the magnitude of force between two charges:

    $Q_1 = -20$ coulomb and $Q_2 = +30$ coulomb when they are 2.0 cm apart in a vacuum? Is the force one of attraction or repulsion?

3. What is the amount of work done (or energy used) to move a mass with a 5 newton force a distance of 4 meters?

4. How much power is used to move a mass with a force of 8 newtons a distance of 3 meters in 4 seconds?

5. How much power is used to move the mass in Problem 3 the 4 meters if it is done in 4 seconds? in 2 seconds?

6. What is the current in a wire that is carrying an average of 18 coulombs in 6 seconds? What would the current be if the same charge moved past the point is 2 seconds?

7. How much charge will pass a point in a circuit that is passing 7 amperes for 3 seconds? What about a current of 3 amperes for 7 seconds?

8. What would be the rate of energy use of a small motor rated at one-fifth horsepower?

9. What is the amount of work done (or energy used) to move a mass with a 6 pound force a distance of 4 feet?

10. State the first law of electrostatics.

11. State Coulomb's Law.

12. Describe the three common subatomic particles.

CHAPTER 2
What is DC Electricity?

Review

Looking back at Chapter 1, the discussion of the forces between charged particles (electrostatics) showed that the force of attraction between unlike charges and the force of repulsion between like charges (in newtons) varies directly according to the amount of each charge (in coulombs) and inversely according to the square of the distance (in meters) between them. Recall also that a charged particle moving in an electric field gains electrical potential energy because work was done in moving it against an electrical force. A measure of potential energy, or "potential" as we will call it, is in units of volts. A potential of one volt is gained when one joule of energy is used in moving one coulomb of charge; therefore, in general, volts are joules of energy per coulomb of charge:

volt = joule/coulomb

To maintain a current (coulombs of charge moving in one second) in a conductor, there must be an electric field. The field is produced by a potential difference across it. Conducting materials differ from one another in the magnitude (or amount) of current resulting from a given electric field produced by a given voltage. Recall that the symbol used to identify voltage is V, E, or EMF.

Resistance – R

Georg Simon Ohm (1789-1854) discovered experimentally that for a fixed voltage, the amount of current through a material depends on the type of material and the physical dimensions of the material. In his honor, the ohm is the unit of electrical resistance (or opposition) offered by a material to a moving charge (an electric current). The upper case Greek letter omega, Ω, is the symbol used to represent the ohm. *A material is said to have a resistance of one ohm if a potential of one volt results in a current of one ampere.*

Georg Simon Ohm determined from his various experiments that the resistance, R, of a conductor is given by the relationship:

$$R = \frac{\rho l}{A}$$

where: R = the resistance in **ohms**
l = the length in **feet**
A = the cross-sectional area in **circular mils**
ρ = the resistivity, which is a parameter that depends on the molecular structure of the sample and its temperature

Example 1. Find the Resistance of a Wire

Find the resistance of 10 feet of copper wire whose diameter is 0.024 inches and whose resistivity is 10.3 ohm-circular mils/foot.

$$R = \frac{\rho l}{A} = \frac{10.3 \text{ ohms-cir mils/ft} \times 10 \text{ ft}}{576 \text{ cir mils}} = 0.18 \text{ ohms}$$

$d = 0.024$ inches $= 24$ mils
$A = d^2 = 24^2$ cir mils

It is important to remember that electrical resistance is present in every electrical circuit, including the components and the interconnecting wires or printed circuit board conductor runs. The amount of resistance offered by the wire or PC board runs is very small compared to the overall circuit resistance; consequently, it is considered to be zero and disregarded in the typical electrical or electronic circuit. An electrical component that intentionally introduces a lumped amount of resistance into a circuit is called a resistor. We will discuss resistors in more detail a little later in this chapter.

Figure 2-1 presents an experiment that displays the principal characteristics of electrical conductors. As shown in *Figure 2-1c,* use a soft-lead pencil to completely fill in the space outlined between point A and B of the first "conductor" shape in *Figure 2-1a.* Make the pencil mark as dark as possible. As shown in *Figure 2-1d,* hold the common lead of an ohmmeter firmly against the "conductor" end that you have formed at point A. Slide the other ohmmeter probe from one end of the figure to the other as you watch the ohmmeter display. Measure the resistance between A and each of the numbered positions and write the resistance under each number. The resistance will probably be between 20,000 and 40,000 ohms per division. Repeat the procedure for the second shape in *Figure 2-1b.* The conductor in *Figure 2-1b* has a larger area than the conductor in *Figure 2-1a;* therefore, its resistance will be smaller than the first conductor.

Electrical Circuits and Laws

An electrical circuit is a path, or a group of interconnecting paths, capable of carrying electrical currents. It is a closed path that contains a voltage source or sources. There are two basic types of electrical circuits — series and parallel. The basic series and parallel circuits may be combined to form more complex networks (another name for the term "circuit"), but these combinational circuits may be simplified and analyzed as the two basic types. We will develop some rules for finding (analyzing) the currents based on the sources available in the basic and complex circuits. Before we do, let's clarify what is meant by a circuit, series circuit, parallel circuit, voltage, voltage drop, current branch, and circuit loop.

Basic Circuit Definitions

Look at *Figure 2-2a*. In this *series* circuit, the interconnection of a flashlight battery, a flashlight lamp and an ammeter to measure current is shown pictorially. It is a circuit because the path from the positive (+) terminal of the battery to the negative (−) terminal is complete. We represent the pictorial circuit with a schematic shown in *Figure 2-2b*. You can see that the series circuit is one complete loop where the current is the same through each component connected in the circuit. In *Figure 2-2c* are shown the symbols used in the schematic and what they represent. Note that the ammeter has resistance, R_M, and the lamp has resistance, R_L. The current, I, in the series circuit is present in one complete current loop. Note also that the voltage of the battery is divided between a voltage drop across the ammeter and a voltage drop across the flashlight lamp. The voltage across each component divides according to the ratio of R_M and R_L to the total resistance.

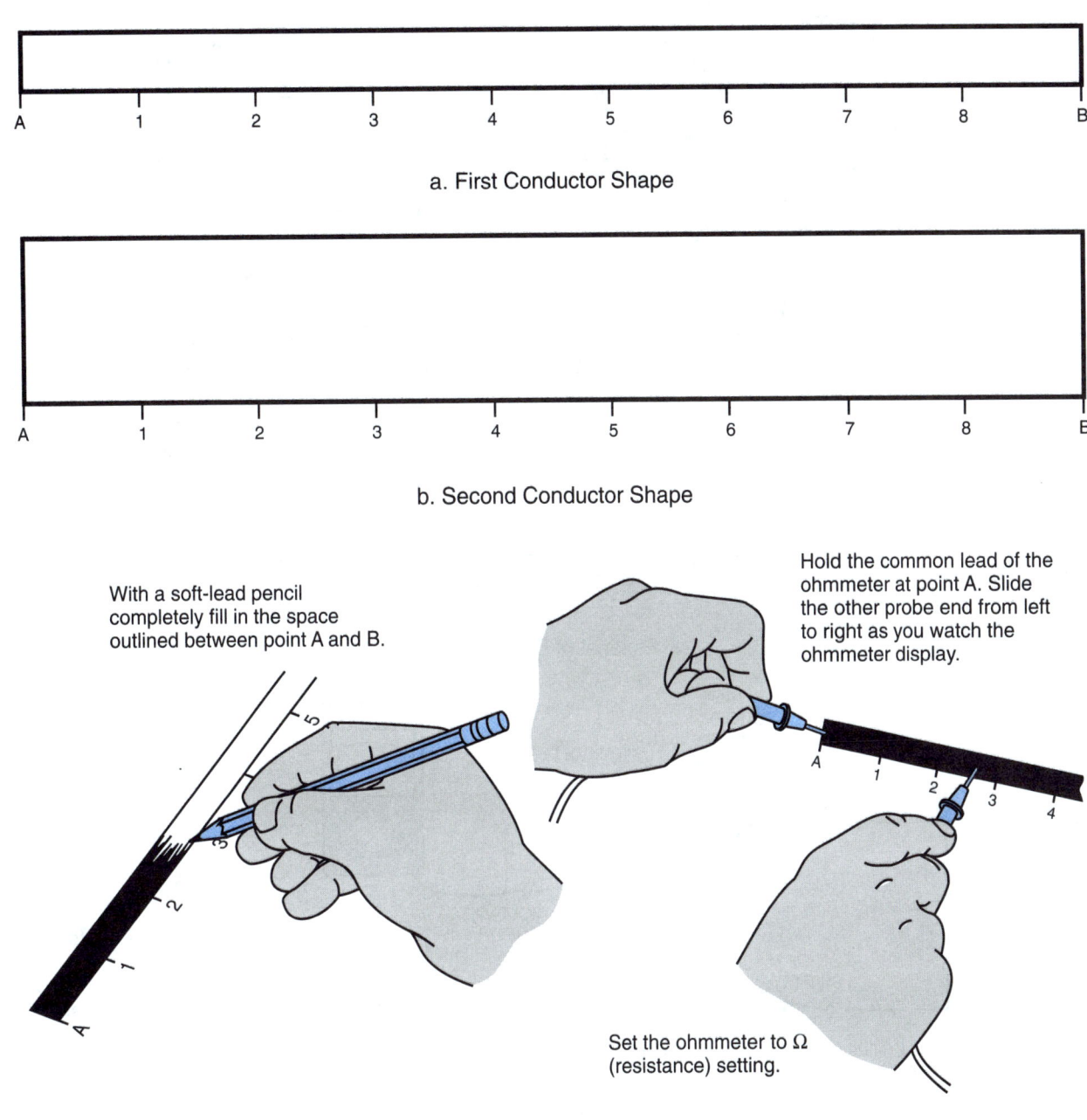

Figure 2-1. Resistance of a pencil mark.

In *Figure 2-2d*, we see the schematic of a *parallel* circuit. Note that the current I_T out of the battery divides into I_2 and I_3 at the current branch shown. Therefore, the current is not the same in every part of the circuit. I_2 is called one current branch, and I_3 is called another current branch. The point where the current branches is usually called a junction. Notice that there are two circuit loops. The source voltage, V, is divided into two voltage drops which are proportional to the current and resistance in each portion of the circuit; one across R_1 and the other across the parallel combination of R_2 and R_3. Next, let's look at two laws that will provide additional tools for analyzing and understanding circuits.

Basic Electronics 2

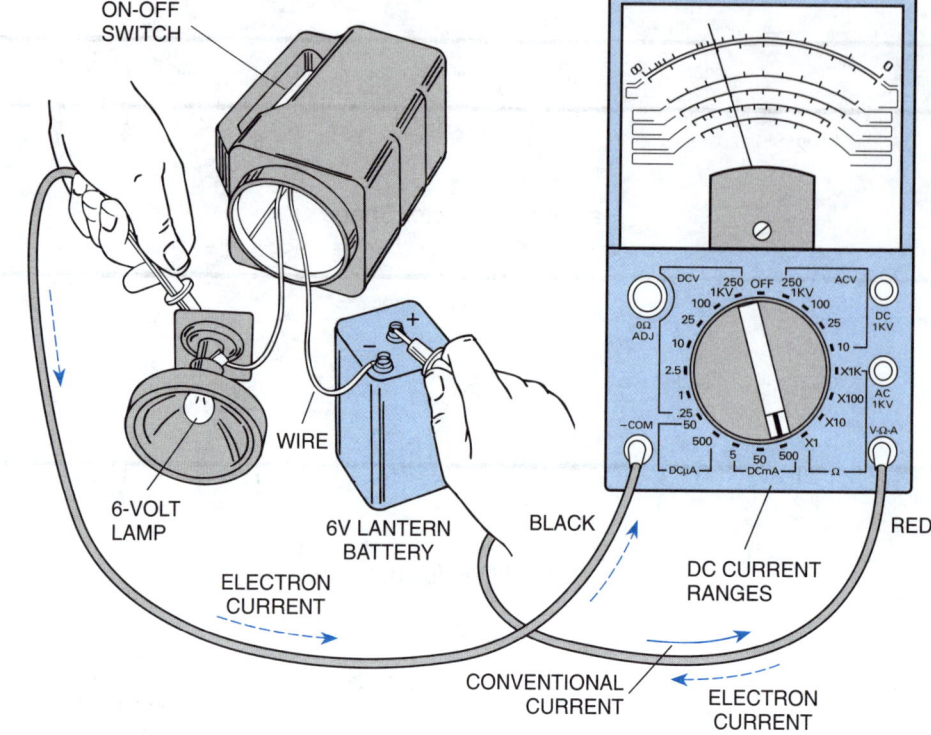

a. Pictorial of Series Circuit

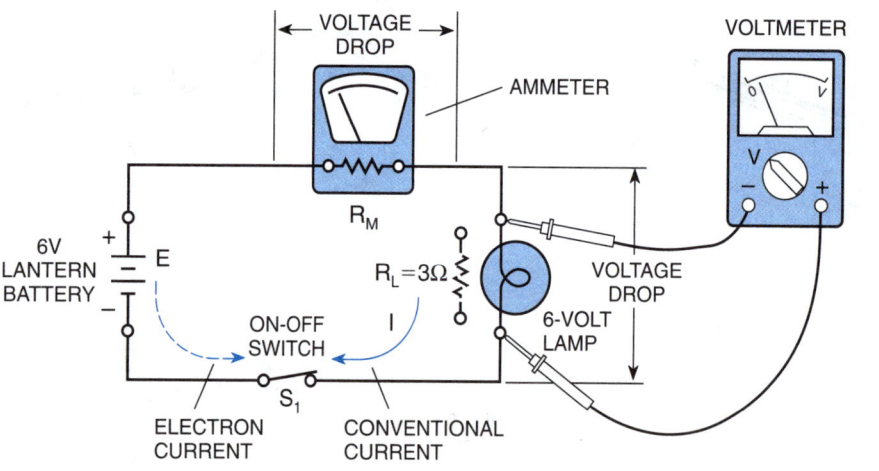

b. Schematic of Series Circuit

c. Symbols Used

I = Current in Amperes
V = Voltage in Volts
E = EMF in Volts

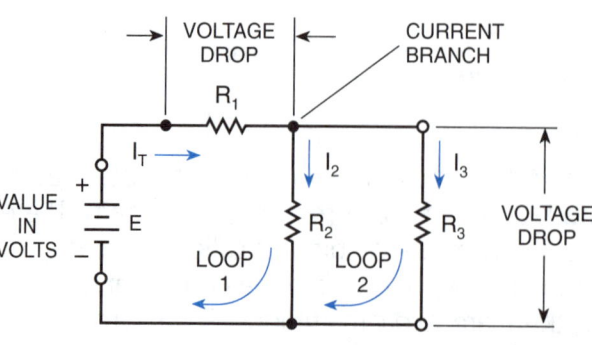

d. Parallel Circuit with Current Branches

Figure 2-2. Basic circuit definitions.

Kirchhoff's Laws

Gustav Robert Kirchhoff (1824-1887), a German physicist, developed two laws for analyzing circuits through experimentation around 1857. The two conclusions which he developed and became his laws are stated as follows:
1. Kirchhoff's Current Law (KCL) – The algebraic sum of the currents at any junction in an electrical circuit is equal to zero. A junction is a point where paths of the circuit come together and are interconnected. More simply stated, all of the current that goes into a junction is equal to all of the current that goes out of that junction. There is no extra and none is lost.
2. Kirchhoff's Voltage Law (KVL) – The algebraic sum of the electromotive forces and the voltage drops around any closed electrical loop is zero. In other words, if we started at a particular point in a closed circuit and went around that circuit adding the individual differences in potential (voltage drops) until all were considered and the starting point was reached, there would be no extra voltage and none would be left unaccounted for.

Through the application of Kirchhoff's laws, a means of determining unknown circuit values is provided. The laws of voltage or current, when applied to complex networks, enable a more comprehensive understanding of the distribution of voltages and currents even without actually measuring them. These two laws, coupled with Ohm's law, which is discussed next, give us the tools to analyze any circuit.

Ohm's Law – Relationship of Voltage, Current and Resistance

One of the most important laws of electricity and electronics was established in the early part of the 19th century by Georg Simon Ohm. It describes the relationship between the three electrical parameters: voltage, current and resistance. Called Ohm's law in his honor, it is stated as follows:

> **OHM'S LAW**
> *The current in an electrical circuit is directly proportional to the voltage and inversely proportional to the resistance.*

Recall that current is a drift or flow of charges, usually carried by electrons, and that voltage is a measure of the potential difference, electromotive force or electrical pressure. The part of Ohm's law that says that the current varies directly with the voltage means that if the resistance value is held fixed, the current change will follow the voltage change. If the EMF is doubled, then the current will double. If the EMF is reduced by one-half, then the current will be reduced by one-half.

Ohm's law also states that current varies inversely with the resistance or opposition. Assuming that the EMF is fixed, this means that as the resistance in a circuit increases, the current decreases proportionally. Likewise, assuming that the EMF is fixed, the lower the resistance, the higher the current.

This relationship can be summarized by a single mathematical equation:

$$\text{Current} = \frac{\text{Electromotive Force}}{\text{Resistance}} \quad \text{or, stated in terms of units: Amperes} = \frac{\text{Volts}}{\text{Ohms}}$$

When electrical parameters are used in mathematical equations or relationships, single letters are used to represent them. Resistance is represented by the letter R. Difference of potential may be represented by either a V for voltage or an E for electromotive force (EMF). Current is represented by the letter I (intensity of charge motion). The equation then becomes:

$$I = \frac{E}{R} \quad \text{or} \quad I = \frac{V}{R}$$

Applying Ohm's Law

Let's again consider the circuit of *Figure 2-2b*. We want to know the current in this series circuit when the lamp has an R_L equal to 3 ohms. The ammeter measures the circuit current and a voltmeter measures the voltage drop across the lamp. Using the basic form of Ohm's law, I = E/R, we see that 6 volts divided by 3 ohms equals 2 amperes. Since the ammeter is a very low resistance compared to the 3-ohm resistor, the voltage drop across it is neglected.

A simple aid for remembering Ohm's law is shown in *Figure 2-3a*. Just cover the letter in the circle that you want to find and read the equation formed by the remaining letters. Since the current in *Figure 2-2b* is unknown, but the voltage and resistance are known, the basic equation to be solved for I is found by using the aid of *Figure 2-3c*. The result is:

$$I = \frac{E}{R}$$

or, in this example, 6 volts divided by 3 ohms equals 2 amperes.

Similarly, knowing the current and resistance, the voltage can be calculated by using the equation shown in *Figure 2-3b*;

E = IR

or, for the same example, 2 amperes times 3 ohms equals 6 volts.

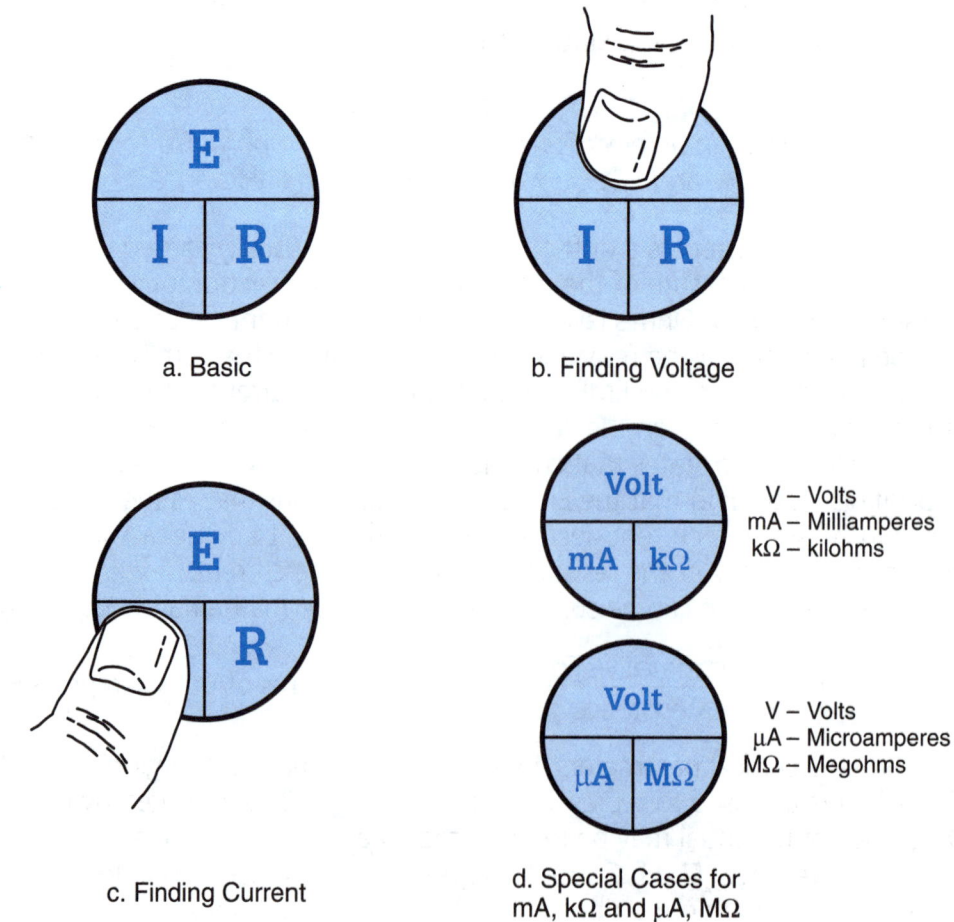

a. Basic

b. Finding Voltage

c. Finding Current

d. Special Cases for mA, kΩ and µA, MΩ

V – Volts
mA – Milliamperes
kΩ – kilohms

V – Volts
µA – Microamperes
MΩ – Megohms

Figure 2-3. Ohm's law circle.

Prefix Shorthand

In electronic circuits, the resistance is typically higher than a few ohms. It is very common that circuit or component resistance is on the order of thousands of ohms (kilohms) or millions of ohms (megaohms shortened to megohms). This high resistance results in lower current, of course. Engineers and technicians use a shorthand of prefixes like kilo and mega. Here are what the prefixes mean:

Symbol

M	mega = millions (multiply by 10^6)
k	kilo = thousands (multiply by 10^3)
m	milli = thousandths (multiply by 10^{-3})
μ	micro = millionths (multiply by 10^{-6})
n	nano = one thousandth of a million (multiply by 10^{-9})
p	pico = one millionth of a million (multiply by 10^{-12})

Here are some examples using the prefixes:

Example 2a. Find Current When V and R Are Known

If a 9-volt battery has a 3000 ohm resistor connected across it, what will be the circuit current?

Since I is unknown, we use:

$$I = \frac{E}{R} = \frac{9V}{3000\Omega} = 0.003 \text{ ampere}$$

The "0.003 ampere" is stated as "three one-thousandths of an ampere."

We can see this better if we arrange our original problem a little differently—like this:

$$I = \frac{9}{3 \times 1000} = \frac{3}{1000} = 3 \times \frac{1}{1000} = \frac{3}{1 \times 10^3} = 3 \times 10^{-3} = 3 \text{ "milli" amperes}$$

Example 2b. Find Current When V and R Are Known

If R is 3,000,000 ohms instead of 3000 ohms, what will be the circuit current?

$$I = \frac{E}{R} = \frac{9V}{3,000,000\Omega} = 0.000003 \text{ ampere} = \text{three one millionths of an ampere}$$

This can also be written as:

$$I = \frac{9}{3 \times 1,000,000} = \frac{3}{1 \times 10^6} = 3 \times 10^{-6} = 3 \text{ "micro" amperes}$$

Engineers and scientists don't like to write a lot of zeros before or after numbers, so they developed the shorthand to substitute for the zeros we discussed previously.

Engineers substitute "milli" for 1×10^{-3}.
Engineers substitute "micro" for 1×10^{-6}.

Therefore, the answer to *Example 2a* can be stated as I = 3 mA which is read as "I equals three milliamperes." And the answer to *Example 2b* can be stated as I = 3 μA which is read as "I equals three microamperes." The symbol m stands for "milli" and the symbol μ stands for "micro."

If the EMF is given in volts (V), rather than kilovolts (kV) or millivolts (mV) for example, a resistance in kilohms (kΩ) will result in a current in milliamperes (mA). A resistance in megohms (MΩ) results in a current in microamperes (µA). *Figure 2-3d* shows the relation between these prefixed units using the Ohm's law circle.

To practice using Ohm's law, let's look at some additional examples using Ohm's law to find an unknown parameter when two are known.

Example 3a. Find Voltage When I and R Are Known

In *Figure 2-4a,* the current is 2.5 mA (2.5 x 10^{-3}A) and the resistance is 3 kΩ (3 × 10^3Ω). To calculate the voltage (in volts), use the form of Ohm's law:

$E = IR$
$E = (2.5 \text{ mA})(3 \text{ k}\Omega) = (2.5 \times 10^{-3})(3 \times 10^3)$
$E = 7.5V$

Example 3b. Find Current When E and R Are Known

In *Figure 2-4b,* the resistance is 2.2 MΩ (2.2 × 10^6Ω) and the EMF across it is 11 volts. To find the current (in microamperes), use the form:

$$I = \frac{E}{R} = \frac{11V}{2.2 \text{ M}\Omega} = \frac{11}{2.2 \times 10^6} = 5 \times \frac{1}{10^6} = 5 \times 10^{-6}A = 5 \text{ µA}$$

Example 3c. Find Resistance When E and I Are Known

In *Figure 2-4c,* the current is shown to be 60 µA (60 × 10^{-6}A) and the voltmeter indicates 18 volts. The resistance (in megohms) can be calculated by using the form:

$$R = \frac{E}{I} = \frac{18V}{60 \text{ µA}} = \frac{18}{60 \times 10^{-6}} = 0.3 \times \frac{1}{10^{-6}} = 0.3 \times 10^6 \Omega = 0.3 \text{ M}\Omega \text{ or } 300 \text{ k}\Omega$$

The fundamental relationship between voltage, current and resistance is described by Ohm's law. It is the most used equation for the design of electrical and electronic circuits. Spend time to understand its meaning and how to use it.

DC and AC Circuits

An electric current may be classified as direct current (dc) or alternating current (ac). Direct current is in one direction only, whereas alternating current changes in direction. The directional quality of electricity is called polarity. In this chapter, we will investigate the parameters of direct current circuits only; that is, circuits where the current is in one direction. Later, we will see that Ohm's law applies to alternating current circuits as well.

Measuring Voltage, Current and Resistance

As circuits are analyzed, used or designed, there is a need to measure voltage, current and resistance. There are two basic types of meters used to make these measurements. One type has a needle that deflects along a scale and indicates the value of a quantity by the position of the needle on the scale. This is an analog meter and is commonly called a multimeter, multitester or VOM (volt-ohm-milliammeter). The other type is a digital meter. Any quantity that is measured appears as a number on a digital display. It is commonly called a DMM (digital multimeter). Both the VOM and the DMM can measure the three fundamental electrical parameters of Ohm's

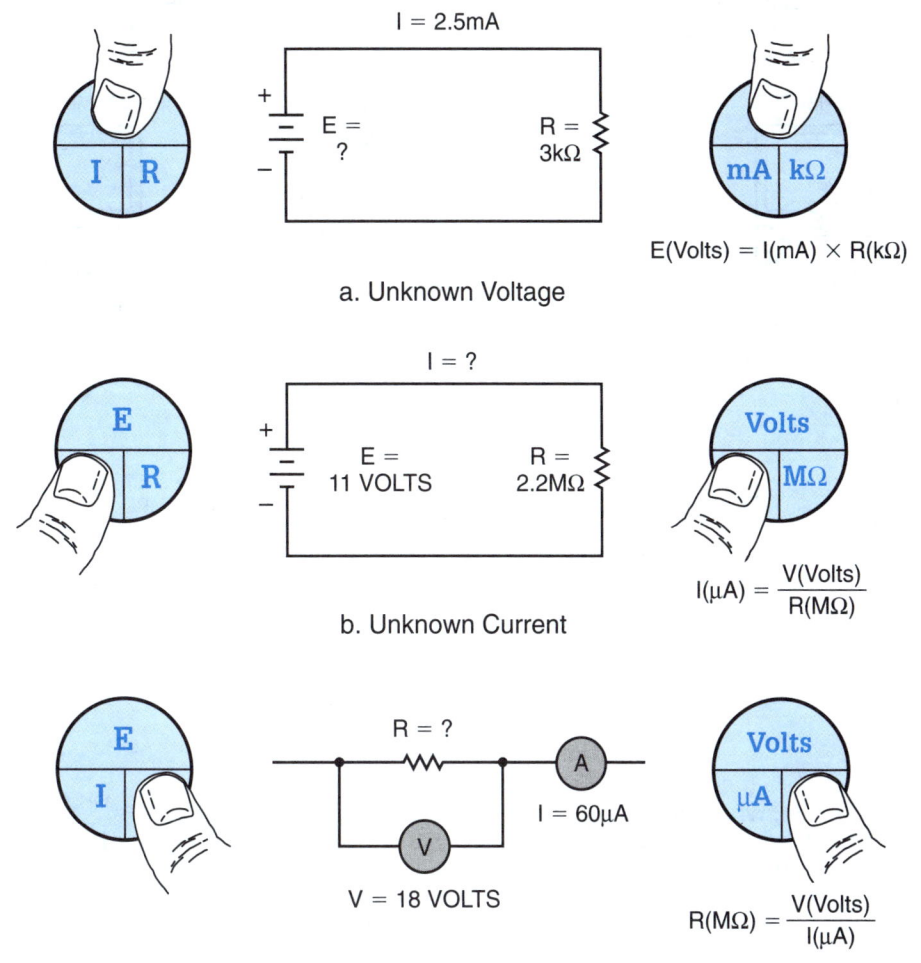

Figure 2-4. Example circuits.

law. They contain, along with other special features, a voltmeter, an ammeter (or a milliammeter or microammeter depending on a selector setting), and an ohmmeter. *Figure 2-5* shows how to connect the three types of meters to make basic measurements. If you desire more complete details on the use of multitesters, visit your Radio Shack store for the book, *Using Your Meter*, (62-2039). Let's now discuss series and parallel circuits in more detail.

Series Circuit and KVL (Kirchhoff's Voltage Law)

As we stated before, a series circuit is one in which the current is the same through all of the components of the circuit; that is, a series circuit has only one path for current. *Figures 2-6a* and *2-6b* show that a series circuit may have one component or many, but there is still only one path for current. This becomes the first of three rules that define a series circuit. The rules are:

The Three Rules of a Series Circuit
1. Current has the same value at any point within a series circuit.
2. The values of the resistances of individual components add up to the total circuit resistance, which is called the equivalent resistance, R_{eq}.
3. Voltage drops across the individual component resistances add up to the total applied voltage.

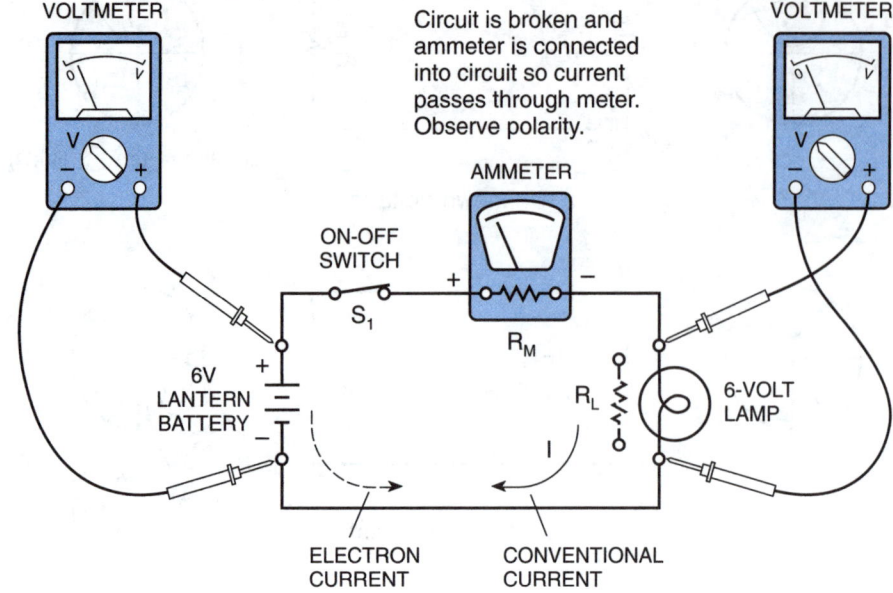

a. Using Voltmeter and Ammeter

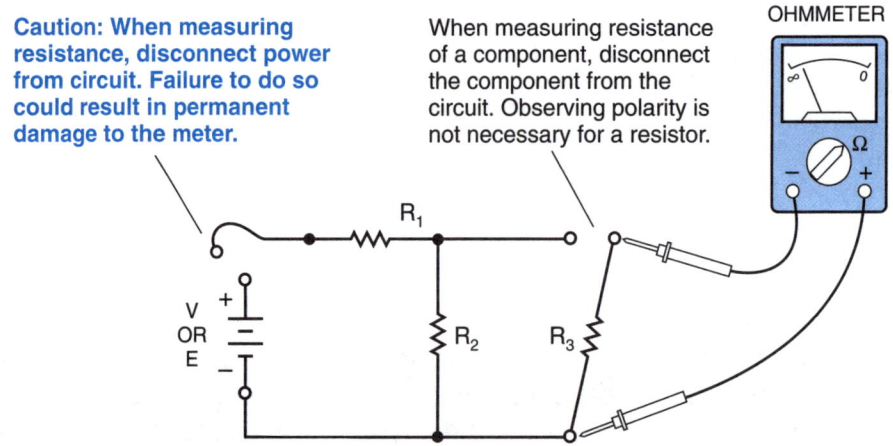

b. Using an Ohmeter to Measure Resistance

Figure 2-5. Measuring with voltmeter, ammeter, and ohmmeter.

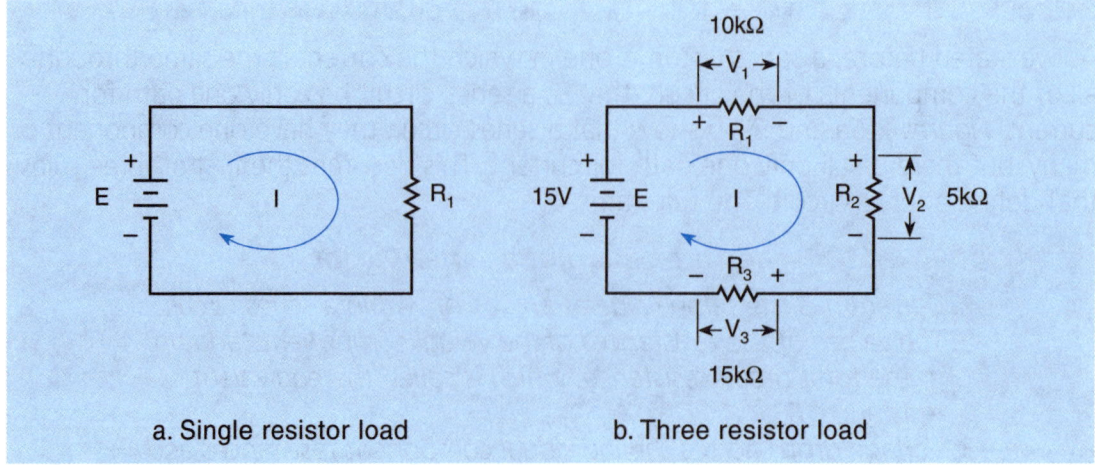

a. Single resistor load

b. Three resistor load

Figure 2-6. Examples of series circuits.

There is a current rule, a resistance rule and a voltage rule, which correspond to the three parameters of Ohm's law. Let's now examine each rule in turn.

The first rule can be examined in the light of Kirchhoff's current law. The amount of current that comes to any point in a circuit is equal to the amount of current that leaves that point because there is no point where current can divide and take more than one path. As a result, there is exactly the same amount of current through every part of the circuit—through every resistor, every component, every device and every wire.

The second rule of a series circuit essentially says that in a series circuit containing several resistors, it appears to the power supply that there is one resistor whose value is the total of all the resistance in the circuit. Once the circuit's total or equivalent resistance is known, Ohm's law can be used to determine the value of current in all parts of the circuit.

The third rule describes the behavior of voltages in a series circuit. Whenever there is current through a resistance, there is a voltage drop across the resistance equal to the current (I) times the resistance (R). This voltage developed across a resistor carrying current is called an IR drop or a voltage drop. In a series circuit, no matter how many resistors the circuit contains, the total of the voltage drops always equals the source, or applied, voltage.

Let's apply the rules to the circuit of *Figure 2-6b*. First, the equivalent resistance is found by adding the values of each of the individual resistors.

$$R_{eq} = R_1 + R_2 + R_3$$
$$R_{eq} = 10 \text{ k}\Omega + 5 \text{ k}\Omega + 15 \text{ k}\Omega = 30 \text{ k}\Omega$$

Once the equivalent resistance, 30 kilohms, is known, the current may be found by Ohm's law:

$$I = \frac{E}{R_{eq}} = \frac{15V}{30 \text{ k}\Omega} = 0.5 \text{ mA}$$

The voltage drop across each of the resistors shown in *Figure 2-6b* may be calculated using Ohm's law by multiplying the current by each resistor value (V_{R1} = 5V; V_{R2} = 2.5V, and V_{R3} = 7.5V). Or each voltage drop could be measured with a voltmeter.

Parallel Circuits and KCL (Kirchhoff's Current Law)

A parallel circuit provides two or more paths or branches for circuit current. There are three rules of a simple parallel circuit similar to those of a series circuit discussed earlier. These rules are:

> ### The Three Rules of a Parallel Circuit
> 1. The same voltage is applied across each individual branch.
> 2. The total current is equal to the sum of the individual branch currents.
> 3. The equivalent (or effective) resistance is equal to the applied voltage divided by the total current, and this value is always less than the smallest resistance contained in any one branch.

An example of a parallel circuit is shown in *Figure 2-7*. Let's apply the rules to this circuit. As you can see from the circuit, the 12 volts of the source E is directly across each branch resistor. Therefore, with the resistor values known, each of the branch currents (in amperes (A)) can be calculated using:

$$I_1 = \frac{E}{R_1} = \frac{12V}{3\Omega} = 4A$$

$$I_2 = \frac{E}{R_2} = \frac{12V}{4\Omega} = 3A$$

$$I_3 = \frac{E}{R_3} = \frac{12V}{6\Omega} = 2A$$

The total current, I_T, is equal to the sum of the branch currents.

$$I_T = I_1 + I_2 + I_3$$
$$I_T = 4A + 3A + 2A = 9A$$

The equivalent resistance of the circuit can be obtained by using Ohm's law:

$$R_{eq} = \frac{E}{I_T} = \frac{12V}{9A} = 1.33\Omega$$

It is smaller than any one of the resistors: R_1, R_2, or R_3.

If you wanted to find the equivalent resistance of a parallel circuit without calculating the individual currents, it may be found for any number of resistors by using the reciprocal method. This method involves taking the reciprocal of each resistor (by dividing the resistor value into 1), adding the reciprocal values together, and taking the reciprocal of the sum to get the equivalent resistance.

$$\frac{1}{R_{eq}} = \frac{1}{R_1} + \frac{1}{R_2} + \frac{1}{R_3} + \ldots + \frac{1}{R_n}$$

For example, let's use the values of the resistors in *Figure 2-7* where $R_1 = 3\Omega$, $R_2 = 4\Omega$, and $R_3 = 6\Omega$:

$$\frac{1}{R_{eq}} = \frac{1}{3\Omega} + \frac{1}{4\Omega} + \frac{1}{6\Omega} = 0.333\Omega + 0.25\Omega + 0.167\Omega = 0.75\Omega$$

$$R_{eq} = \frac{1}{0.75\Omega} = 1.33\Omega$$

Notice that this is the same value for R_{eq} as we calculated above.

For two resistors in parallel, the product over the sum of the resistors gives the equivalent resistance of the pair. For *Figure 2-7*, to arrive at the equivalent resistance, we could combine R1 and R2 as an equivalent of two parallel resistors, and then combine that equivalent with R3 as two parallel resistors to arrive at the final equivalent resistance. The calculations are shown in *Figure 2-7*.

For same-valued resistors in parallel, simply divide the value of one of the resistors by the number of like resistors that are connected in parallel. For example, four 60-ohm resistors in parallel have an equivalent resistance of 60/4 = 15 ohms. In other words, the current drawn by the four 60-ohm resistors in parallel would be the same as one 15-ohm resistor. They are equivalent. Other examples are shown in *Figure 2-7*.

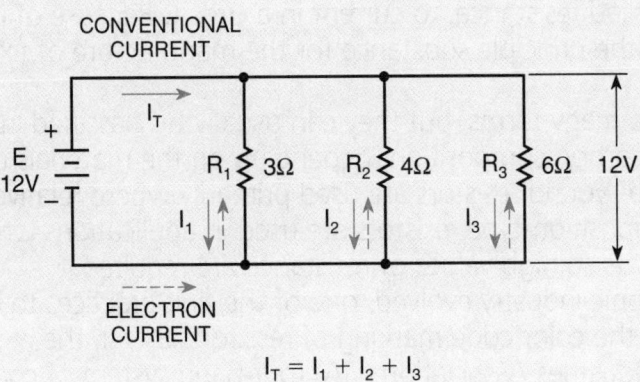

Using two parallel resistor method for equivalent circuit resistance:

$$R_{eq(1-2)} = \frac{R_1 \times R_2}{R_1 + R_2} = \frac{3\Omega \times 4\Omega}{3\Omega + 4\Omega} = \frac{12\Omega}{7\Omega} = 1.71\Omega$$

$$R_{eq} = \frac{R_{eq(1-2)} \times R_3}{R_{eq(1-2)} + R_3} = \frac{1.71\Omega \times 6\Omega}{1.71\Omega + 6\Omega} = \frac{10.29\Omega}{7.71\Omega} = 1.33\Omega$$

If R_1, R_2 and R_3 had all been equal in value, the resultant equivalent resistance would have been:

$$R_{eq} = \frac{R_1}{3} \quad \text{Proof:} \quad \frac{1}{R_{eq}} = \frac{1}{R_1} + \frac{1}{R_1} + \frac{1}{R_1} = \frac{3}{R_1} \quad \therefore R_{eq} = \frac{R_1}{3}$$

If R_1 and R_2 were equal to 4 ohms each and R_3 were equal to 8 ohms, then the equivalent resistance, R_{eq}, could be calculated as:

$$R_{eq(1-2)} = \frac{4\Omega}{2} = 2\Omega$$

$$R_{eq} = \frac{R_{eq(1-2)} \times R_3}{R_{eq(1-2)} + R_3} = \frac{2\Omega \times 8\Omega}{2\Omega + 8\Omega} = \frac{16\Omega}{10\Omega} = 1.6\Omega$$

Figure 2-7. A parallel circuit with three branches.

Combination of Series and Parallel Circuits

The term series-parallel is applied to a circuit that is composed of both series components and parallel branches. All of the facts and rules presented previously concerning the individual series and parallel networks are applicable in solving for the unknown circuit parameters of a combination circuit.

The solving of an apparent complex resistive circuit is not as difficult as it may appear if a logical approach is used. Examine the circuit carefully, then determine the path or paths that current may take through the network to return to the source. Redraw the circuit. These first steps will convert a possible maze into a group of orderly series and parallel connected components. The effective, or equivalent, resistance may then be determined, or additional values of voltage, current or resistance may be found to give a clearer picture of the complete circuit.

The ability to analyze and troubleshoot circuits quickly and accurately requires experience and constant use of the fundamental laws of circuits.

Now that we know a little about circuits, let's talk about the most common components found in typical electronic circuits.

Resistors

Resistors are electrical circuit components that are designed to introduce opposition, which we have called resistance, to current in a circuit. Because of its high resistivity, carbon is used as the principle substance for the manufacture of fixed and variable resistors.

Resistors take many forms, but they can usually be classified into two groups, wire-wound and carbon composition, depending on the materials used for their construction. Wire-wound resistors are used primarily where relatively high currents are handled. Composition-type resistors are used in applications where there are rather small currents so high values of resistance are required.

As the electronic industry evolved, one of the first practices to become standardized was the color code marking of resistors. By far, the most popular system is the Electronic Industries Association (EIA) four-band system of coding. Refer to the Appendix where we have included a color code chart. In this system, the colors are in the form of rings, or bands, at one end of the resistor. The resistance value in ohms can be determined by reading the color bands beginning at the end to which the bands are closer. The correlation between color and number has gained universal acceptance as the standard code for all components using color code for value description.

What Is a Capacitor?

In its simplest form, a capacitor consists of two conductors separated by an insulator (dielectric) as shown in *Figure 2-8a*. (The term "condenser" for a capacitor is obsolete.) The symbol for a capacitor is shown in *Figure 2-8b* and the construction of a common paper capacitor is shown in *Figure 2-8c*. The curved plate of the symbol indicates the outside foil of the paper capacitor and should be connected to the circuit point nearer to ground or the common reference point.

Charging a Capacitor

Assume that the switch S in *Figure 2-9* has been in position 1 for a very long time. The capacitor is not charged and the two plates are neutral; that is, they have equal positive and negative charges. If a potential difference from the battery E is applied

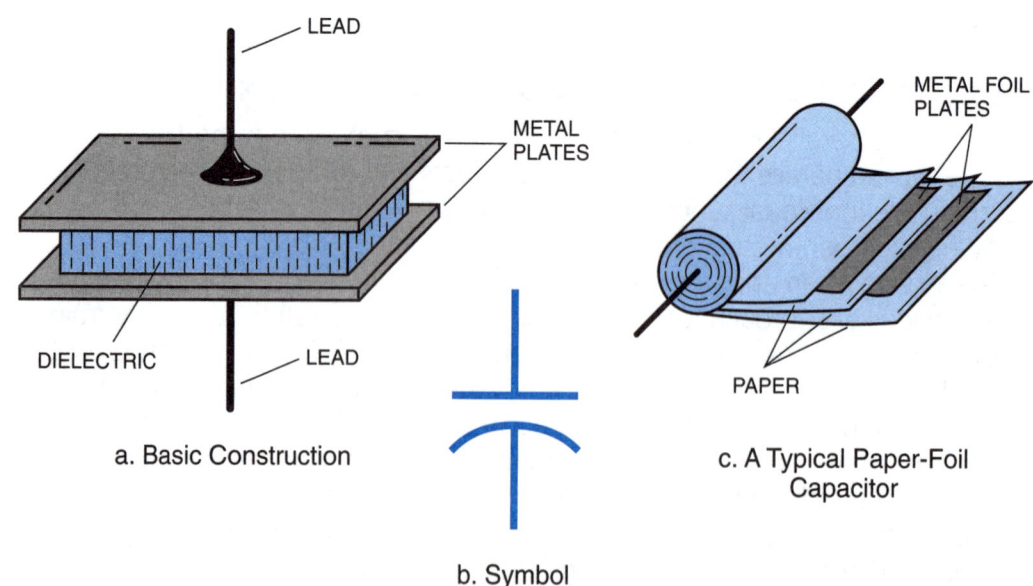

Figure 2-8. Typical construction and schematic symbol for capacitors.

to the plates of the capacitor by throwing the switch, S, to position 2, the capacitor, C, becomes charged over a period of time. The length of time depends on the value of capacitor, C, and resistor, R. The charging action may be described as the positive terminal of the battery extracting electrons from the upper plate of the capacitor, leaving it positively charged. At the same time, the negative terminal of the battery forces electrons onto the bottom plate. As electrons accumulate on the bottom plate of the capacitor and others depart from the top plate, a difference of potential develops across the capacitor. Each electron forced on the bottom plate makes that plate more negative, while each electron removed from the top plate causes it to become more positive. The buildup of the capacitor's voltage is exponential as shown in *Figure 2-11*. When the capacitor is fully charged, the voltage across the capacitor equals the voltage of E and there is no more current in the circuit. Therefore, in this dc circuit, there is current only until the capacitor charges and then no more until the voltage E is changed or the switch position is changed.

Discharging a Capacitor

After the switch in *Figure 2-9* has been in position 2 long enough to fully charge the capacitor, the switch is thrown to position 1. Immediately current carries electrons from one plate of the capacitor to the other to equalize the charge stored in the capacitor. The process continues until the voltage across the capacitor is zero and the current is zero. The circuit remains in this state until the switch is changed to position 2.

Capacitance is measured in **farads**. The abbreviation for farad is F. A one-farad capacitor stores one coulomb (6.28 billion billion electrons) of charge when a potential of 1 volt is applied across the terminals of the capacitor. The farad is a very large unit of measurement of capacitance. More common is the millifarad (mF), microfarad (µF), nanofarad (nF) and picofarad (pF). A given capacitor has the same value of capacitance, no matter in what circuit it is connected.

The value of capacitance depends on three factors:
1. The area A of the plates that face each other.
2. The distance s between the plates.
3. The dielectric constant of the material between the plates.

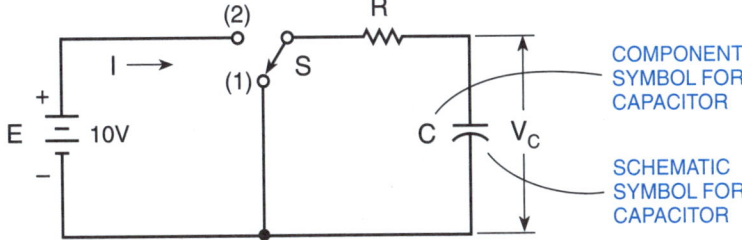

Figure 2-9. Charging and discharging a capacitor in a circuit.

What Is Inductance and an Inductor?

The property of an electric circuit that tends to oppose any change in an electric current is inductance. A component specifically designed to have a known value of inductance is called an inductor.

Basically all inductors are made by winding a length of conductor around a core which is made either of magnetic material or of insulated material. When a magnetic core is not used, the inductor is said to have an air core. The physical characteristics, or geometry, of both the core and the windings around the core affect the amount of inductance produced. *Table 2-1* illustrates these factors. More turns, better magnetic core material, larger core cross-sectional area, and shorter coil length—all increase the inductance.

Table 2-1. Inductance Variables

Physical Characteristics	Lower Inductance	Higher Inductance
Number of Turns	Smaller Number	Larger Number
Core Material	Air	Soft Iron
Core Cross-Section Area	Small Area	Larger Area
Core Length	Long Core	Short Core
Turns Spacing	Wide Spacing	Narrow Spacing

Current Increasing Through an Inductor

Refer to *Figure 2-10* which is a circuit similar to *Figure 2-9* for the capacitor. Let's assume the switch has been in position 1 for a long time. When the switch is thrown to position 2, the current I wants to increase; however, the current I through the inductance L sets up a counter EMF in the inductance which opposes the buildup of current just as the built-up voltage on the capacitor in *Figure 2-9* slowed down the transfer of charge onto the capacitor. The rise of current follows an exponential curve as shown in *Figure 2-11* that depends on the value of inductance L in henries and the value of resistance R in ohms. The current finally rises to a value equal to I = E/R, assuming that R is much larger than the resistance of the coil which forms L.

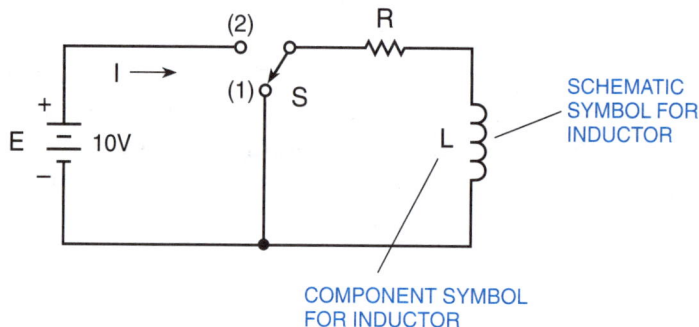

Figure 2-10. Increasing and decreasing current in an inductance in a circuit

Current Decreasing Through an Inductor

Now suppose that, after the current in the circuit of *Figure 2-10* is at its maximum value, the switch is thrown to position 1. The current through the inductance L begins to decrease, but the inductance doesn't want the current to change, so it generates a counter EMF to try to keep the current going in the same direction. The fall or decrease of the current is also shown in *Figure 2-11*. The shape of the exponential curve is similar to the increasing current curve, and again, depends on the values of L in henries and R in ohms.

Time Constant Principles

Figure 2-11 shows the curves for both the charging and discharging voltage of the capacitor in *Figure 2-9*, and the increasing and decreasing current in the inductor in *Figure 2-10*. The curves show how the capacitor voltage changes and the inductance current changes with time. Both curves depend on the time constant of the respective circuit.

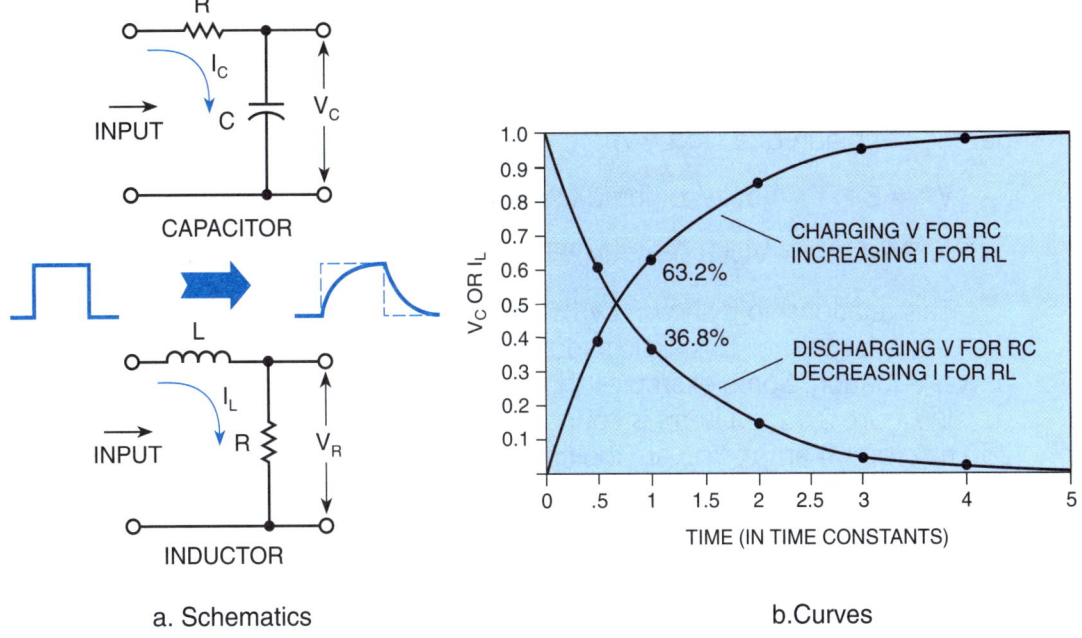

Figure 2-11. RC (RxC) and RL (L/R) time constants determine how voltage V_C and current I_L change in capacitor and inductor circuits.

The time constant for the capacitor circuit (Figure 2-9) is:

$\tau = RC$ where: τ = RC time constant in **seconds**
R = Resistance in **ohms**
C = Capacitance in **farads**

The time constant for the inductor circuit (Figure 2-10) is:

$\tau = \dfrac{L}{R}$ where: $\tau = \dfrac{L}{R}$ time constant in **seconds**
R = Resistance in **ohms**
L = Inductance in **henries**

Note that the voltage across the capacitor and the current through the inductor increases to 63.2% of the final value in one time constant, and that after five time constants, the voltage and current are at their final values. A mirror image of the curves occurs on discharge of the capacitor, or decrease of current through the inductor, and the same time constants apply.

With the exception of the resistor, the capacitor is probably used more than any other electronic component. It is used in filtering (frequency selective) circuits, coupling circuits and timing circuits, just to name a few broad categories. The next is the inductor, followed closely by a combination of the inductor and the capacitor. In all of these applications, it is the RC/RL timing concept that makes the capacitor and inductor so useful in many cases. We will see this concept applied again and again.

Power – How Fast We Use Energy (P = IE)

The concepts of power and energy are often confusing. Basically, power is how fast work is done; that is, how fast energy is used. The basic unit of electrical power is the watt, which is a joule of energy used per second. Recall from Chapter 1 that a volt is a joule of energy per coulomb of charge. Also, an ampere is a coulomb of charge per second. Let's consider the work done by an electromotive force moving charge through resistance. The faster that charge is forced through a resistance, the more power that is dissipated by the resistance; that is, the more heat energy per second that is dissipated into the air around the resistor. Power dissipation in resistors and other devices is an important consideration in electronic equipment. A resistance in a circuit consumes power according to the voltage applied to it and the current through it. In fact, if we multiply the definitions of a volt and an ampere together, we get the definition of a watt (W):

$$W = E \times I = (joule/coulomb)(coulomb/sec) = joule/sec$$

$$\therefore WATT(S) = VOLT(S) \times AMPERE(S)$$

This relationship is shown by the power circle in *Figure 2-12a,* which is similar to the Ohm's law circle. This circle should help you remember the various forms of the power equation. Some examples should make the power concept clearer.

In *Figure 2-12a,* a lamp is connected across 12 volts and "draws" a current of 250 mA (or 1/4 amp). You are looking for power with voltage and current given. The power dissipated by the lamp is calculated as:

$$P = E \times I = 12V \times 0.250A = 3W$$

The 40-ohm resistor in *Figure 2-12b* is connected to a 20-volt voltage source (battery or power supply). To use the power equation given in the circle, we must first know the current. Using Ohm's law:

$$I = \frac{20V}{40\Omega} = 0.5A \qquad \text{then} \qquad P = E \times I = 20V \times 0.5A = 10W$$

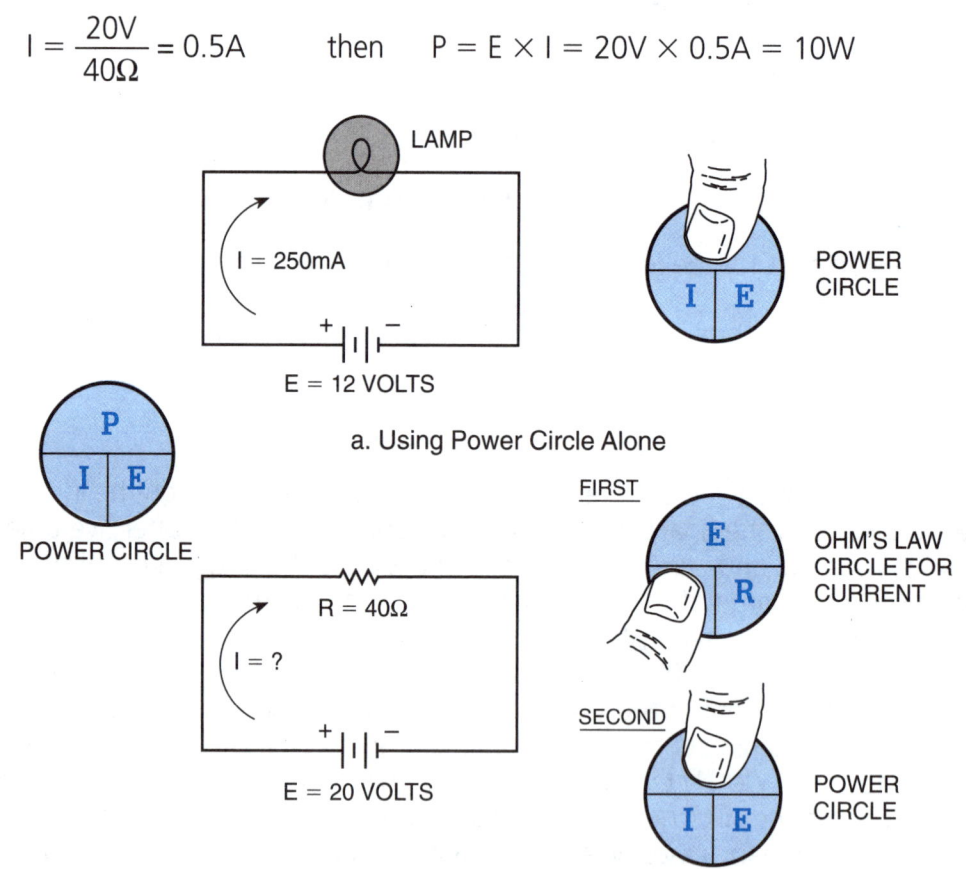

Figure 2-12. Power can be calculated by using the power circle alone or in combination with the Ohm's law circle.

Summary

We have considered direct current (dc) electricity and some of its properties. In the next chapter, we'll look at alternating current (ac) and see how it differs from dc.

Quiz for Chapter 2

1. A wire is replaced by one that is twice as long and one-half the diameter. Its resistance will be:
   a. one-fourth
   b. four times
   c. eight times
   d. the same

2. All of the current that goes out of a junction or point in a circuit must equal to the:
   a. voltages around the circuit
   b. resistance of the circuit
   c. sum or total current from the power source
   d. current going into that junction

3. For the current in a circuit to increase, the voltage must _____ and/or the resistance must _____ .
   a. increase, decrease
   b. decrease, increase
   c. increase, increase
   d. decrease, decrease

4. The directional quality of electricity is called _____.
   a. current
   b. amplitude
   c. phase
   d. polarity

5. A circuit where the current is in one direction is called a _____ circuit.
   a. DC
   b. AC
   c. AM
   d. FM

6. How fast energy is used is a measure of _____ .
   a. impedance
   b. power
   c. work
   d. potential

7. The basic unit of electrical power is the:
   a. volt
   b. ampere
   c. ohm
   d. watt

8. A dc motor rated at 220 volts and 660 watts should draw a current of:
   a. 1 amp
   b. 2 amps
   c. 3 amps
   d. 4 amps

9. A series circuit is one in which the current is the:
   a. product of the power and the voltage
   b. sum of the resistors divided into the applied voltage
   c. sum of the resistors divided by the applied voltage
   d. same through each part

10. A series circuit has four 100-ohm resistors. Its equivalent resistance is:
    a. 400 ohms
    b. 200 ohms
    c. 50 ohms
    d. 25 ohms

11. A parallel circuit has four 100-ohm resistors. Its equivalent resistance is:
    a. 400 ohms
    b. 200 ohms
    c. 50 ohms
    d. 25 ohms

12. An RC circuit with values of R = 200 kΩ and C = 8 μF has a time constant of:
    a. 0.4 second
    b. 1.6 seconds
    c. 4.0 seconds
    d. 16 seconds

Answers:
1 c, 2 d, 3 a, 4 d, 5 a, 6 b, 7 d, 8 c, 9 d, 10 a, 11 d, 12 b

Questions and Problems for Chapter 2

1. Find the resistance of 50 feet of copper wire whose diameter is 0.016 inches and whose resistivity is 18.4 ohm circular mils/foot.

2. If five wires come together at a terminal block and four of them are found to have the following currents, find the current and direction in the fifth wire:

    #1 wire 2.5 amps into the terminal, #2 wire 6 amps into the terminal, #3 wire 1.4 amps away from the terminal, and #4 wire 3.2 amps away from the terminal. #5 wire=?

3. Calculate the current through an 8-ohm resistor connected across a 24-volt battery.

4. The current in a circuit is seen to be 5 mA and the resistance 4 kΩ ohms. Calculate the voltage across the circuit.

5. If the current is known to be 60 µA through a transistor and a voltmeter reads 8 volts across it, what will the resistance of the transistor be calculated to be?

6. A lamp is connected across 120 volts and "draws" a current of 500 mA (or 1/2 amp). Calculate the power dissipated by the lamp.

7. Find the equivalent resistance of four resistors connected in series. The values of the resistors are 1.2 kΩ, 2.4 kΩ. 3.7 kΩ and 4.1 kΩ.

8. Find the equivalent resistance of four 2 kΩ resistors connected in series.

9. Find the equivalent resistance of the same four 2 kΩ resistors of problem 8 if they are connected in parallel.

10. Find the equivalent resistance of four resistors connected in parallel. The values of the resistors are 2 kΩ, 4 kΩ, 5 kΩ and 8 kΩ.

11. A resistor that has the color bands Red, Violet, Orange, Gold has a resistance and tolerance of _____ .

12. How much time does it take to charge a 0.5 µF capacitor to the value of the 20-volt supply if it is charging through a 1 MΩ resistor?

CHAPTER 3
What Is AC Electricity?

Introduction

Alternating current (ac), as the term implies, is electric current first in one direction for a period of time and then in the reverse direction for a period of time. In fact, even though you may not know it, the power distribution lines that supply power to your home are ac electricity lines, and the voltage across these lines is increasing from zero to a maximum positive value and then reversing and increasing to a maximum negative value at 60 times a second.

What Is DC and What Is AC, Really?

Direct current (dc) is a flow of charge that can change in magnitude but not in polarity (the directional quality). Alternating current, on the other hand, is a flow of charge that can change in polarity as well as in magnitude. The term "magnitude" refers to the value of the current.

Figure 3-1 shows circuits with both direct current and alternating current examples. In *Figure 3-1a,* the current is dc. As the voltage is varied, the current will change in magnitude but not in direction. The circuit of *Figure 3-1b* is an alternating current example. It has the capability of producing current through the resistor in either direction. As the switch (S) is changed from voltage E_a (positive) to voltage E_b (negative), the direction of the current reverses. This arrangement produces a steady magnitude of current, first in one direction for a period of time (as determined by how long S is in position A); and then in the reverse direction for a period of time (while S is in position B). The magnitude of current in each direction is 3 amperes—a steady value at all times. The direction depends on the switch position.

The circuit of *Figure 3-1c* is also capable of varying both the magnitude and the direction of the current, this time through a lamp. The position of the sliding contact on the variable resistor determines the magnitude and direction of the current through the lamp at any particular time. Suppose we attach a small motor running at a constant speed to the sliding contact. As the contact rotates with a steady velocity, it produces a particular kind of alternating current through the lamp. Let's put the contact at the zero voltage point A, and then start the motor. The current that is produced through the lamp increases in magnitude at a constant rate in one direction to its positive maximum, then decreases down to zero as the contact on the variable resistor goes through the center zero voltage point B. As the sliding contact continues, the current increases to its negative maximum. It then reduces to zero and starts again in a positive direction. Since the current changes polarity, it is ac. If we measure the voltage across the lamp and plot the value of the voltage against time as the contact rotates, the plot of the voltage looks like *Figure 3-1d.*

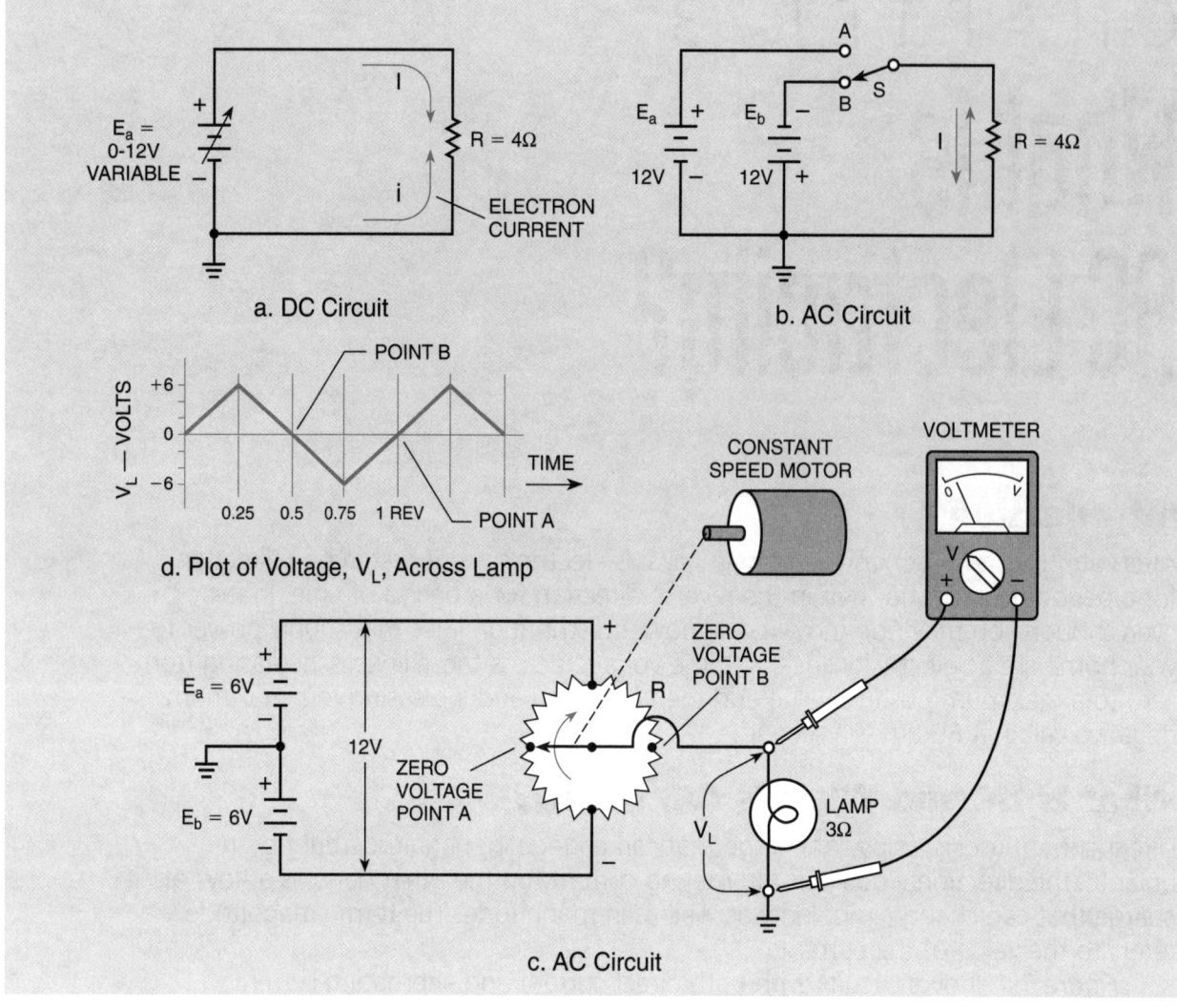

Figure 3-1. Examples of DC and AC circuits.

Waveforms

A very valuable instrument for studying ac electricity is the oscilloscope. *Figure 3-2a* shows an example of one, and variations of an ac voltage with time on its screen. The squiggly line seen on the screen of the oscilloscope when an ac circuit voltage is measured is a graph (picture) of the voltage variation plotted on the vertical axis versus time on the horizontal axis. This graph is called a *waveform*. The time it takes to complete one cycle of the ac repeating waveform is time, T, as shown in *Figure 3-2b*. The period of time represented on the graph is determined by the setting of the TIME/DIV (or SEC/DIV) control. This control sets the speed of the movement of the dot from left to right on the screen; that is, the rate of change of the time base generator's voltage that sweeps the electron beam dot that forms the waveform across the screen. Note that a constant dc voltage, as shown in *Figure 3-2b*, is a straight line on the screen.

In order for the waveform to be stationary on the screen, it must be triggered. The trigger starts the time base generator moving the dot. It must begin at the same relative point on the voltage waveform plotted on the vertical axis each time it sweeps (travels from left to right). The oscilloscope is calibrated so that the waveform presented on the screen can be visually analyzed and its parameters measured. Determining a signal's voltage at any instant of time is the job that the oscilloscope does best.

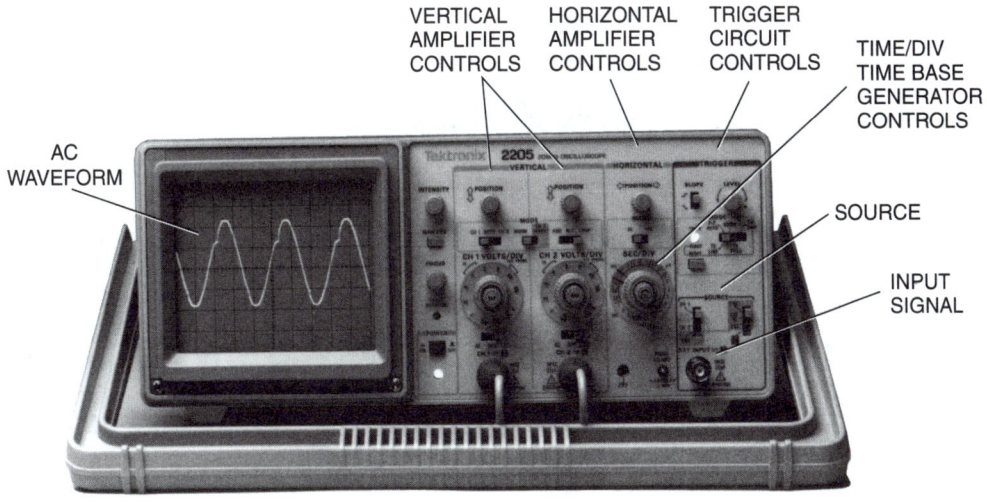

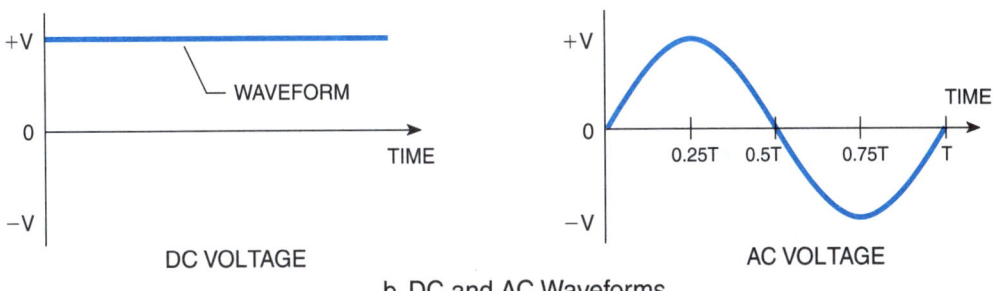

b. DC and AC Waveforms

Figure 3-2. An oscilloscope is an instrument that plots how ac voltages or currents change with time. DC voltages that do not change in magnitude appear as a straight line (constant value.) Courtesy of Tektronix, Inc.

The Sine Wave

The most fundamental of waveshapes, the sine wave, is the voltage function that is produced in the alternator of an automobile battery charging system as well as on the power lines from a power company's generating station. Before we look at those voltages, let's first consider the sine wave produced in mechanical applications in order to gain a better understanding of the function and the shape of the waveform.

Mechanical Analogies That Produce Sine Waves

Suppose that a lead ball hanging at the end of a spiral steel spring, as illustrated in *Figure 3-3a,* is pulled down a distance S below its equilibrium position. When the ball is released, the restoring force produces an acceleration proportional to the distance S but opposite in direction. This type of motion is called simple harmonic motion (SHM). There are other vibrating bodies that produce simple harmonic motion. Another, as shown in *Figure 3-3b,* is a flat spring with one end clamped in a vise that has a lead ball attached to its free end. If the ball is pulled sideways bending the spring and is then released, its motion is approximately SHM.

A third common example of approximate SHM is the motion of a pendulum, shown in *Figure 3-3c.* A simple pendulum consists of a heavy bob at the end of a very light string. Both with the flat spring and the pendulum, the displacement must be small or an error will exist because the circular arc of the weight deviates from linear motion.

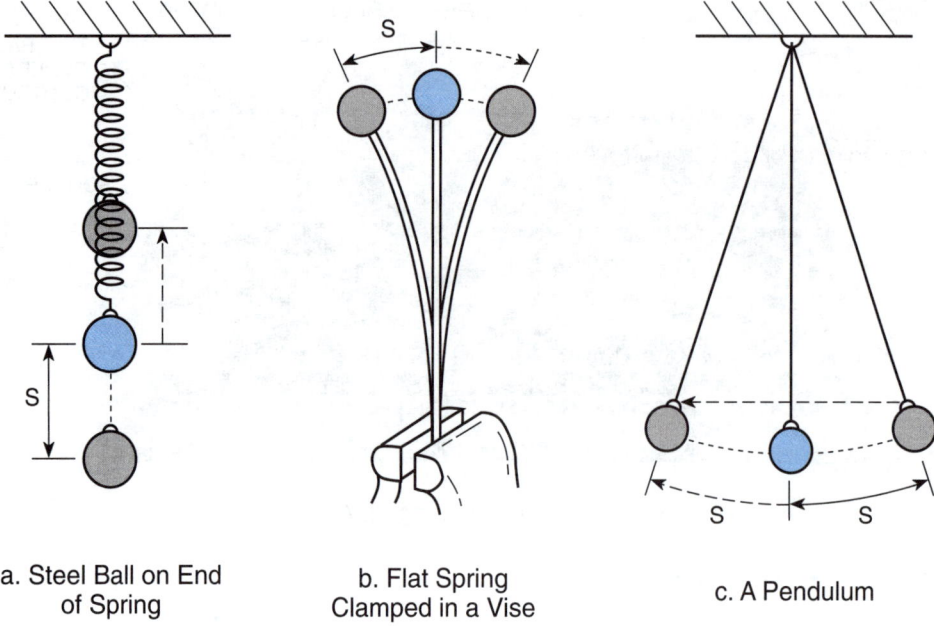

a. Steel Ball on End of Spring

b. Flat Spring Clamped in a Vise

c. A Pendulum

Figure 3-3. Mechanical simple harmonic motions that produce sine waves.

Electrical Sine Waves Are Similar to Displacement Sine Waves

Now, what does all this have to do with a sine wave? In each of the above examples, if we had a method of plotting a graph of the position of the ball as a function of time, the shape of the graph would be a sine wave, such as is shown in *Figure 3-4*. There is a maximum positive (plus peak) and maximum negative (minus peak) displacement. Shown is one cycle which takes time T as it starts at zero, goes through plus peak, back through zero to its minus peak, and back to zero again. This cycle then repeats itself.

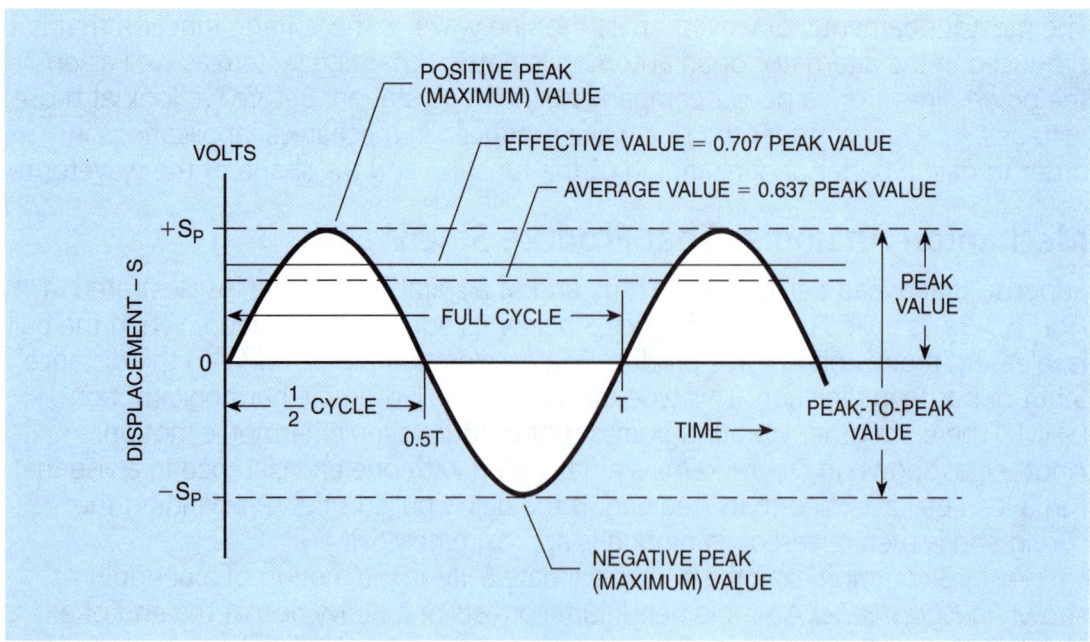

Figure 3-4. Sine wave describes displacement of objects which have simple harmonic motion.

As we said previously, such a sine wave of ac voltage is produced by an automotive alternator or power company generator. *Figure 3-5* shows how this is done by a loop conductor rotating in a magnetic field. This is a simplified diagram of what happens in an alternator or power company generator. Notice that when the loop conductor moves parallel to the magnetic field, zero voltage is generated, and when it cuts across the field perpendicular to the lines, then the maximum value of voltage is generated. This occurs when the loop is moving through the 90° and 270° positions. At 90°, the voltage is at its maximum positive value; at 270°, the voltage is at its maximum negative value. One complete revolution around the circle is a cycle. The number of cycles per second (cps) is the frequency of the sine wave, which is stated in the unit hertz (Hz). The time per cycle, T in seconds, is the period of the wave. To find the frequency of a wave, the oscilloscope can be used to measure its period and then the reciprocal of the period gives the frequency. For example, if the waveform in *Figure 3-4* has a period, T, of 2 milliseconds (2 ms or 0.002 second), the frequency of the waveform is 500 Hz. Here's the math:

$$f = \frac{1}{T} = \frac{1}{0.002} = 500 \text{ Hz} \quad \text{or}$$

$$f = \frac{1}{2 \times 10^{-3}} = 0.5 \times 10^3 \text{ Hz} = 0.5 \text{ kHz}$$

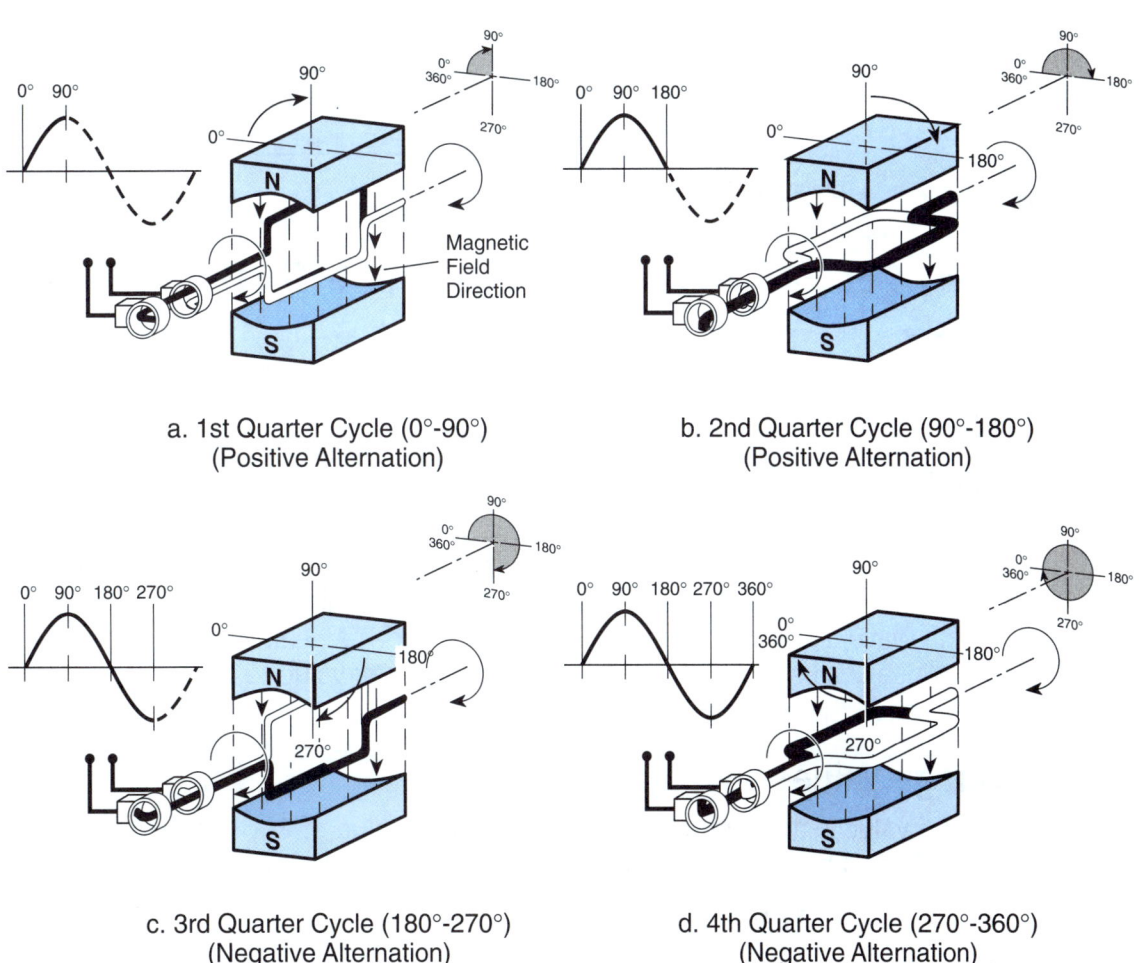

a. 1st Quarter Cycle (0°-90°)
(Positive Alternation)

b. 2nd Quarter Cycle (90°-180°)
(Positive Alternation)

c. 3rd Quarter Cycle (180°-270°)
(Negative Alternation)

d. 4th Quarter Cycle (270°-360°)
(Negative Alternation)

Figure 3-5. A sine wave of voltage is generated as a loop of wire is rotated in a constant magnetic field.

Example 1. Determining Frequency of Oscilloscope Waveform

In the oscilloscope waveform shown, the vertical voltage deflection is 1V/division and the horizontal time rate is 0.001 s/division. What is the frequency of the waveform?

Since each time division is 0.001 second, one cycle of the waveform takes 0.004 second. Therefore, T = 0.004s and the frequency is:

$$f = \frac{1}{T} = \frac{1}{0.004} = 250 \text{ Hz}$$

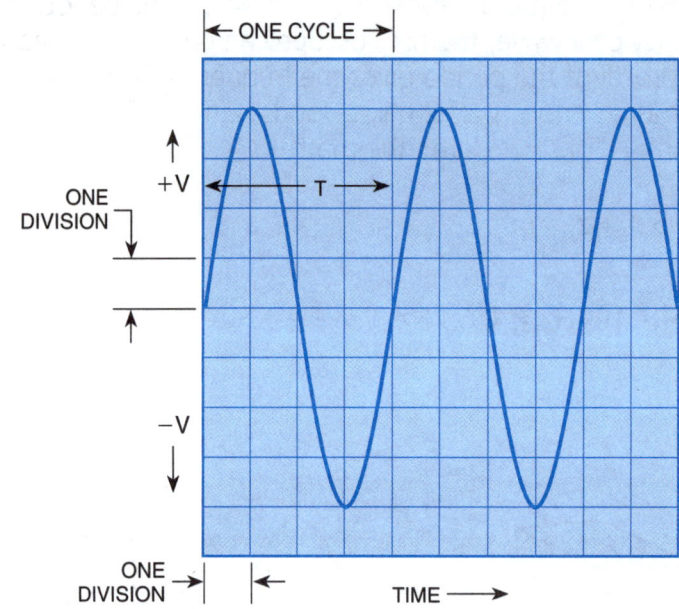

Peak, Peak-to-Peak, Average and Effective (RMS) Values

Figure 3-4 summarizes the four values of an ac voltage. Each value is explained in the following paragraphs.

Peak and Peak-to-Peak Values

Since the instantaneous value of the sine wave continuously changes, it is necessary to be specific when describing the value of the waveform. There are four values used to describe an ac voltage or current. The first is the peak value. The maximum value of a sine wave, either positive or negative, is called the peak value. The second is the peak-to-peak value. The value from the positive peak value to the negative peak value is called the peak-to-peak value.

Average Value

The third value is the average value. The average value of any shape of waveform is its area divided by its length. The average value, V_{avg}, of a sine wave voltage is:

$$V_{avg} = 0.637 \, V_p \qquad \text{where } V_p = \text{peak value}$$

Effective (RMS) Value

When there is a direct current through a resistor, a certain amount of power is dissipated by the resistor in the form of heat. Heat is also produced if there is an alternating current through the same resistor. An alternating current that will produce the same amount of heat in a specific resistance as a direct current is said to have the same *effective* value. The effective value of any waveform may be determined by a mathematical process known as the root-mean-square (RMS) method. We will not go through the mathematical procedure used, but just state that for a sine wave, the *effective or RMS value,* V_{rms}, is equal to the peak value, V_p, divided by the square root of 2. In equation form, this is:

$$V_{rms} = \frac{V_p}{\sqrt{2}} = \frac{V_p}{1.414}$$

$$V_{rms} = 0.707\, V_p$$

The effective value is the most commonly used. If a notation is not included with an ac value, it is usually assumed to be the effective or RMS value.

Most ac voltmeters and ammeters are calibrated to read the effective or RMS values of a sine wave. It is important to remember that if the waveshape differs from a sine wave, the effective (heating) value will differ from 0.707 times the peak and conventional meters will not indicate the value accurately. However, a special type of meter, called a "true RMS" voltmeter will accurately indicate the effective value regardless of the waveshape.

Example 2. Determining Values of AC Voltages

For the waveform in *Example 1*, what are the V_{peak}, $V_{peak-to-peak}$, and V_{rms} values for the voltage waveform?

Since the vertical deflection is 1V/division:

$V_{peak} = 4$ volts
$V_{peak-to-peak} = 8$ volts
$V_{rms} = 0.707\, V_{peak} = 2.83$ volts

Phase Relationships – Angle Measurement

As shown in *Figure 3-6a,* an ac value can be represented as a vector of length A from origin 0 rotating through angles from 0° to 360°. The vector is described as A sin θ, where θ is the angle from the 0° axis that the vector is rotated through. The magnitude of the sine wave at angle θ is equal to A sin θ. Therefore, as shown in *Figure 3-6,* the value of the sine wave can be plotted against the angle of rotation. When these points are connected, the result is the sine wave waveform shown in *Figure 3-6b.* If the vector has rotated by an angle θ = 30°, then the magnitude of the sine wave at 30° is A sin 30°. Since sin 30° = 0.5, then the magnitude of the sine wave is 0.5A. In like fashion, the magnitude of the sine wave can be determined at any angle of rotation by multiplying the sine of the angle by the vector magnitude.

Also, since the ac value of the sine wave can be represented by A sin θ, different ac quantities will have different angles. Therefore, they are at different points in the cycle. Thus, one voltage may be 20 sin 30° and another voltage may be 30 sin 60°. The angle θ is referred to as the phase angle, and the differences in the angle θ between quantities shows that the ac quantities are "*out-of phase*". The first vector is at 30° in its cycle and the second vector is at 60° in its cycle.

The concept of phase is seldom used in reference to a single sinusoidal voltage or current. The consideration of phase is very important, however, when two or more sine waves of voltage or current are present, especially in the same circuit. This becomes very important when capacitors and inductors are present in ac circuits. The

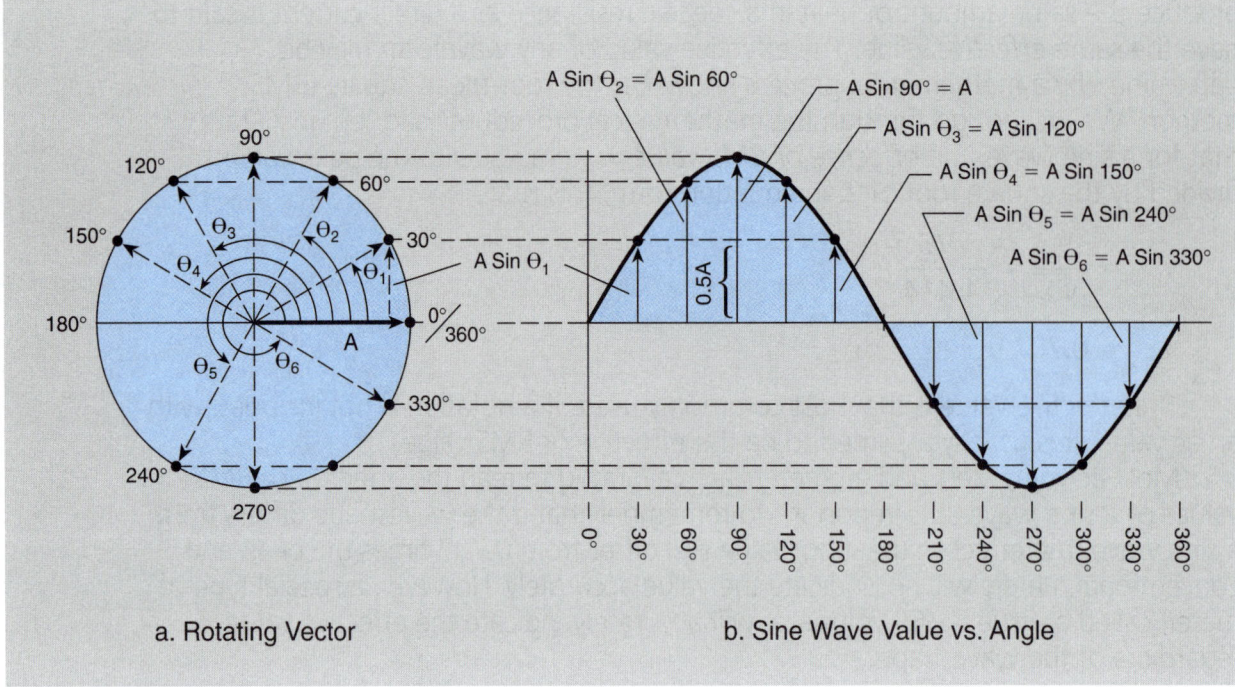

a. Rotating Vector   b. Sine Wave Value vs. Angle

Figure 3-6. Sine wave generated by the vector A sin θ.

capacitors and inductors cause circuit currents and voltages to be at different phase angles (different angles in their cycles), and, thus, the phase angle is an important part of the value of the voltage or current. Let's examine this in more detail.

AC Characteristics of Capacitors

Although a changing current does not pass through a capacitor's dielectric, it certainly appears to do just that. The reason for this is seen in the very definition of capacitance, q = Cv. If the voltage across the plates of a capacitor is continuously changing, the charge on the plates varies. Thus, the alternating current appears, in effect, to be through the dielectric.

When a capacitor is placed in an ac circuit, the circuit current will be reduced. The electrical property that causes this decrease in current is an opposition to current (an impedance) known as *capacitive reactance*.

As we saw in Chapter 2 *(Figure 2-11)*, even though a voltage is applied to a capacitor and there is current to charge a capacitor, the voltage across the capacitor takes some time before it reaches its final value. Because of the ability of a capacitor to oppose a change in voltage by charging and discharging, the vector representing the current and the vector representing the voltage applied to a pure capacitance are 90° out of phase, with the voltage lagging the current. We have plotted the two waveforms against time in *Figure 3-8a* so that we can compare them to the waveforms of inductors and resistors also shown in *Figure 3-8*.

Capacitive Reactance

The opposition to ac offered by a capacitance; that is, capacitive reactance, X_C, measured in ohms, is:

$$X_C = \frac{1}{2\pi f C}$$

where: X_C = capacitive reactance in **ohms**
f = frequency in **Hz**
C = capacitance in **farads**
2π = 6.28 — number of radians in 360° (one cycle)

In an ac circuit, a capacitor is just as effective as a resistor for controlling the current. Ohm's law again can be used for the basic relationship between current, voltage and capacitive reactance in a purely capacitive circuit. This relationship is expressed as:

$$I = \frac{E}{X_C}$$ where X_C is substituted for R

> ### Example 3. Determining Capacitive Reactance
> What is the reactance of a 2 μF capacitor at 400 Hz? (1μF = 1 × 10⁻⁶F)
>
> $$X_C = \frac{1}{2\pi fC} = \frac{1}{6.28 \times 400 \times 0.000002} = 199\Omega$$

Look again at the equation for capacitive reactance. X_C varies inversely with the frequency of the applied voltage and the amount of capacitance. As frequency increases, X_C decreases; as capacitance increases, X_C decreases. As frequency decreases, X_C increases; as capacitance decreases, X_C increases. Some examples are shown in *Figure 3-7*. In *Figure 3-7a*, all voltages with frequencies lower than f_C will be opposed with high capacitive reactance which will prevent those frequencies from getting to the load. In other words, the capacitive reactance is much larger than the load resistance. In *Figure 3-7b*, all voltages with frequencies above f_C will be bypassed around the load ("shorted out") by the low capacitive reactance of C in parallel with the load. In other words, the capacitive reactance is much smaller than the load resistance.

AC Characteristics of Inductance

The property of an electric circuit that tends to oppose any change in an electric current is inductance. The magnetic field *established by a steady or non-time varying current does not react* upon the electric circuit which produces it. However, whenever the circuit *current changes in magnitude* the magnetically stored energy *does react* upon the electric circuit, demanding additional energy from the system if

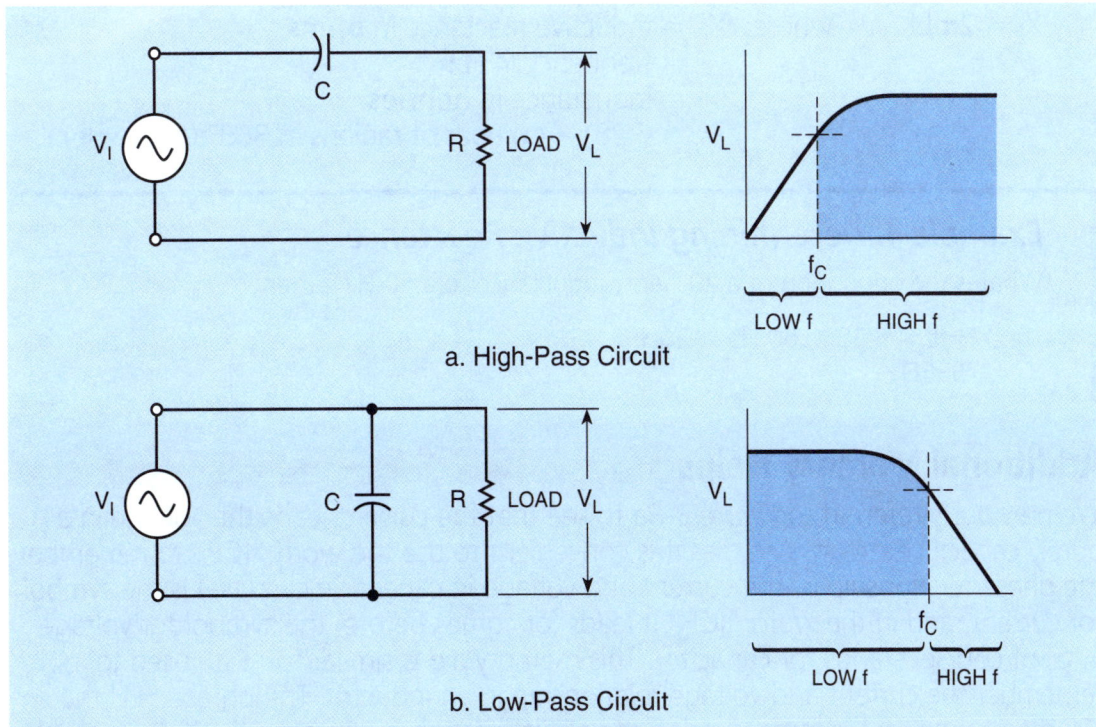

Figure 3-7. Effect of capacitive reactance for frequency control.

the current is increasing and returning energy to the system if the current is decreasing.

A circuit having pure resistance (if such a thing were possible) would have the alternating current through it and the voltage across it rising and falling together. This is illustrated in *Figure 3-8b,* which shows the sine waves for current and voltage in a purely resistive circuit having an ac source. The current and voltage do not have the same amplitude but they are *in phase*.

As we saw in Chapter 2, in a dc circuit containing inductance, the current took time to rise to its maximum value even though the full applied voltage was immediately at maximum. This was due to the counter EMF (CEMF) developed in the inductor. In the case of a circuit having inductance, the opposing force of the CEMF is enough to keep the current from remaining in phase with the applied voltage. *Figure 3-8c* shows the waveforms for a purely inductive ac circuit for a complete cycle. The vector representing the current lags the vector representing the voltage in a purely inductive circuit by 90 degrees.

A simple aid for remembering the relationship of voltage and current in an inductive circuit is the word "ELI." Since E is the symbol for voltage, L is the symbol for inductance and I is the symbol for current, the word "ELI" indicates that current comes after (lags) voltage in an inductor.

Inductive Reactance

When the current through an inductor continuously reverses itself, as in the case of a sine wave ac source, the inertia effect of the CEMF is greater than with dc. The greater the amount of inductance (L), the greater the opposition from this inertia effect. Also, the faster the reversal of current, or frequency (f), the greater this inertial opposition. This opposing force which an inductor presents to the alternating current cannot be called resistance since it is not the result of friction within a conductor. The name given to it is *inductive reactance* because it is the reaction of the inductor to alternating current. Inductive reactance is measured in ohms and its symbol is X_L.

Inductive reactance increases directly with an increase of frequency and with an increase of inductance. Its value is calculated from:

$$X_L = 2\pi f L$$

where: X_L = inductive reactance in **ohms**
f = frequency in **Hz**
L = inductance in **henries**
2π = 6.28 — number of radians in 360° (one cycle)

Example 4. Determining Inductive Reactance

What is the opposition of a 20-henry inductance to a 60-Hz current?

$X_L = 2\pi f L = 6.28 \times 60 \text{ Hz} \times 20 \text{H}$
$X_L = 7536 \Omega$

Additional Memory Helps

We previously referred to *Figure 3-8a* to see that the current leads the voltage in a purely capacitive circuit. At times it is convenient to use the word "ICE" to remember the phase relationship of the current and voltage in capacitive circuits. I is the symbol for current, and in the word "ICE" it leads (or comes before) the symbol for voltage E. C, of course, stands for capacitor. This memory aid is similar to "ELI" used to remember the current and voltage relationship in an inductor. The phrase "ELI the ICE man" is helpful in remembering the phase relationship in both the inductor and capacitor circuits.

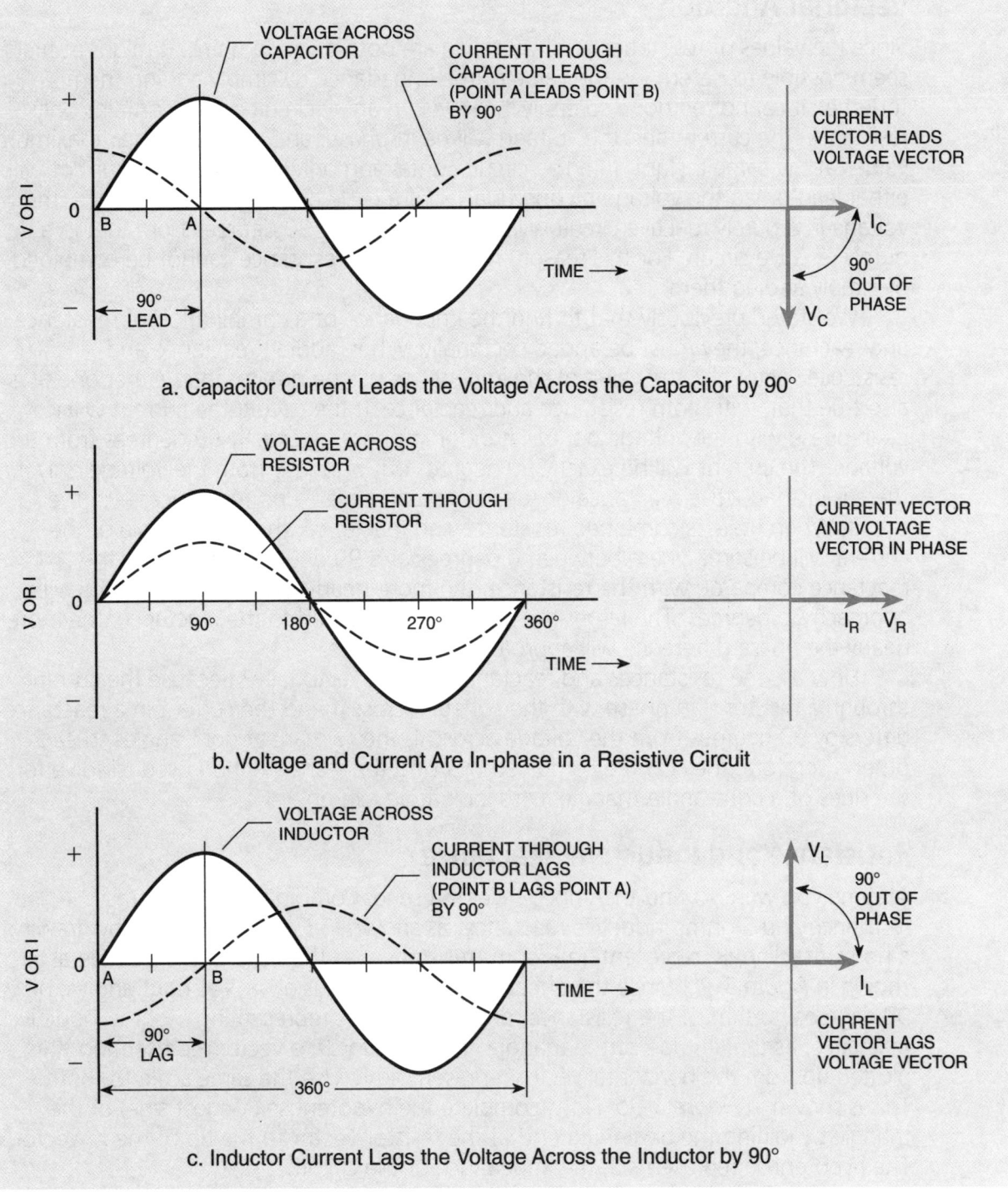

Figure 3-8. Phase shift – The difference in the angles of the vectors representing voltage and current in a circuit containing capacitance, resistance or inductance.

Impedance and Vectorial Addition

We have seen that the opposition that capacitance or inductance offers to an alternating current is called capacitive reactance or inductive reactance, respectively. Since practical ac circuits that contain reactance also contain resistance, the two combine to oppose current. This combined opposition by the resistance and the reactance is called the *impedance* and is represented by the symbol Z. Combining all impedances then gives the total opposition to circuit current and provides the total impedance of an ac circuit.

Vectorial Addition

Since the values of resistance and reactance are both given in ohms, it might at first seem possible to determine the value of the impedance by simply adding them together. It cannot be done so easily, however. In an ac circuit which contains only resistance, the current and the voltage will be in phase, and will reach their maximum values at the same instant. In an ac circuit containing only reactance, the current will either lead or lag the voltage by one-quarter of a cycle (90 degrees). Therefore, the voltage in a purely reactive circuit will differ in phase by 90 degrees from that in a purely resistive circuit. For this reason, reactance and resistance cannot be combined by simply adding them.

We stated previously that to find the impedance of a combination of resistance and reactance, they must be added vectorially. When adding reactance and resistance vectorially, the value of the impedance will be greater than either one. It is also true that, with both resistance and reactance in the circuit, the current will not be in phase with the voltage nor will it differ in phase by exactly 90 degrees from the voltage. The current will be exactly 90 degrees out of phase from the voltage only if there is just inductive reactance or capacitive reactance — no resistance — in the circuit. When there is combined resistance and reactance, the phase angle of the current will be somewhere between 0 degrees and 90 degrees. The larger the reactance compared with the resistance, the more nearly the phase difference will approach 90 degrees. The larger the resistance compared to the reactance, the more nearly the phase difference will approach zero degrees.

How are the resistances and reactances added vectorially? Because the current through a resistor is in phase with the voltage across it and the current in a reactance differs by 90 degrees from the voltage across it, the two are at right angles to each other. Therefore, they can be combined by using the same method used to solve for the sides of a right-angle triangle. Let's look at an example.

Resistance and Inductive Reactance

Assume you want to find the impedance of a series combination of 8 ohms resistance and 5 ohms inductive reactance, as shown in *Figure 3-9a*. Start by drawing a horizontal line, R, representing 8 ohms resistance, as the base of the triangle as shown in *Figure 3-9b*. Since the effect of the reactance is always at right angles, or 90 degrees, to that of the resistance, draw the line X_L, representing 5 ohms inductive reactance, as the altitude of the triangle. In other words, a vector representing X_L is plotted up from the horizontal vector representing R. Use the same scale for both. This is shown in *Figure 3-9b*. Now, complete the hypotenuse (longest side) of the triangle by joining the beginning end of the resistor vector to the tip of the X_L vector. The hypotenuse represents Z, the impedance of the circuit.

One of the right-triangle rules used is the Pythagorean theorem. It states that for a right triangle:

$$(\text{hypotenuse})^2 = (\text{base})^2 + (\text{altitude})^2$$

Applied to impedance, this becomes:

$$(\text{impedance})^2 = (\text{resistance})^2 + (\text{reactance})^2$$

Substituting symbols:

$$Z^2 = R^2 + X_L^2$$

therefore:

$$Z = \sqrt{R^2 + X_L^2}$$

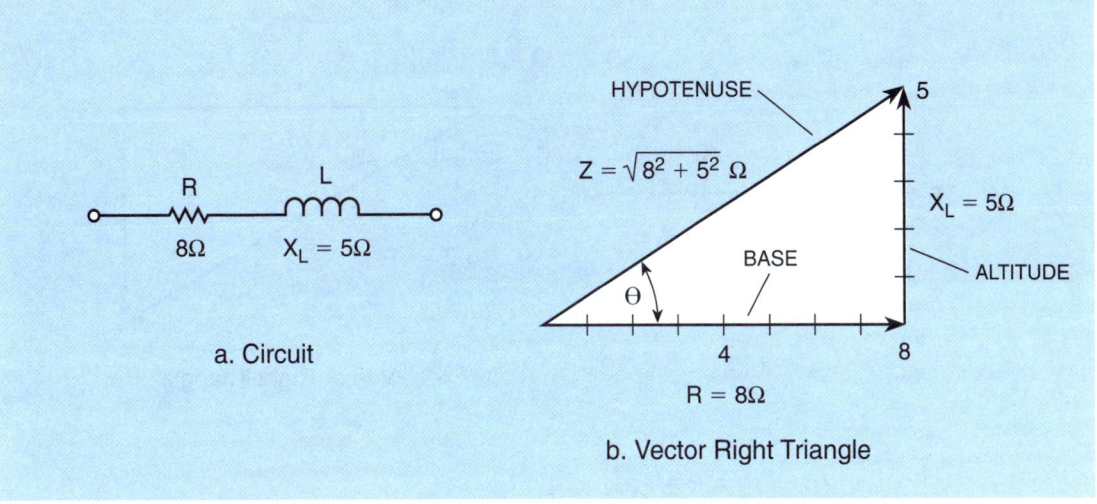

Figure 3-9. Solving for Z in a R and X_L circuit.

Applying this equation to our example where we are given:

R = 8 Ω, X_L = 5 Ω

We find Z by:

$Z = \sqrt{8^2 + 5^2} = \sqrt{64 + 25} = \sqrt{89} = 9.43 Ω$

Note that the impedance in ohms is more than either the resistance or the reactance alone.

Resistance and Capacitive Reactance

When dealing with capacitive reactance instead of inductive reactance, as in the previous example, it is customary to draw the line representing the capacitive reactance in a downward direction. This is shown in *Figure 3-10*. The line is drawn downward for the capacitive reactance to indicate that it acts in a direction opposite to inductive reactance, which is drawn upward. In a series circuit containing capacitive reactance, the same Pythagorean theroem is used and the equation for finding the impedance becomes:

$Z = \sqrt{R^2 + X_C^2}$

In both *Figures 3-9* and *3-10*, we have concentrated on solving for the magnitude of Z and have not been concerned about the angle θ, which is the phase angle for Z. To find θ, we know from trigonometry that:

$\tan θ = \dfrac{\text{side opposite}}{\text{side adjacent}} = \dfrac{X_L}{R}$

For *Figure 3-9*:

$\tan θ = \dfrac{5}{8}$ therefore, θ = 32°

For *Figure 3-10*:

$\tan θ = \dfrac{-X_C}{R}$ therefore, θ would be a negative angle.

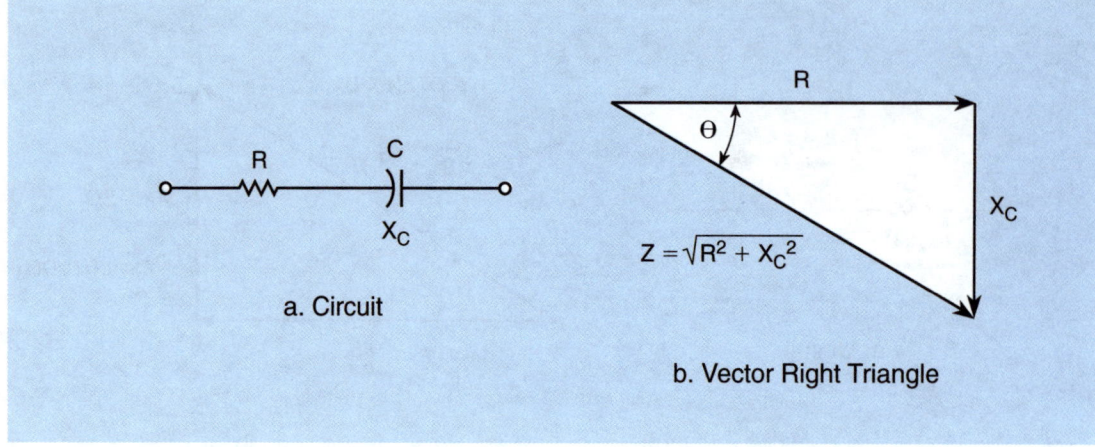

Figure 3-10. Solving for Z in a R and X_C circuit.

Example 5. Determining Impedance

For a series circuit with a 30Ω resistor and a capacitor that has a capacitive reactance of 40Ω at 3981 Hz, calculate the (a) impedance, (b) angle θ, and (c) value of the capacitor.

a. $Z = \sqrt{R^2 + X_C^2} = \sqrt{30^2 + 40^2} = \sqrt{900 + 1600} = \sqrt{2500} = 50\Omega$

b. $\tan\theta = \dfrac{-X_C}{R} = \dfrac{-40}{30} = 1.33 \quad \therefore \theta = 53°$

c. $X_C = \dfrac{1}{2\pi fC}$ and $C = \dfrac{1}{2\pi fX_C} = \dfrac{1}{6.28 \times 3.981 \times 10^3 \times 4 \times 10^1}$

$C = \dfrac{1}{100 \times 10^4} = \dfrac{1}{1 \times 10^6} = 1 \times 10^{-6} = 1\,\mu F$

Up to this point, our circuits have used either inductance or capacitance — not both in the same circuit. In the remainder of this chapter, we will be concerned with the combination of inductance, capacitance and resistance in ac circuits.

AC Circuits with R, L and C

To explain the various properties that exist within ac circuits, the series RLC circuit will be used. *Figure 3-11a* is the schematic diagram of the series RLC circuit. The symbol shown in *Figure 3-11a* that is marked E is the general symbol used to indicate an ac voltage source.

The effect of inductive reactance is to cause the current to lag the voltage. The effect of capacitive reactance is to cause the current to lead the voltage. Since inductive reactance and capacitive reactance are exactly opposite in their effects, which we demonstrated in *Figures 3-9* and *3-10* by plotting the vectors for X_L and X_C in the opposite directions, what will be the result when the two are combined? It is not difficult to see that the net effect is a tendency to cancel each other, with the combined effect then equal to the difference between their values. This is shown in *Figure 3-11b*. This resultant is called "equivalent reactance." It is represented by the symbol X and expressed by the equation:

$X = X_L - X_C$

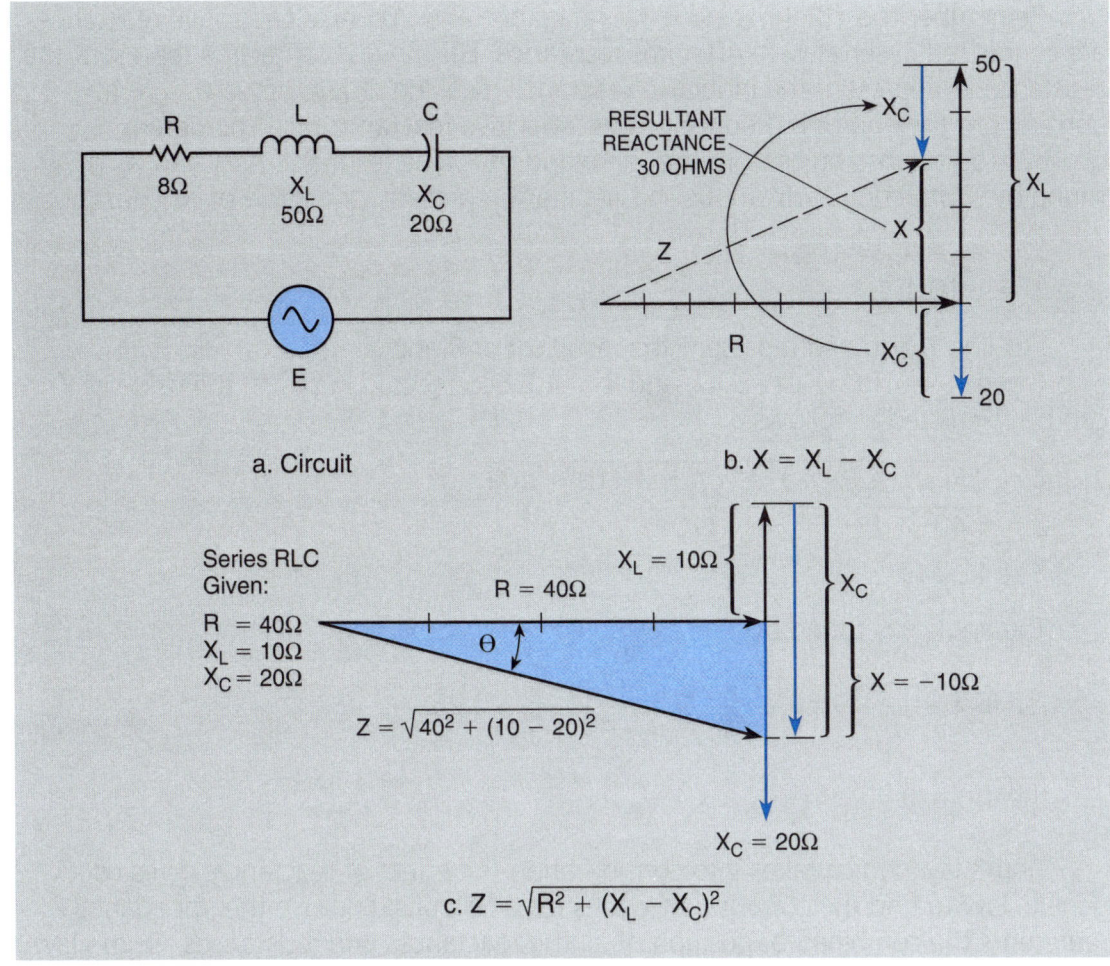

Figure 3-11. Solving for Z in a RLC series circuit.

Thus, if a circuit contains 50 ohms of inductive reactance and 20 ohms of capacitive reactance in series, the net reactance of X, is:

50 ohms − 20 ohms = 30 ohms of reactance.

As shown in *Figure 3-11*, this is like plotting the X_C vector from the tip of the X_L vector.

In this series RLC circuit, the equation for the impedance containing R, X_L, and X_C then becomes:

$$Z = \sqrt{R^2 + (X_L - X_C)^2}$$

NOTE: This equation and the previous one can be used to calculate Z only if the resistance and reactance are connected in series.

Remember that Z also has an angle θ, and:

$$\tan \theta = \frac{X_L - X_C}{R}$$

Trigonometry tables are used to look up θ in the tangent tables.

Another Series RLC Circuit

To make sure we understand vector addition clearly, let's solve another RLC series circuit given in *Figure 3-11c*. This time the capacitive reactance is greater than the inductive reactance. We will find the impedance and its angle.

Remember the 10 ohms inductive reactance and 20 ohms capacitive reactance are connected in series with 40 ohms resistance. Let the horizontal line represent the resistance R. Represent the inductive reactance by a line drawn upward at right angles from the end of R. Represent the capacitive reactance by a line drawn downward at right angles from the same end of R. The resultant of X_L and X_C is found by subtracting X_C from X_L. This resultant represents the value of X. Thus:

$$X = X_L - X_C = 10\Omega - 20\Omega$$
$$X = -10\Omega$$

The line Z will then represent the resultant of R and X.

Given $X_L = 10\Omega$, $X_C = 20\Omega$, and $R = 40\Omega$ connected in series, the value of Z can be calculated as follows:

$$Z = \sqrt{R^2 + (X_L - X_C)^2} = \sqrt{40^2 + (10 - 20)^2}$$
$$Z = \sqrt{40^2 + (-10)^2} = \sqrt{1700}$$
$$Z = 41.2\ \Omega$$

The angle θ is found by:

$$\tan\theta = \frac{X_L - X_C}{R} = \frac{10 - 20}{40} = \frac{-10}{40} = -0.25$$

therefore, $\theta = -14°$.

Ohm's law, which takes into consideration the effect of reactance, gives us a general law to find the current that works for ac circuits. Because the impedance Z represents the combined opposition of all the reactances and resistances, Ohm's law for ac is:

$$I = \frac{E\ \sqrt{\theta}}{Z\ \sqrt{\theta}}$$

This general Ohm's law applies to alternating current in any circuit. The magnitude of any one of the I, E or Z values may be found from the equation without considering the angles if the other values are known, just as for dc circuits. However, if the phase angle is required, the phase angle of each of the values must be taken into account as indicated in the general equation.

Other Circuit Combinations

Of course, parallel ac circuits, and combination series and parallel ac circuits, with any combination of reactance and resistance are encountered in working systems. The mathematics of these solutions are somewhat beyond our goals in this text. Some of you may want to pursue study of the more complex circuits after you finish this book.

Power In an AC Circuit

In a direct current circuit, the power is equal to the voltage times the current, or:

$$P = E \times I$$

If a voltage of 100 volts applied to a circuit produces a current of 10 amperes, the power is 1000 watts. This is also true in an ac circuit when the current and voltage are in phase; that is, when the circuit is effectively resistive. However, if the ac circuit contains reactance, the current will lead or lag the voltage by a certain amount (the phase angle). When the current is out of phase with the voltage, the power indicated by the product of the applied voltage and the total current gives only what is known as the "apparent power." The "true power" depends upon the phase angle between the current and voltage. The phase angle θ is determined as before from the resultant reactance and resistance:

$$\tan \theta = \frac{X}{R}$$

θ = the angle that has a tangent of (X/R)

When an alternating voltage is impressed across a capacitor, power is taken from the source and stored in the capacitor as the voltage increases from zero to its maximum value. Then, as the impressed voltage decreases from its maximum value to zero, the capacitor discharges and returns the power to the source. Likewise, as the current through an inductor increases from its zero value to its maximum value, the field around the inductor builds up to a maximum. When the current decreases from maximum to zero, the field collapses and returns the power to the source. Therefore, no power is used up in either case, since the power alternately flows to and from the source. This power that is returned to the source by the reactive components in the circuit in called "reactive power." Apparent power, reactive power and true power form a right triangle as shown in *Figure 3-12*.

As shown in *Figure 3-12*, true power can be found as follows:

true power = apparent power × cos θ

The cos θ is called the power factor (PF), thus, PF = cos θ. The power factor is a number, represented as a decimal or a percentage, that represents the portion of the apparent power dissipated in a circuit. From trigonometry, the easiest way to find the power factor is to find the cosine of the phase angle θ. As shown in *Figure 3-12*:

$$\cos \theta = \frac{\text{true power}}{\text{apparent power}}$$

Therefore, you do not need to use trigonometry to find the power factor. Since the power dissipated in a circuit is true power, and apparent power is E x I, then:

$$PF = \frac{\text{true power}}{\text{apparent power}}$$

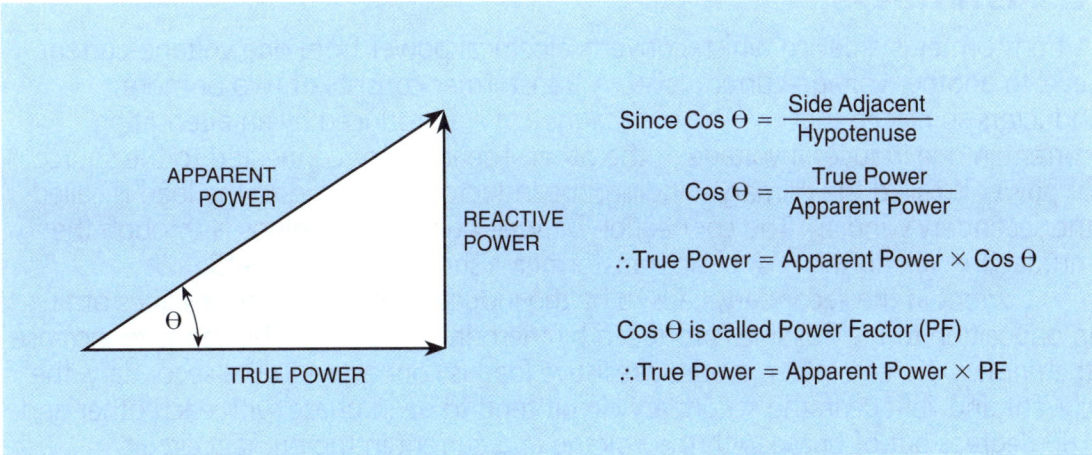

Figure 3-12. The AC power right triangle.

Example 6. Determining Power Factor

What is the power factor when a circuit has 100 VAC applied and 20 amperes of current and has a true power dissipation of 1800 watts?

$$PF = \frac{\text{true power}}{\text{apparent power}} = \frac{1800}{100 \times 20} = 0.9$$

Resonance

A special condition of an RLC circuit, called resonance, occurs at a particular frequency. The definition of resonance for an RLC circuit is the non-zero (not dc) frequency such that the circuit current and voltage are in phase. In a series circuit, this means that inductive reactance and capacitive reactance are equal and cancel. Therefore, the impedance is equal to just the resistance of the series circuit. In a parallel circuit, the vector sum of the currents in the inductor and the capacitor have to be equal so that they cancel. This means that the current going into a parallel circuit at resonance is minimum, resulting in a high equivalent impedance at the resonant frequency. We will see series and parallel resonant circuits in many applications such as filters and tuned circuits.

Example 7. Determining Resonant Frequency

What is the resonant frequency of a series RLC circuit that has a 0.5 µF capacitor in series with a 20 mH inductor?

Since

$$X_L = 2\pi f L \quad \text{and} \quad X_C = \frac{1}{2\pi f C} \quad \text{and} \quad X_L = X_C \text{ at resonance}$$

then

$$2\pi f L = \frac{1}{2\pi f C} \quad \text{and} \quad f^2 = \frac{1}{(2\pi)^2 LC}$$

Therefore, for a series circuit, the resonance frequency f_r is:

$$f_r = \frac{1}{2\pi\sqrt{LC}} = \frac{1}{2\pi\sqrt{(0.5 \times 10^{-6})(20 \times 10^{-3})}} = 1590 \text{ Hz}$$

Transformers

A transformer is a device which converts electrical power from one voltage-current level to another voltage-current level. A transformer consists of two or more inductors so placed that the varying magnetic field produced by an alternating current in one induces a voltage in the other. The inductor connected to the source of power is called the primary winding; the inductor connected to the load is called the secondary winding. The connection between the two windings is through the mutual flux linking them. A basic transformer is shown in *Figure 3-13*.

Current in the secondary, caused by the induced voltage, produces lines of flux in opposition to the flux lines produced by the primary current. This tends to increase the primary current. When a purely resistive load is connected to the secondary, the current and voltage in the secondary circuit tend to be in phase with each other and 180 degrees out of phase with the voltage and current in the primary circuit.

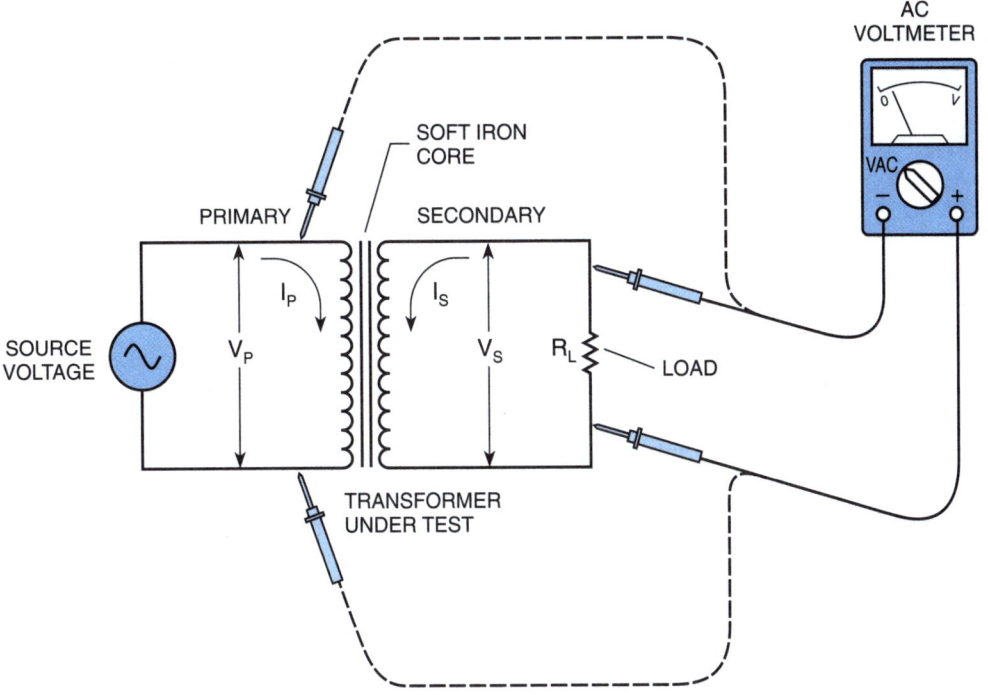

Figure 3-13. A power transformer.

The core around which the primary and secondary are wound may be iron for low frequencies, as in the case of power and audio transformers. Primary and secondary windings on an air core may be employed for transformers that couple energy in higher frequency circuits.

Transformer Losses and Efficiency

Losses occur in transformers because of flux leakage, copper losses, hysteresis, flux saturation of the core and eddy currents. However, many practical transformers have efficiencies above 95 percent. Transformers are rated according to voltage, current, power and frequency. For maximum efficiency and protection of transformers, they should never be operated so as to exceed any of the ratings.

If the transformer were ideal, there would be no power loss from the primary to secondary and 100% of the source power would be delivered to the load. Since voltage times current equals power, the power relationship is given by:

$V_P \times I_P = V_S \times I_S$ where: V_P = primary voltage in **volts**
I_P = primary current in **amperes**
V_S = secondary voltage in **volts**
I_S = secondary current in **amperes**

Turns Ratio

In an ideal transformer, the ratio of primary voltage, V_P, to the voltage induced in the secondary, V_S, is the same as the ratio of the number of turns in the primary, N_P, to the number of turns in the secondary, N_S. The following equation expresses the relationship:

$$\frac{V_P}{V_S} = \frac{N_P}{N_S}$$

The turns ratio of a transformer is:

$$\frac{N_S}{N_P}$$

the ratio of the secondary turns to the primary turns, which is equal to the ratio of the secondary voltage to the primary voltage.

The turns ratio of a transformer can be measured using the circuit of *Figure 3-13*. Apply a small ac voltage to the primary. Measure V_P with an ac voltmeter. Now measure the secondary voltage V_S with the same voltmeter. The turns ratio is equal to the secondary voltage divided by the primary voltage.

$$\frac{N_S}{N_P} = \frac{V_S}{V_P}$$

Isolation, Step-down, and Step-up Transformers

If the number of turns in the primary and the secondary are equal, then the voltages appearing across the primary and secondary are equal. This type of transformer with a one-to-one turns ratio is called an *isolation transformer*. If a lower voltage appears across the secondary than across the primary, it is called a step-down transformer. The turns ratio would be less than 1. However, if a higher voltage appears across the secondary than across the primary, it is called a step-up transformer. The turns ratio would be greater than 1. According to the primary and secondary power relationship equation given previously, the secondary current will be stepped down if the secondary voltage is stepped up; and if the secondary voltage is stepped down, the secondary current is stepped up.

Transformers are useful in providing electrical isolation between the primary and the secondary, since there is no electrical connection between the two windings. Energy is transferred completely by magnetic coupling. We will see capacitors, inductors, as well as transformers, used in the many electronic circuits.

Summary

Now that we know about dc and ac electricity, we understand the basics and we can move on to investigate the operation of diodes and transistors so that we can understand more detailed electronic circuits. Diodes and transistors, combined with resistors and capacitors, especially in integrated circuits, have contributed to the greatest advancement in electronics since the beginning of time.

Quiz for Chapter 3

1. A waveform on an oscilloscope screen that varies positively and negatively around an axis is:
   a. a dc voltage
   b. an ac voltage
   c. a constant voltage
   d. in phase

2. When a wire loop is caused to rotate in a magnetic field created by a single pair of magnets, every time it rotates it produces:
   a. two cycles (one for each pole)
   b. a sine wave cycle
   c. a smooth DC current
   d. a square wave of current

3. The average value of a full cycle of a sine wave is equal to:
   a. 0.707 times the peak value
   b. 0.637 times the peak value
   c. 2.0 times the peak value
   d. zero

4. Most ac voltmeters and ammeters are calibrated to read:
   a. effective RMS values of a sine wave
   b. average values of a sine wave
   c. twice the dc value
   d. 1/2 the dc value

5. If the frequency of a voltage applied to a capacitor is increased, the current through the capacitor will:
   a. become dc
   b. lower in frequency
   c. increase because its reactance decreases
   d. decrease because its reactance increases

6. The total opposition of reactance and resistance is called:
   a. equivalent resistance
   b. reluctance
   c. conductance
   d. impedance

7. The current is said to _____ the voltage in a purely inductive circuit.
   a. line up with
   b. lead
   c. lag behind
   d. cause

8. If a sine wave voltage and current reach their maximum positive value at different instants of time:
   a. they may be out of phase
   b. they may not be ac
   c. they are in phase
   d. they are 360 degrees out of phase

9. The impedance of a circuit is _____ either the resistance or the reactance alone.
   a. more than
   b. equal to
   c. less than
   d. twice

10. When the current is out of phase with the voltage, their product gives what is known as the:
    a. true power
    b. reactive power
    c. AC power
    d. apparent power

11. In a series circuit, if inductive reactance and capacitive reactance are equal and cancel, the circuit is:
    a. unstable
    b. saturated
    c. at resonance
    d. matched

12. A device which converts electrical power from one voltage-current level to another voltage-current level is a:
    a. transistor
    b. capacitor
    c. diode
    d. transformer

Answers:
1 b, 2 b, 3 b, 4 a, 5 c, 6 d, 7 c, 8 a, 9 a, 10 d, 11 c, 12 d

Questions and Problems for Chapter 3

1. What is the frequency of a sine wave that has a period of 0.05 seconds?

2. The voltage at the wall outlet is found to be 120 volts RMS. What is its peak value? What is its peak-to-peak value?

3. An ac circuit has a sine wave input of 50 volts peak. What is the average value of the output voltage?

4. What would a peak-to-peak sine wave voltage of 100 volts measure on a common ac voltmeter?

5. What is the reactance of an 8 µF capacitor at 200 Hz?

6. Find the opposition offered to 60 Hz current by a 0.5 H inductor.

7. A 0.4 µF capacitor is to be used in a speaker crossover circuit in series with a 25Ω resistor. What is the impedance of the circuit at 10 kHz?

8. A circuit contains 80 ohms of inductive reactance and 40 ohms of capacitive reactance in series with 30 ohms of resistance. What is the impedance of the circuit?

9. If the circuit in problem 8 is connected to the 120 volt wall outlet, what would the current be through the resistor?

10. A 0.8 µF capacitor in series with a 60 mH inductor has a resonant frequency of _____ Hz.

11. What turns ratio is needed in a transformer to deliver 28 volts to a circuit from the 120 volt power line?

12. If the primary voltage for a given transformer is 220 volts and the load requires 60 volts at 2 amperes, what will be the primary current? (Assume negligible transformer losses.)

CHAPTER 4
Diodes and Transistors – How They Work

Now that we are familiar with electricity, and how dc electricity has current in one direction only, and that the current changes direction continuously in cycles in ac electricity, we should be able to understand how semiconductor diodes and transistors operate. The operation of all semiconductor devices is based on the same principles. To begin to understand these principles, we will start with a discussion of diodes and the P-N junction.

Diodes: How the Simplest Semiconductor Devices Work

The simplest semiconductor device is a diode. As shown in *Figure 4-1,* it is made up of a junction of N and P semiconductor material. We will be concerned only with silicon diodes. Diodes are also made from other semiconductor materials, such as germanium and gallium arsenide. Diodes made from these materials work essentially the same way.

Rectification Is a Form of Switching

It is easy to see that a semiconductor diode is basically an electrically-controlled switch. As an example, consider the silicon diode as shown in *Figure 4-1*. The working part of the diode is a specially processed piece of silicon that has two regions—an anode (explained later as a P-type semiconductor region), and a cathode (explained later as an N-type semiconductor region).

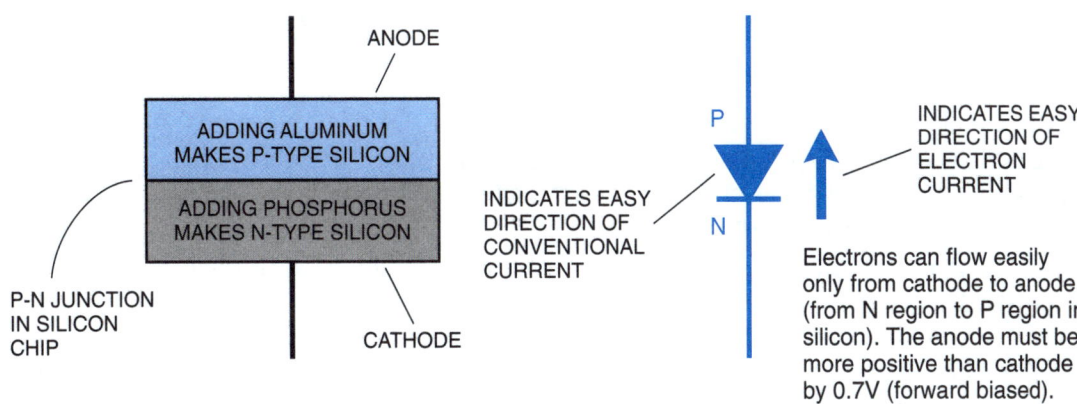

a. Physical Construction of a Silicon Diode   b. Schematic Symbol

Figure 4-1. The P-N junction in a diode chip acts as a one-way valve for electrons.

The diode acts as a one-way valve for current. The diode allows no flow of electrons (turns off) when it is reverse biased, but it allows an easy flow of electrons (turns on) when it is forward biased. Forward bias is when the anode is more positive than the cathode and above the threshold voltage of about 0.7 volt. Reverse bias means the cathode is more positive than the anode, trying to cause current in the reverse direction. Thus, *a diode is basically a form of automatic switch*—when forward biased, current is allowed; when reverse biased, no current is allowed.

The switching occurs in response to an electrical signal (the voltage bias across the diode), and it can happen very rapidly. The process of allowing current in one direction and allowing no current in the other direction is called *rectification*. Some diodes can rectify ac at frequencies measured in gigahertz, which means billions of cycles per second.

The Junction Between P-Type and N-Type Silicon Rectifies Current

Let's talk more about the piece of silicon in *Figure 4-1* with its two regions called the anode and the cathode that form the diode. *Electrons* pass easily from the cathode to the anode. (Electron current, which is indicated by the separate arrow, is opposite from the conventional current direction, which is indicated by the arrowhead in the schematic symbol). Electron current from anode to cathode is blocked.

In studying semiconductor devices, it is easier to understand the electronic operation when we think about the flow of electrons instead of conventional current. Something about the anode and cathode regions of the silicon chip allows electrons to flow from cathode to anode, but not the other way.

The Different P and N Regions

What is different about the P anode and N cathode regions of silicon? To begin with, the basic material from which the silicon chip is made is a single crystal silicon. That means all the atoms inside it line up in the same rows and layers all through the chip without any interruptions. However, some modifications have been made to the single crystal silicon crystal for each region. The anode region has a few aluminum atoms mixed in with the silicon. As a result, for reasons we will see later, the anode material is called *P-type silicon*. In like fashion, the cathode region of the crystal has a few phosphorus atoms scattered here and there. This type of material is called *N-type silicon*.

The place where the two types of silicon meet inside the crystal is called the *P-N junction*. What we will find out in this chapter is how a P-N junction acts as a one-way valve for electrons. This will help us understand how transistors work.

Each Silicon Atom Is Connected to Four Others by Covalent Bonds

Silicon is a chemical element; that is, silicon is one of the basic elements which are combined to make other substances. The rocks and soil of the earth probably contain more silicon than any other element. A grain of sand, for instance, is a quartz crystal which is made of silicon and oxygen. Pure silicon is obtained from sand by separating the silicon from the oxygen. Silicon is used to make semiconductor devices because of the special ways in which electrons flow among the atoms of a silicon crystal. These ways depend on how the atoms are connected together.

If we could look inside a piece of silicon crystal with a microscope with super magnification, we would see silicon atoms arranged in very even rows and layers. As shown in *Figure 4-2a,* each atom would look like a fuzzy, cloudy ball, with four fuzzy extensions that connect it to four other atoms. The ball part of an atom is called the atom's *core*. The extensions stretching between atoms are called *covalent bonds*. This particular arrangement of atoms in a crystal is called the *lattice*. This same model could represent any semiconductor crystal; for example, germanium, silicon or carbon. Yes, carbon, in the form of diamond, can be used as a semiconductor material that operates at extremely high temperatures.

Figure 4-2b shows the usual way of drawing a diagram of silicon atoms and covalent bonds in a crystal. The circles represent atoms or atomic cores, and the lines represent covalent bonds. The atoms are placed in a square pattern, with each one connected to four neighbors. In explaining how electrons flow through the crystal, *we can pretend that the crystal is a single flat layer* of silicon atoms arranged in squares instead of a three-dimensional model with the bonds in the X, Y and Z directions.

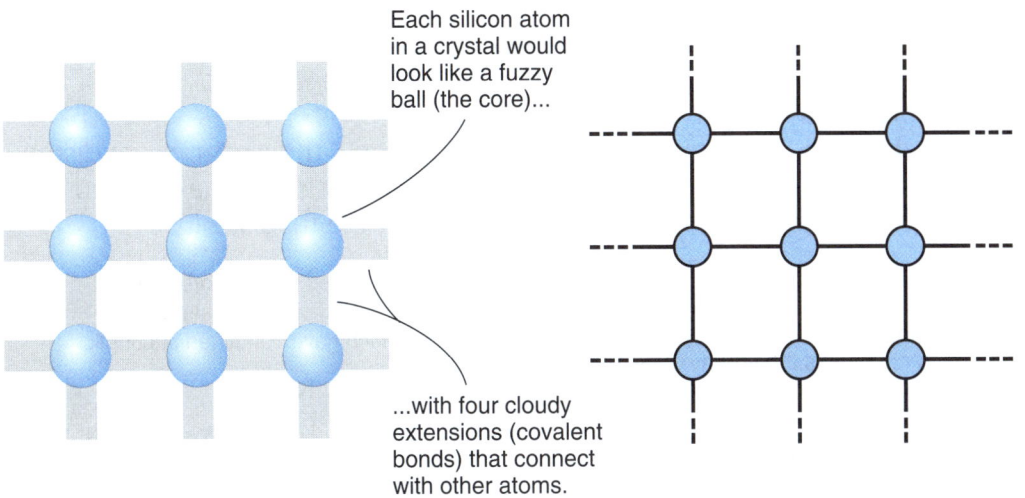

a. Atoms with Covalent Bonds   b. Usual Diagram of Silicon Crystal

Figure 4-2. Each atom of a silicon crystal is connected to four other atoms by covalent bonds in an orderly arrangement called the diamond lattice. For simplicity, we can use a simple, flat diagram with atoms in a square pattern.

An Atom's Electrons Are Arranged In Shells

To understand how electrons flow in a semiconductor crystal, we have to see how covalent bonds work. Those bonds are a result of the way the electrons of each atom are arranged.

As stated in Chapter 1, an atom consists of a tiny, positively charged nucleus surrounded by a swarm of negatively charged electrons. The speeding electrons are held in orbits around the nucleus by electrostatic attraction. The nucleus receives its positive charge from positively charged protons. Each chemical element (hydrogen, oxygen, and silicon, for example) has a different number of positive protons in its nucleus. In a normal atom, there are just as many negatively-charged electrons as there are positively-charged protons. So the entire atom is neutral in charge.

Now, as shown in Chapter 1, an atom's electrons do not orbit just anywhere at random around the nucleus. Instead, the orbits in all atoms follow a certain plan. That plan is determined by a set of rules from physics called quantum mechanics.

Electron Orbits

Let's review the general plan for electron orbits in all atoms started in Chapter 1. We repeat the customary model shown in *Figure 4-3*. The actual orbits of electrons may be oval-shaped, and they do not all lie in the same plane. But in this model, *each electron's orbit is a circle* that represents the electron's average distance from the nucleus.

In *Figures 4-3a* and *4-3b,* the radius or height of a circular orbit also represents the *energy* that the electron has in its actual orbit. *The greater the height of the orbit above the nucleus, the more energy the electron has*. This is because it takes energy to raise an electron to a higher orbit against the electrostatic pull of the nucleus.

The plan for all atoms is that *the electrons are permitted to have only certain amounts of energy*. In this model, that means orbits with certain heights. *The permitted orbits are grouped in separate layers called shells*. In *Figure 4-3b,* the shells are shown as thick, shaded circles. The shells (K, L, M, N, O) and their contained subshells were shown in *Figure 1-2b.*

There are empty gaps between the shells. That means it takes a certain amount of energy to raise an electron from a permitted orbit in one shell or subshell to a permitted orbit in the next higher shell or subshell.

Look at the silicon atom in *Figure 4-4b*. Notice how the electrons are distributed in the shells. The silicon atom has 14 protons and 14 electrons. The first and second shells are full, and there are four valence electrons in the third shell.

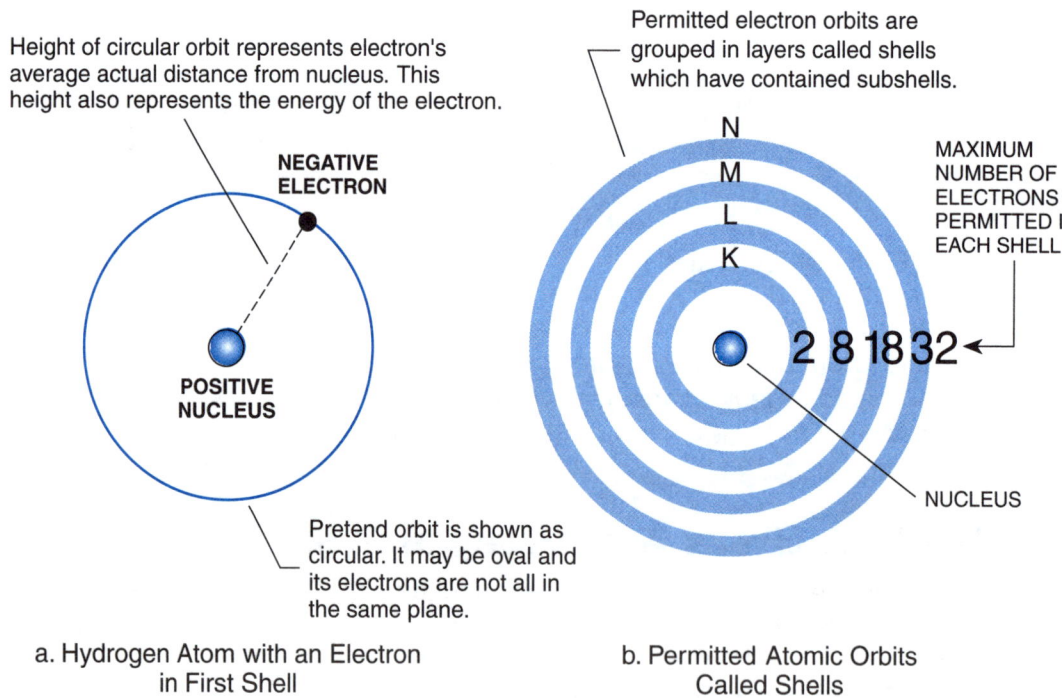

a. Hydrogen Atom with an Electron in First Shell

b. Permitted Atomic Orbits Called Shells

Figure 4-3. The electrons of an atom are permitted to have only certain energies (orbit heights), grouped in layers called shells. All atoms follow the same plan of shells and subshells dictated by physical laws.

Atoms Prefer to Have Full Valence Shells

What does all this have to do with covalent bonds and the flow of electrons in a semiconductor crystal? Well, *an atom uses the valence electrons of its outermost shell or subshell to form bonds* with other atoms. These bonds determine whether the material conducts current, and if so, how. The bonds also determine the kind of chemical reactions in which the element engages.

Example 1. Determining Distribution of Orbiting Electrons

If an oxygen atom has an atomic number of 8 and an arsenic atom has an atomic number of 33, how are the orbiting electrons distributed around the nucleus?

The atomic number identifies the number of positive protons in the nucleus, and, as a result, the number of orbiting electrons permitted in the following shells:

Shell		K	L	M	N	O	
Subshells	s	2	2	2	2	2	
	p		6	6	6	6	
	d			10	10	10	
	f				14	14	
Maximum Orbiting Electrons Permitted		2	8	18	32	32	
Oxygen has 8 Orbiting Electrons		2	6				Oxygen has 2 electrons in first (K) shell and 6 in second (L) shell.
Arsenic has 33 Orbiting Electrons		2	8	18	5		Arsenic has 2 electrons in first (K) shell, 8 in second (L) shell, 18 in third (M) shell, and 5 in fourth (N) shell.

The reason that the outer electrons form bonds is that *atoms prefer to have the outer subshell or shell full of electrons*. The other shells down below, if there are any, are already full. An atom forms bonds only if the outer shell is not filled with its maximum permitted number of electrons. This has to do with the fact that *when a shell or subshell is filled, its electrons are held especially tightly* by the nucleus. That is why elements like helium (atomic number 2), neon (atomic number 10) and argon (atomic number 18, shown in *Figure 4-4c)* are very stable elements. Their outer shells or subshells are full and they don't particularly want to react with other atoms.

Valence Electrons and the Atom's Core

Look at the sodium atom and the chlorine atom in *Figure 4-4*. There is a special name for an atom's outer subshell or shell if it is not full. It is called the atom's *valence shell*. "Valence" means the number of bonds the atom forms. For instance, the valence of silicon atoms in a crystal is four, because every atom forms four bonds. As mentioned previously, the electrons in the valence shell are called the atom's *valence electrons*.

The rest of the atom, consisting of filled shells and the nucleus, is what is called the *core*. Remember, shells filled with electrons don't have anything to do with bonds, chemical reactions, or current, so they can be considered as separate from the valence electrons. The core has a positive charge equal to the number of electrons in the valence shell.

For example, look again at the sodium atom in *Figure 4-4*. Its outermost shell contains only one electron. Since this shell is not filled to its capacity, it is the atom's valence shell. The nucleus and the filled first and second shells are the core of the sodium atom. The core has a positive charge of plus one, which is balanced by the negative charge of the single valence electron. The chlorine atom, on the other hand, has a core with a positive charge of plus seven because it has seven electrons in its valence shell.

When nearly all the orbits of an atom's valence shell are empty, the atom easily gives up the few electrons in that shell. So the single electron in a sodium atom's valence shell is not bound (tied) very tightly to the core.

The chlorine atom's valence shell has seven of the eight electrons that it desires, so a chlorine atom can easily grab and hold one extra electron in its valence shell. If a chlorine atom bumps against a sodium atom, *the chlorine atom steals the sodium atom's single valence electron*. In this way, the chlorine atom achieves a full outer shell, and the sodium atom ends up with an outer shell completely void of electrons.

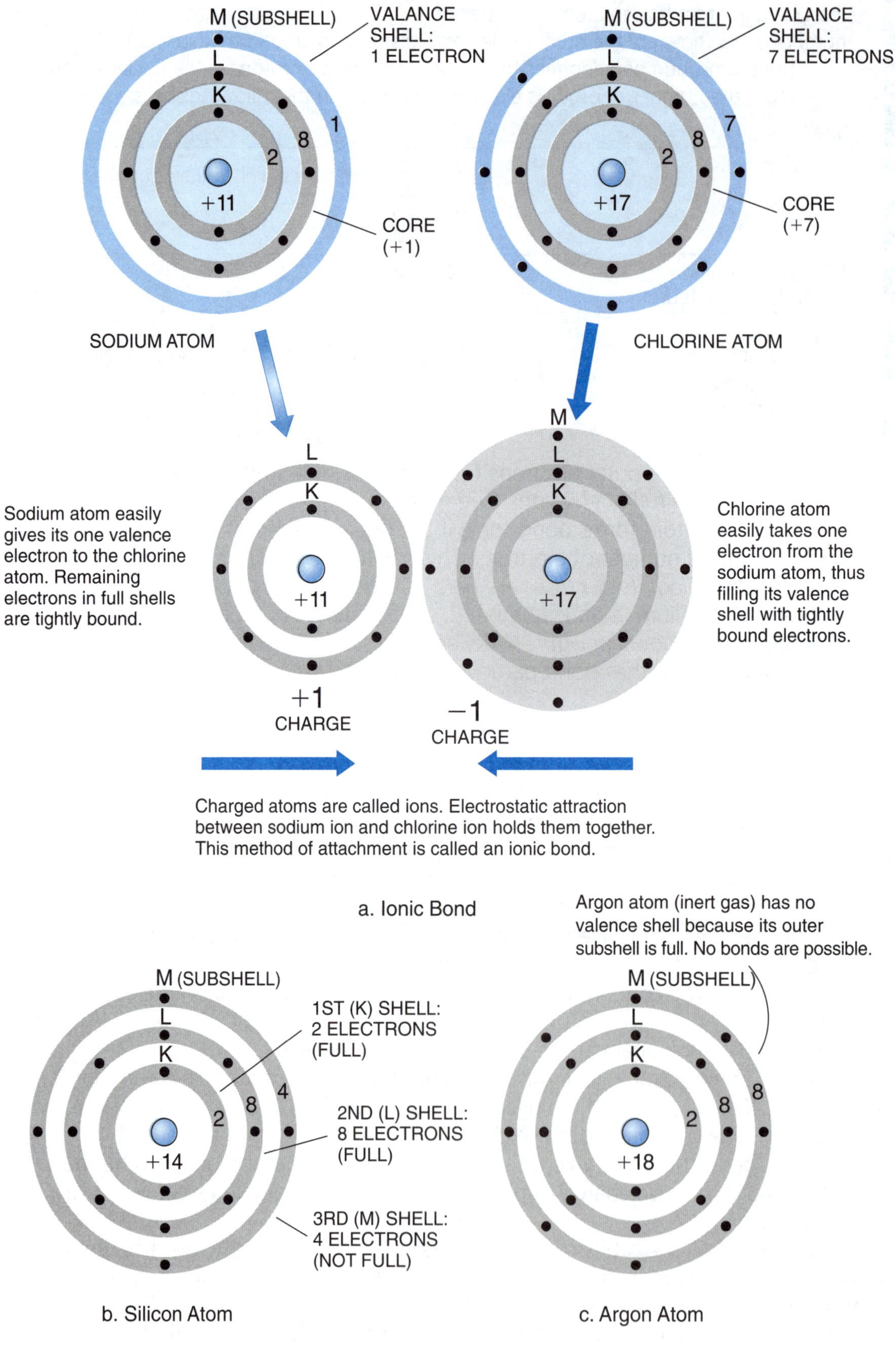

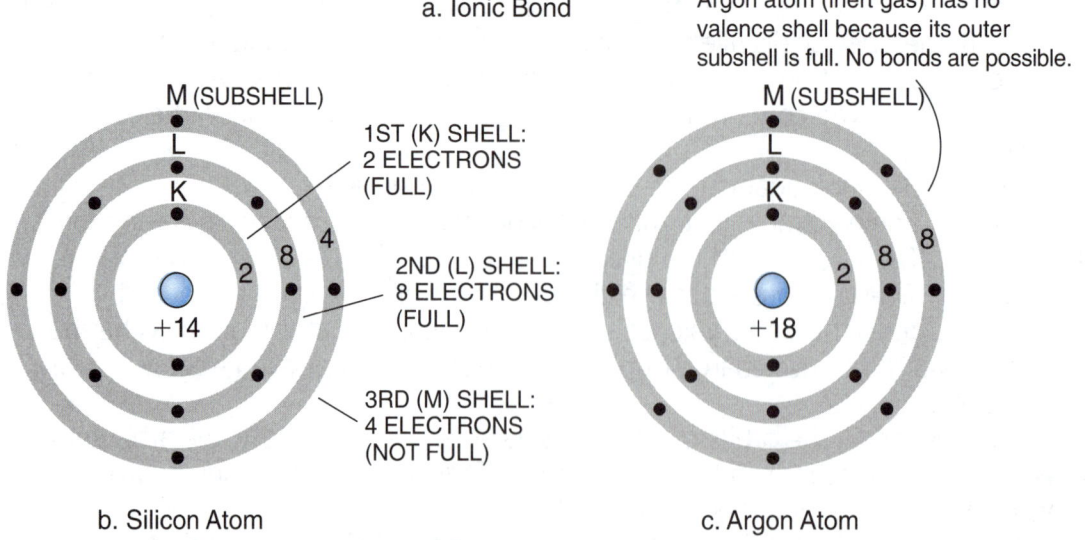

Figure 4-4. Atoms form bonds because they prefer to have full outer subshells or shells. An unfilled outer subshell or shell is called the valence shell.

The result is a type of bond between the two atoms. The sodium atom has a single positive charge because an electron is missing. And the chlorine atom has a negative charge because it has one extra electron. Because of their opposite charges, the two atoms attract each other and stick together. Charged atoms are called *ions*, and this kind of bond between atoms is called an *ionic bond*.

Example 2. Determining Which Atoms Are Good Conductors

Which of the following atoms will be good conductors and which will be good insulators of electricity?

Atom		Atomic No.	Shell					
			K	L	M	N	O	
			Orbiting Electrons					
Na	Sodium	11	2	8	1			Good Conductor, one electron in outer subshell
Ca	Calcium	20	2	8	8	2		Good Conductor, two electrons in outer subshell
Kr	Krypton	36	2	8	18	8		Insulator, outer subshell filled
Xe	Xenon	54	2	8	18	18	8	Insulator, outer subshell filled

A Covalent Bond Consists of Two Shared Valence Electrons

Now we can see how silicon atoms form covalent bonds. Look at the silicon atom in *Figure 4-5a*. Its valence shell is in the third (M) shell, which has four electrons, but desires four more electrons. The core of a silicon atom is the nucleus and the filled first and second shells. The core has a charge of plus four, which is balanced by the four negative electrons in the valence shell.

A silicon atom forms bonds because it tries to get four more electrons to fill up the valence shell. But it does not do this by grabbing additional electrons, as a chlorine atom does. Instead, *the atom shares its four valence electrons* with four neighboring atoms in the silicon crystal.

Figure 4-5b shows how this sharing creates a covalent bond between two silicon atoms. *A covalent bond consists of two valence electrons orbiting around the cores of both atoms*. Each atom provides one of the electrons. These two electrons occupy two of the eight permitted orbits in the valence shells of both atoms. Each electron occupies a permitted orbit in both atoms.

In this way, a covalent bond gives each atom one of the four additional electrons that it desires to fill up its valence shell. But neither atom has a charge, as in the case of sodium and chlorine. This is because neither atom has actually given up an electron. Each atom still owns four valence electrons—no more, and no less.

The two orbiting electrons hold the two cores together. Meantime, the positively charged cores hold each other apart due to their electrostatic repulsion. Since a silicon atom has four valence electrons and room for four more, *Figure 4-6c* shows how an atom forms four bonds with neighboring atoms in a silicon crystal. Thus, all eight permitted orbits in every atom are occupied by electrons. As in any full subshell or shell, these electrons are held very tightly in their orbits.

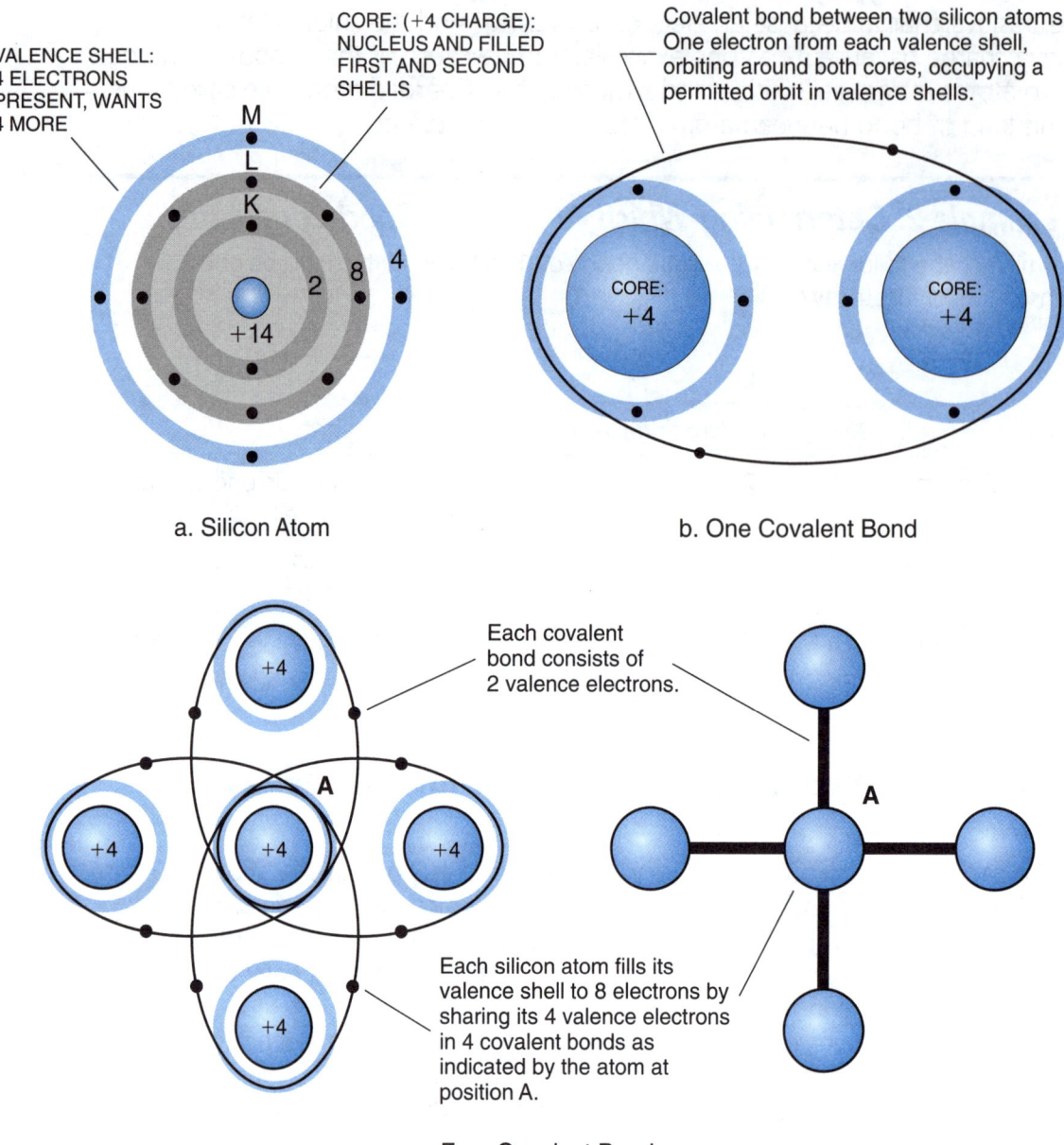

Figure 4-5. How covalent bonds hold silicon atoms to each other in a crystal

A Phosphorus Atom's Fifth Valence Electron Gets Squeezed Out

Now we come to the N-type and P-type silicon that we mentioned at the beginning of the chapter. In order for a material to conduct current, it must contain tremendous numbers of electrons that are able to move from one atom to another. The movable electrons are called *carriers* of current. But in a pure silicon crystal at low temperatures, all the electrons are tied down in their orbits. Therefore, *pure silicon, or any other semiconductor material, is an insulator*. It is useless as a material for electronic devices.

This is where the aluminum and phosphorus come in, as we mentioned earlier. Adding a tiny amount of either element is called *doping* the silicon. The added material is called the *dopant* material. First, let's see what happens when we dope a piece of silicon crystal with phosphorus, producing N-type silicon.

As shown in *Figure 4-6a,* a phosphorus atom has 15 electrons and protons, one more than a silicon atom has. So the valence shell of a phosphorus atom has *five electrons instead of four*. The valence shell lacks three electrons to be full.

As shown in *Figure 4-6b,* each phosphorus atom takes the place of a silicon atom in the crystal. It is surrounded by four silicon atoms, each one eager to form a covalent bond with it. So the phosphorus atom does just that. Its four neighbors grab four of its five valence electrons and force the phosphorus atom to form four covalent bonds.

But that leaves no room in the phosphorus atom's valence shell for the fifth electron. The atom's valence shell is filled with four electrons of its own and four shared with its neighboring silicon atoms. *So the fifth valence electron is literally squeezed out of the valence shell and becomes a "free" electron.* This electron is not held very tightly by the positive charge of its atom's core. The electron is much like the single valence electron of a sodium atom that we discussed earlier. Each phosphorus atom contributes one free electron to conduct current. The atom would easily turn loose of the electron and be satisfied with its valence shell, which has been nicely filled due to the four covalent bonds.

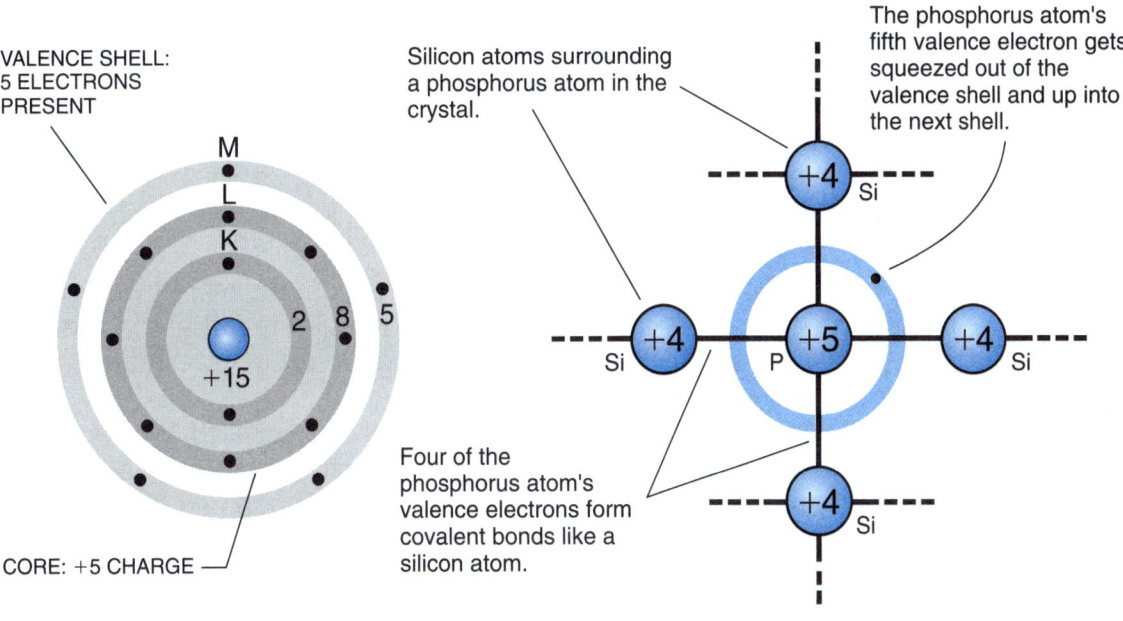

a. Phosphorus Atom   b. Phosphorus Atoms Take Place of Silicon Atoms

Figure 4-6. A phosphorus atom in a silicon crystal takes the place of a silicon atom, forming four covalent bonds. But it has one more electron, which pops up into the next shell of the phosphorus atom.

"Free" Electrons Drift to Form Current

So the electron can easily move over to a similar orbit of a neighboring silicon atom. In fact, the atoms are so close together that their outer shells actually blend together. Thus, the electron is what we call a *free electron*. It carries with it a negative charge. It gives silicon atoms a negative charge as it orbits in the outer shell, and the phosphorus atom that it leaves behind is left with a positive charge. But very quickly, a drifting electron comes into its outer shell and equalizes the charge.

As shown in *Figure 4-7,* when an electric field is applied to the N-type crystal due to a voltage difference across it, the free electrons drift mainly in the direction they are pushed by that field. Thus, the free electrons act as negatively charged carriers of electric current.

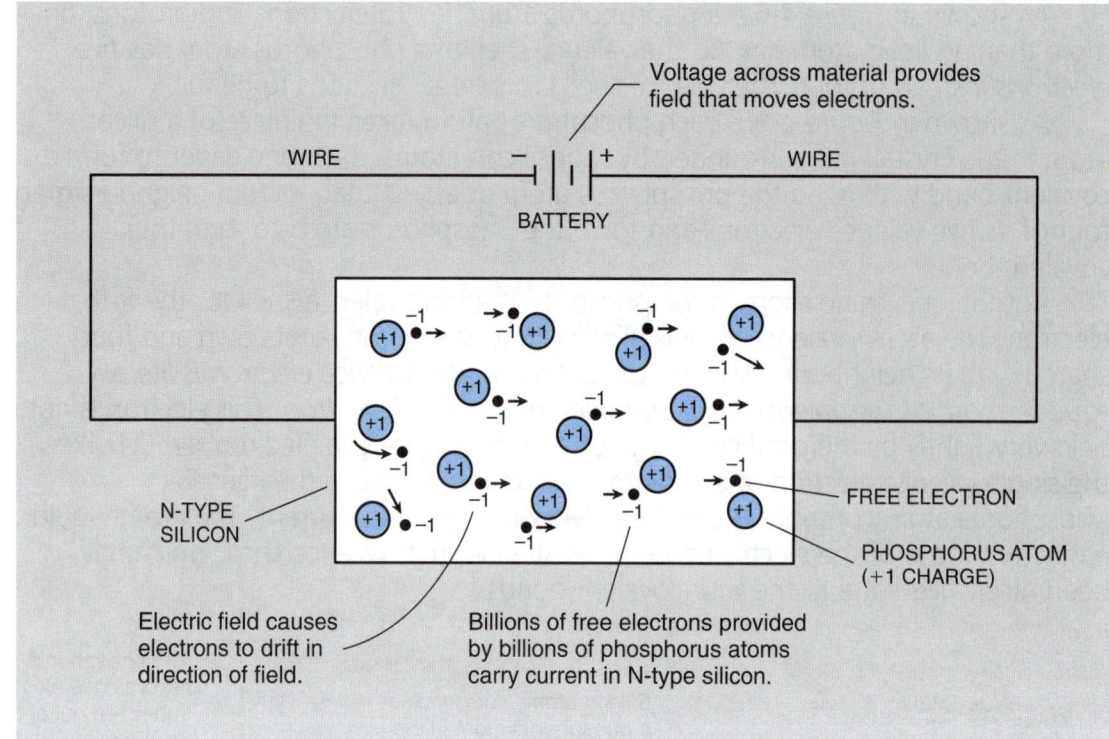

Figure 4-7. How "free" electrons flow in N-type silicon

N-Type Silicon

The more phosphorus atoms there are in a piece of silicon, the more free electrons there are, and the lower the electrical resistance of the silicon. But silicon used in semiconductor devices is never doped heavily enough to make its resistance as low as metals such as copper.

So now we know why silicon doped with phosphorus is called *N-type silicon*. The "N" stands for the *negative* charge of the free electrons that serve as *current carriers*. Notice that this *does not mean the N-type crystal has a negative charge*. It is normally *neutral in charge*, because there are still just as many protons in the nuclei as there are orbiting electrons.

An Aluminum Atom Causes a Positive Hole Among the Valence Electrons

Now let's see what happens when we dope a piece of silicon with aluminum instead of phosphorus. Remember, that makes P-type silicon. As shown in *Figure 4-8a,* an aluminum atom has only 13 electrons and protons, which is one less than a silicon atom has. So, *an aluminum atom has only three electrons in its valence shell*.

Like a phosphorus atom, an aluminum atom takes the place of a silicon atom in the crystal. Surrounded by four silicon atoms that are offering electrons to share, the aluminum atom desperately wants to form four covalent bonds because it wants to fill its valence shell. However, the aluminum atom has only three valence electrons to share, so it can form only three bonds, with three of its four neighbors.

After forming those three bonds, the aluminum atom acts much like the chlorine atom that we studied earlier. It needs only one more electron to be able to form a fourth bond. That would let it finish filling the eight available orbits of its valence shell. *The aluminum atom easily steals one electron from the valence shell of a silicon atom next door*. We can think of this as happening by the flipping of a covalent bond, as shown in the snapshots of *Figure 4-8b* and *4-8c*.

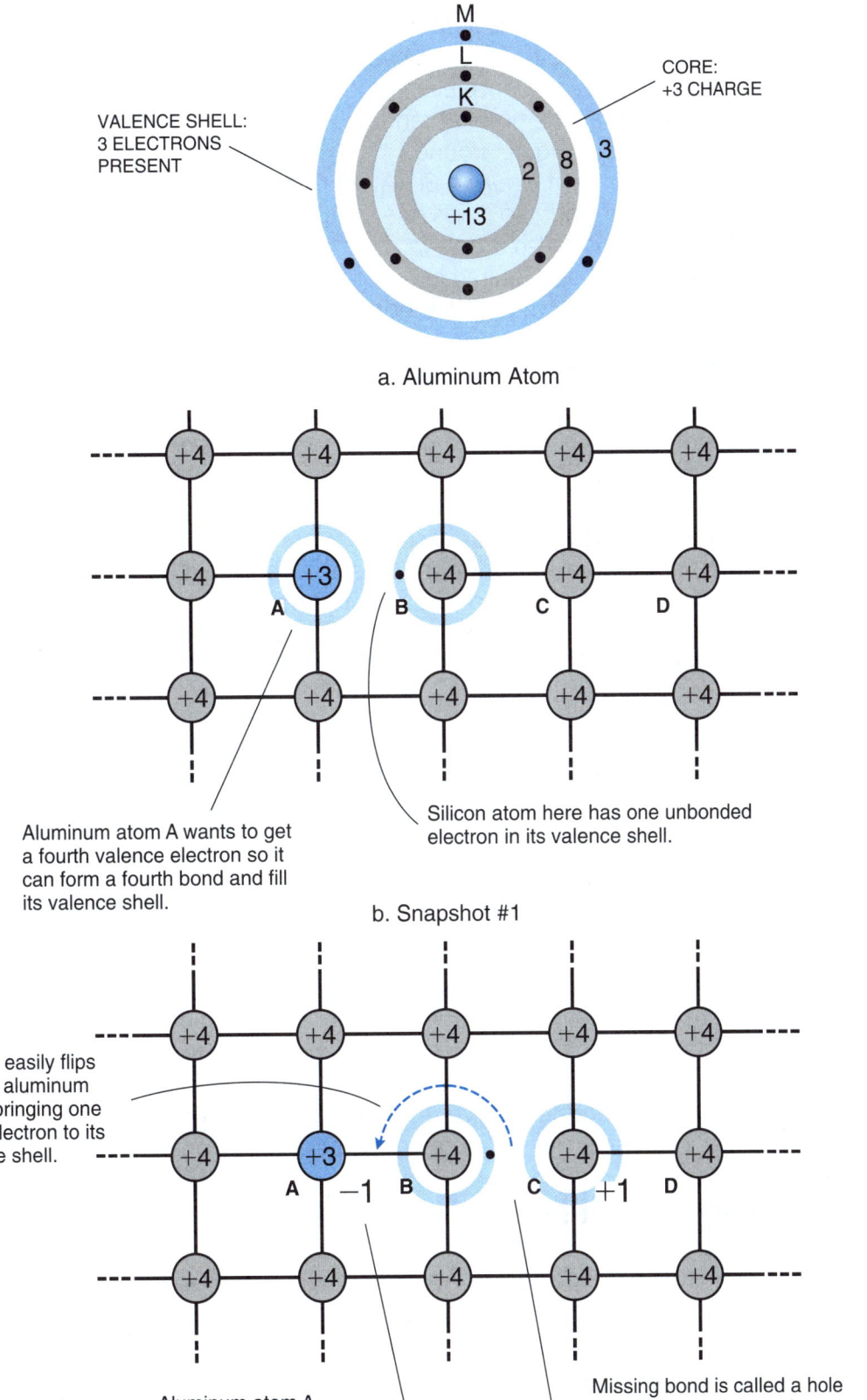

Figure 4-8. An aluminum atom takes the place of a silicon atom in the crystal. It steals an electron to form a fourth bond, creating a hole in the crystal lattice.

In the first snapshot of *Figure 4-8b,* the aluminum atom (Atom A) has only the three bonds that it can form originally. In the second snapshot an instant later, a bond between Atoms B and C has flipped over to the left, between Atoms B and A. Along with the bond comes a valence electron that did belong to Atom C. This sort of bond-flipping goes on all the time, at random. It is partly a result of the heat in the crystal, making all the atoms jiggle around.

In the second snapshot of *Figure 4-8c,* Atom C is missing one negative electron as well as one of its four bonds, so it has one positive charge. *This missing bond, with one positive charge, is called a hole in the crystal lattice.* Meantime, the aluminum atom (Atom A) has acquired one negative charge with the extra electron that it desired to form the fourth covalent bond.

Holes Act As Positive Carriers of Current

The remarkable thing about a hole is that *it easily drifts around in the crystal*. And whenever a hole moves one step in one direction, *a valence electron slips one step in the opposite direction*. Actually, the movement of an electron to form a bond produces a missing bond (a hole) in another location. As electrons move in one direction, holes move in the opposite direction.

For instance, in the snapshot of *Figure 4-8c,* the bond that linked Atoms B and C has flipped to the left, filling the hole between Atoms A and B of the first snapshot. A valence electron has moved to the left, and the hole (positive charge) has moved to the right, to Atom C.

P-Type Silicon

Thus, holes (missing bonds) provide a way for valence electrons to flow and form an electric current, but in doing so, *holes act as positively charged carriers of current*. Thus, the doped material is called a P-type silicon. The "P" stands for the *positive* charge of the holes (missing bonds) that serve as *current carriers*. When an electric field moves valence electrons to the left as in the sequence of snapshots 1 and 2, holes with a positive charge move to the right. It is as though the holes were actually some sort of positively charged objects that move.

Each aluminum atom in a piece of silicon gains a covalent bond with its resultant negative charge that stays with the aluminum atom and contributes a hole with a neighboring silicon atom. The more holes that this piece of P-type silicon has, the more easily it carries current. As with N-type silicon, the more heavily the material is doped, the lower its electrical resistance. As with the N-type, the P-type silicon is *neutral in charge*, because it has no more protons than electrons.

P-Type Silicon with Applied Electric Field

Figure 4-9 shows how holes carry current through a piece of P-type silicon. This is similar to *Figure 4-7* for N-type silicon. An electric field is applied to the P-type crystal by connecting a voltage difference across it. The billions of holes in the doped crystal rush across it in the opposite direction from the flow of electrons in the wires.

As valence electrons flow out into the wire at one end, *new holes are created there*. And as holes arrive at the other wire, *incoming electrons fill up the holes at that end*. Positive holes are created at one end of the crystal just as fast as they are filled at the other end. So the total number of holes remains constant, and the crystal remains neutral in charge.

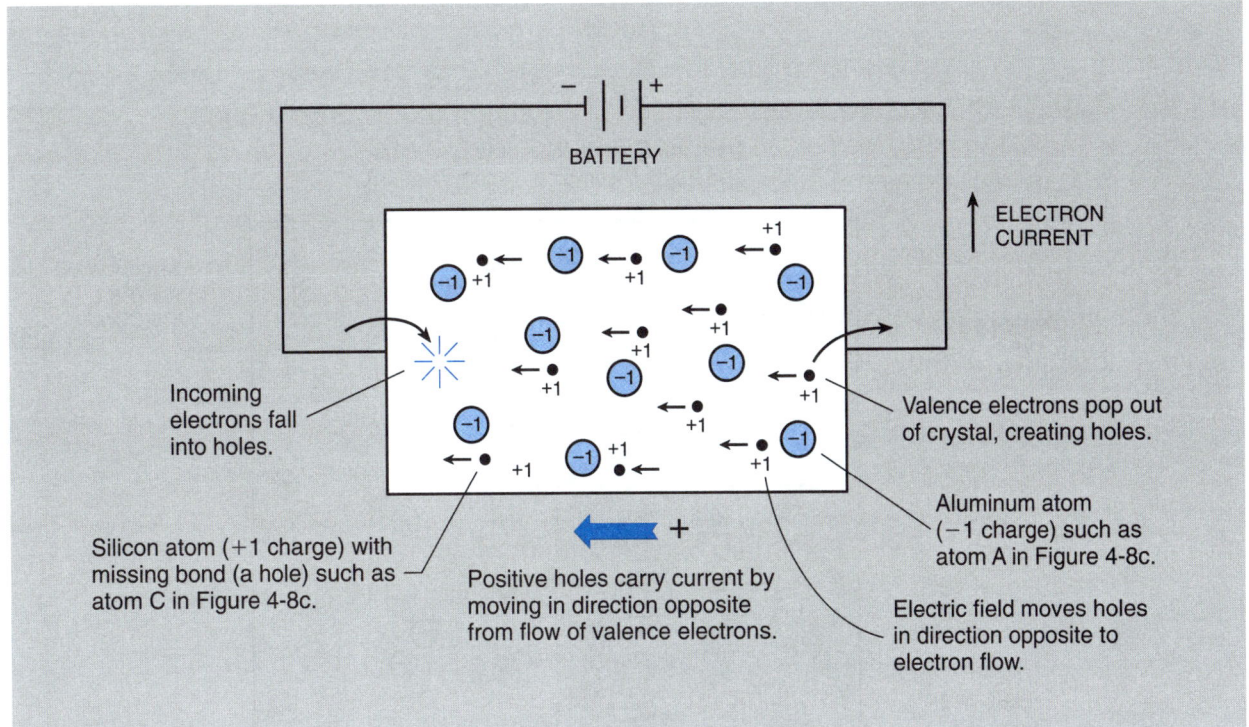

Figure 4-9. How valence electrons flow in P-type silicon, by positive holes moving the other way

Combining P-Type and N-Type Materials at a Junction

Figure 4-10 shows a P-N junction where the two types of semiconductor materials are joined. Recall that N-type semiconductors have phosphorous atoms with a "free" electron in its outer ring as the carrier of current. When the free electron leaves the atom, the atom will have a positive charge. For the P-type semiconductors, aluminum atoms have a missing bond (a hole). When an electron fills the hole, the atom has a negative charge.

When the junction is formed, and with no external field due to an applied voltage, right at the junction the "free" electrons of the phosphorous atoms combine with the holes of the aluminum atoms. This is called *recombination*. The atoms on each side of the junction *cannot move* because they are fixed in the crystal, so they form a potential barrier on each side of the junction—a positive potential on the N-type side and a negative potential on the P-type side. As the recombination continues, the potential region expands. The region is called the *depletion zone* because all free electrons and holes recombine and *all carriers are depleted* from the zone. As shown in *Figure 4-10*, the resultant negative potential barrier prevents any electrons from the N-type semiconductor from entering the depletion zone, and the positive potential barrier prevents any holes from the P-type semiconductor from entering the depletion zone. The result is a stabilized depletion zone and no more movement of carriers.

Forward Bias Pushes Holes and Free Electrons Together

Let's gradually apply a bias voltage across the diode chip as shown in *Figure 4-11* in what is called the forward direction. That is, we gradually make the cathode (N region) more and more negative in voltage with respect to the anode (P region). The electric field produced draws the negative electrons from the N-type cathode and the positive holes from the P-type anode toward the junction.

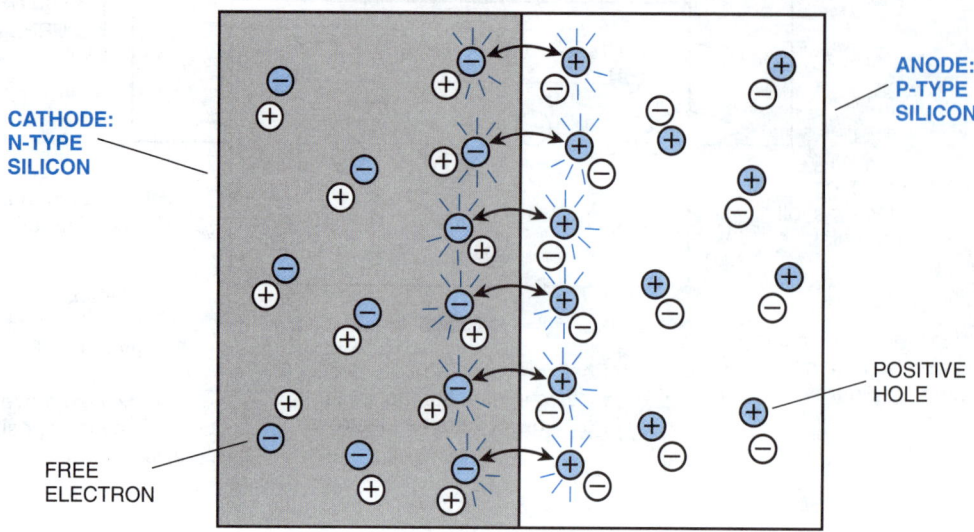

a. Recombination at Junction

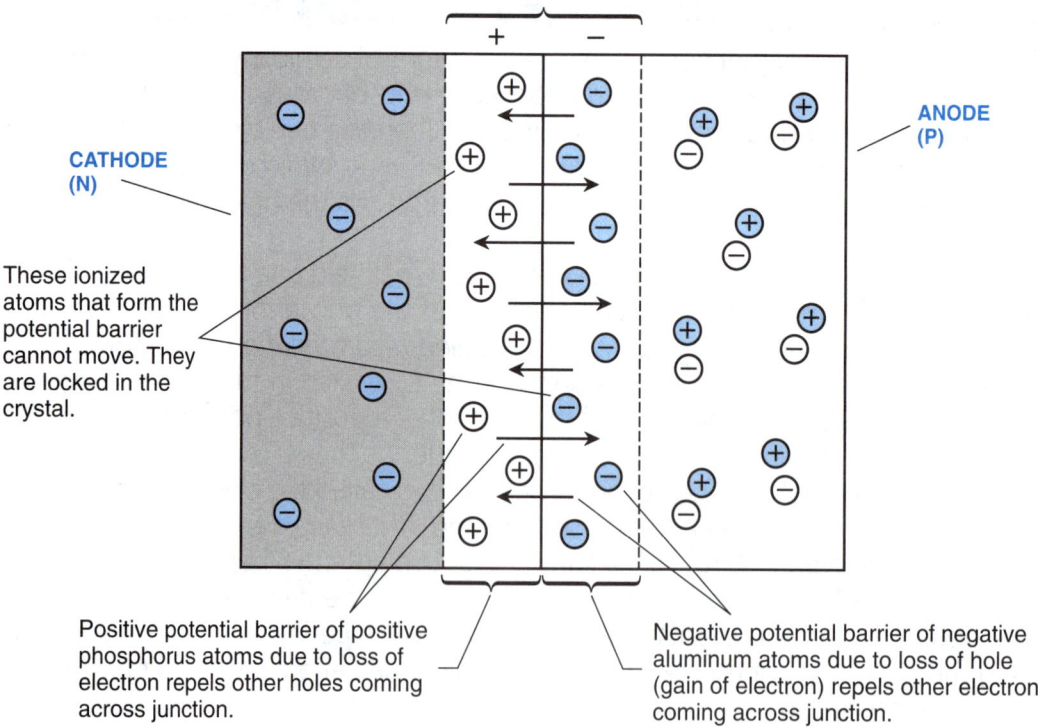

b. Depletion Zone

Figure 4-10. Free electrons and holes recombine near the P-N junction. This creates a depletion zone where there are no carriers and where potentials prevent other carriers from coming into the zone.

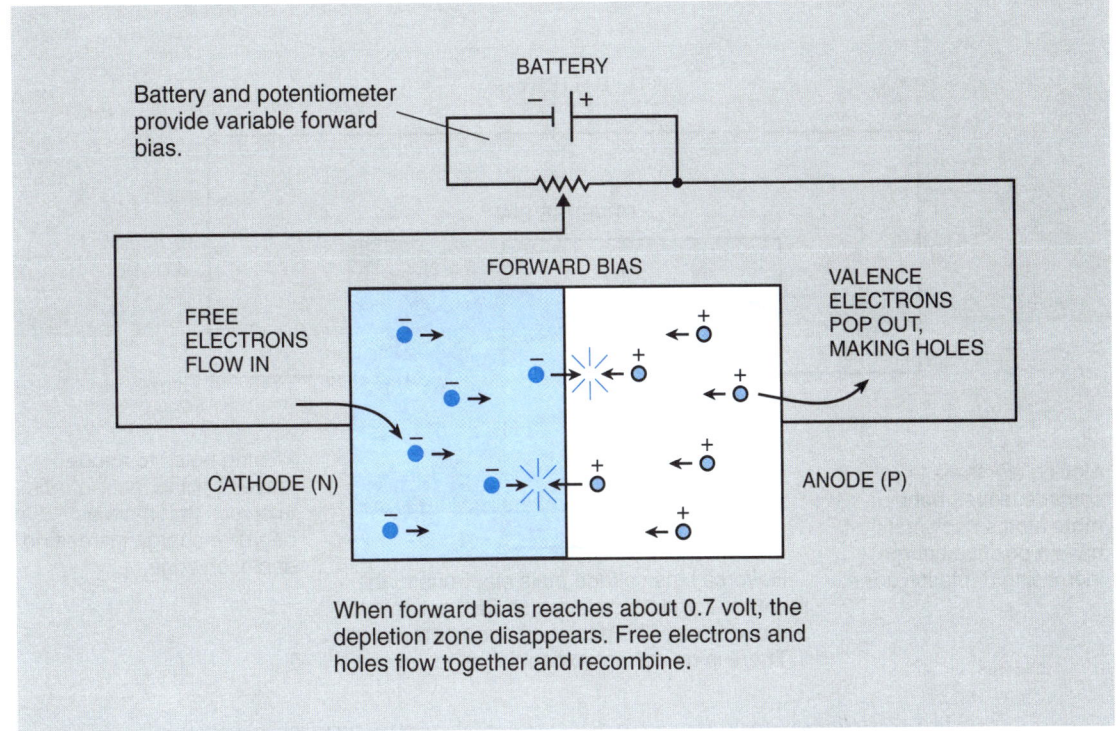

Figure 4-11. How a diode conducts current in the forward direction when bias reaches about 0.7 volt.

At first, we have free electrons and holes looking at each other from opposite sides of the depletion zone. But as the voltage is increased, the applied voltage field overcomes the potential barrier of the depletion zone. That is, as the voltage is increased, the depletion zone becomes thinner and thinner, but there is still no current. The depletion zone is like an insulating barrier between the cathode and anode regions, because no carriers can cross it until the potential barrier is overcome.

Finally, when the voltage difference (the forward bias) reaches about 0.7 volt, *the depletion zone thins down to nothing and disappears*. Then there is nothing to keep free electrons and holes from crossing the junction and recombining. They recombine happily, and there is forward current through the diode. When the forward bias increases above 0.7 volt, the current increases rapidly. Small changes in voltage cause large changes in current.

This forward bias of about 0.7 volt is called the *threshold voltage* of a silicon P-N junction diode. The threshold voltage is the same for all silicon diodes of this type. It is determined by the amount of energy that is required to raise a silicon valence electron to the conduction shell.

Reverse Bias Pulls Holes and Free Electrons Apart

Now consider what happens at the P-N junction when a reverse bias is applied across the diode as shown in *Figure 4-12*. In the P region, holes move toward the anode terminal, away from the junction; in the N region, free electrons move toward the cathode, away from the junction. Electrons are pulled from the phosphorous atoms and holes from the aluminum atoms. As a result, the depletion zone becomes wider instead of thinner. It still acts as an insulator. There is no way for great numbers of electrons to flow across the depletion zone, because there are no carriers there. This is how a diode acts as a one-way valve, turning off when reverse voltage bias tries to cause current in the reverse direction.

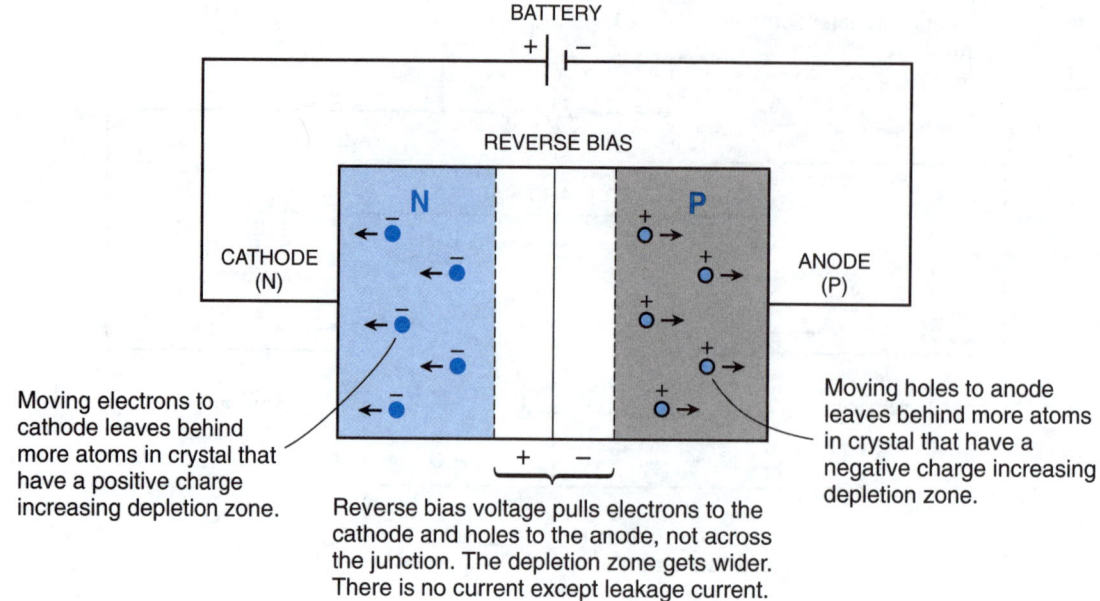

Figure 4-12. Why a diode does not conduct current in the reverse direction.

Example 3. Plotting Current Against Voltage for P-N Junctions

Plot the current against voltage for a silicon P-N junction both for forward bias and reverse bias.

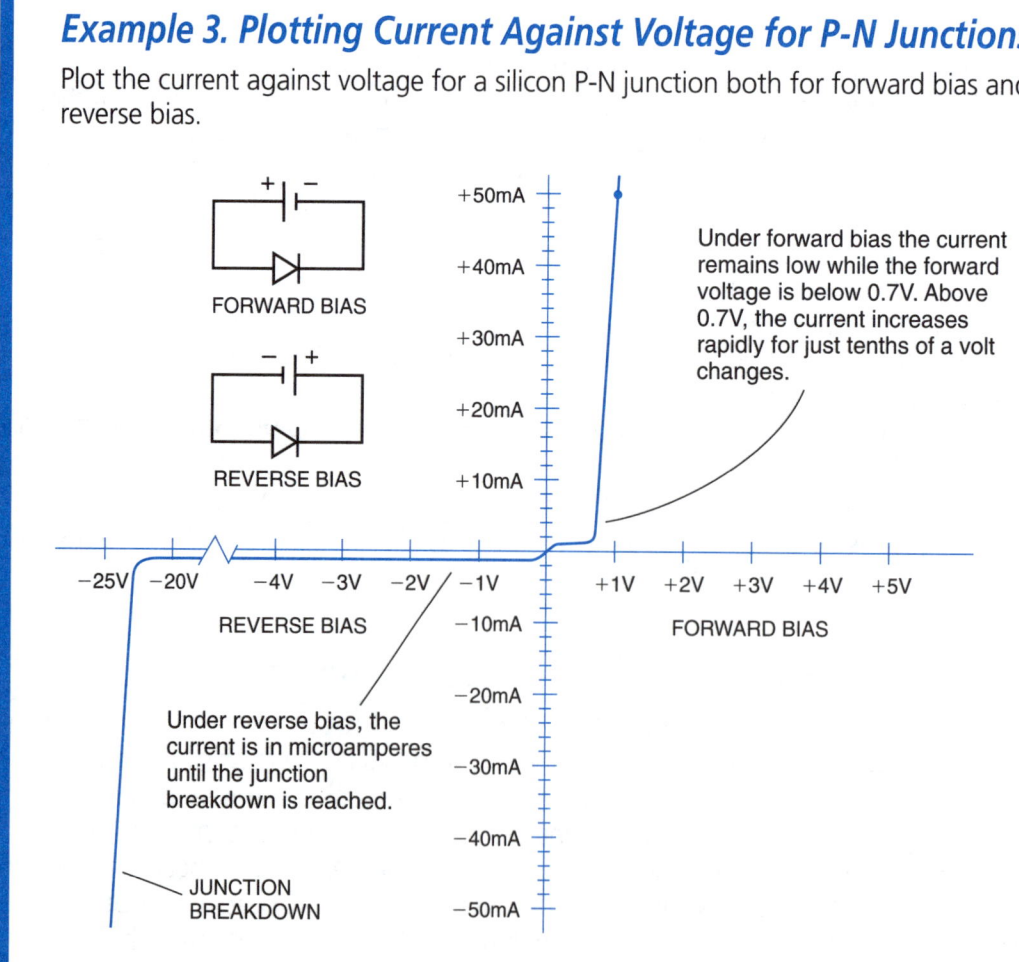

All Semiconductors Devices Operate by Means of Holes and Free Electrons

Now you have seen the basic concepts of semiconductor action illustrated in the simplest form of semiconductor device. It is the special behavior of holes and free electrons that makes all semiconductor devices work. Next, we will see how these special carriers of current allow a transistor to act as a switch or an amplifier.

Typical Discrete NPN Transistors Have Wide, Flat Regions

First, let's talk about how typical NPN transistors are made. Let's look at *Figure 4-13*. The NPN transistor schematic symbol of *Figure 4-13a* suggests that the three silicon regions are segments of a bar as shown in *Figure 4-13b*. That is actually how some of the first transistors were made, about the year 1948. The bar was a single crystal of semiconductor material doped into the different regions as the transistor was pulled from a liquid. It was not silicon initially, but a similar element which was easier to work with, called germanium. The three connections of the transistor are emitter, base and collector.

A more modern structure is shown in *Figure 4-13c*. Here the doping is diffused into single crystal material to modify the selected material into N-type or P-type material. The areas to be modified are selected by photographic techniques in steps so the operations are performed in proper sequence. The diffusions are made in special furnaces.

In an NPN transistor, the emitter and collector regions of the crystal are N-type, like the cathode of a junction diode. And the base region in the middle is P-type, like the anode of a junction diode. A typical transistor mounted in a package is shown in *Figure 4-13d*. It is used for power amplifier and high-current applications.

How an NPN Transistor Works

To understand how a transistor works, let's look at the circuit shown in *Figure 4-14*. In the circuit, the power supply or source of electrons is a set of batteries supplying three volts. Note that the transistor emitter is at –3 volts. One of the light bulb terminals and one of the battery terminals are at ground or 0 (zero) volts. The negative voltage at the emitter will supply free electrons to the emitter which flow through the transistor, the lamp, and to ground, which is the other side of the battery.

The transistor controls current to the lamp, in proportion to a smaller current of electrons allowed to flow out of the base. This small base current is controlled by the variable resistor (potentiometer) R2. R1 is in the circuit to limit the maximum positive voltage on the base with respect to the emitter. The base voltage will not be able to get more positive than –2 volts.

Both P-N Junctions Reverse Biased

As we begin the imaginary experiment, the potentiometer wiper contact is turned all the way to the end which is connected to –3 volts. This keeps the base at the same voltage as the emitter, and the P-N junction of base and emitter is shorted. *The transistor can be represented by two P-N junctions in the chip which act as two junction diodes facing opposite directions.* Since the base is at –3 volts, the collector-base P-N junction is reverse biased. With the emitter-base junction shorted, neither junction has any forward bias so the free electrons and holes are kept apart by a depletion zone at each junction. There is no current through either junction, so there is no current from emitter to collector. As a result, the transistor at this time is like a switch which is turned off.

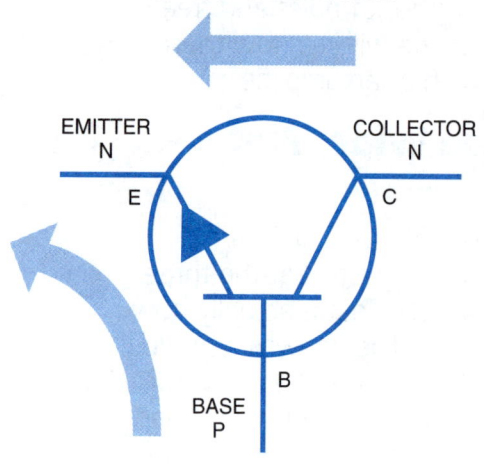
a. Schematic Symbol of NPN Transistor

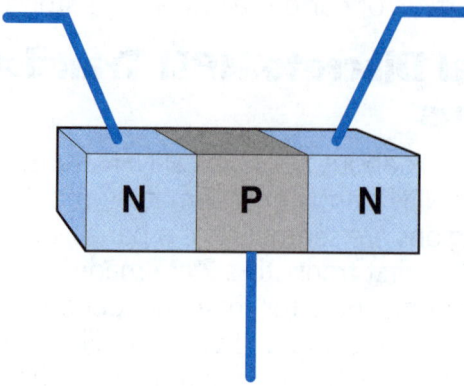
b. Silicon Configuration Suggested by the Symbol

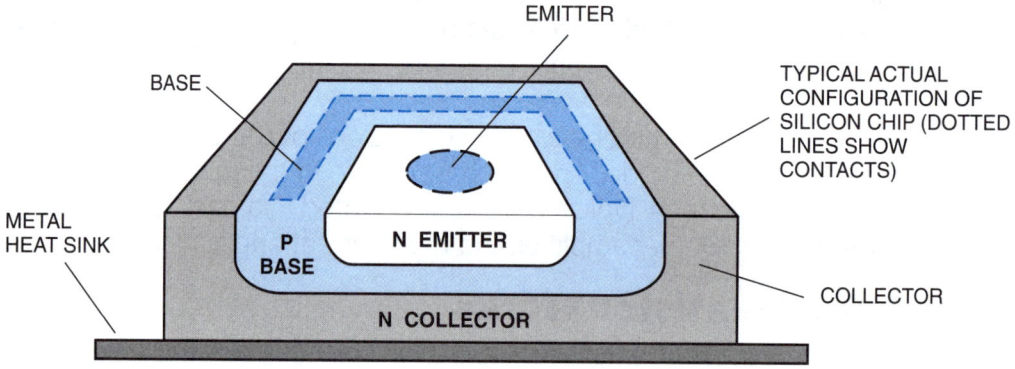
c. Diffused Sandwich Construction

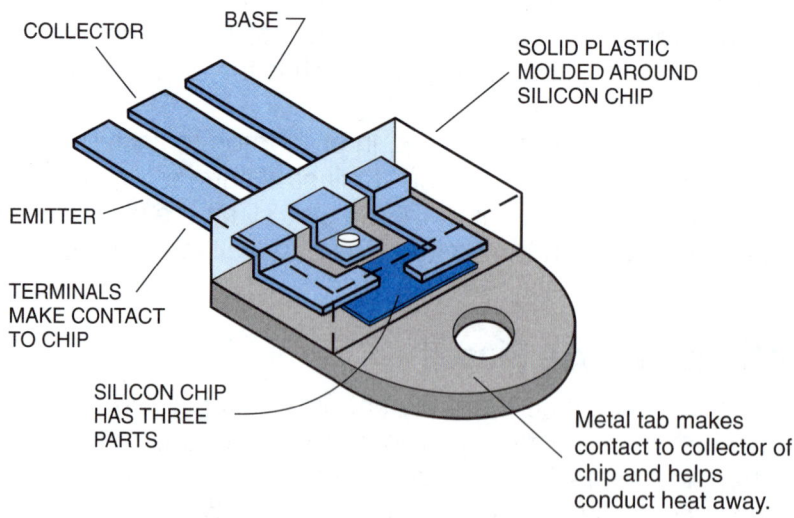
d. Transistor in Package

Figure 4-13. The three parts of an NPN transistor crystal are usually thin, flat layers in a stack, with the collector on the bottom.

Forward Biasing the Base-Emitter Junction

Let's gradually turn the potentiometer to move the wiper contact away from the −3 volt terminal. This produces a smaller and smaller negative voltage on the base so that the base becomes more positive with respect to the emitter. This increases the forward bias across the emitter-base diode junction so the depletion zone gets narrower. Remember, the emitter is N-type and the base is P-type. When the base-emitter forward bias reaches about 0.6 volt (base at −2.4 volts), the depletion zone across the base-emitter junction is about to vanish. But in the transistor's construction, we have something more than just two diodes. T*he P-type base region is actually very thin, and it is doped much more lightly* than the emitter, so that holes are scarce in the base region.

As a result, when the forward bias voltage is above 0.7 volt (base above −2.3 volts) as shown in *Figure 4-14, most of the free electrons from the emitter dash right through the base region to the collector*. The collector-base diode is reverse-biased and has a wide depletion zone. In fact, when a transistor is used to amplify signals in this circuit, the collector-base junction is always reverse biased. The output of the transistor is the collector and the current from the emitter ends up there. The only thing in the base to stop the free electrons from the emitter is a few holes on the near side of the collector-base depletion zone. Since holes are in the base region, some electrons will combine with the holes.

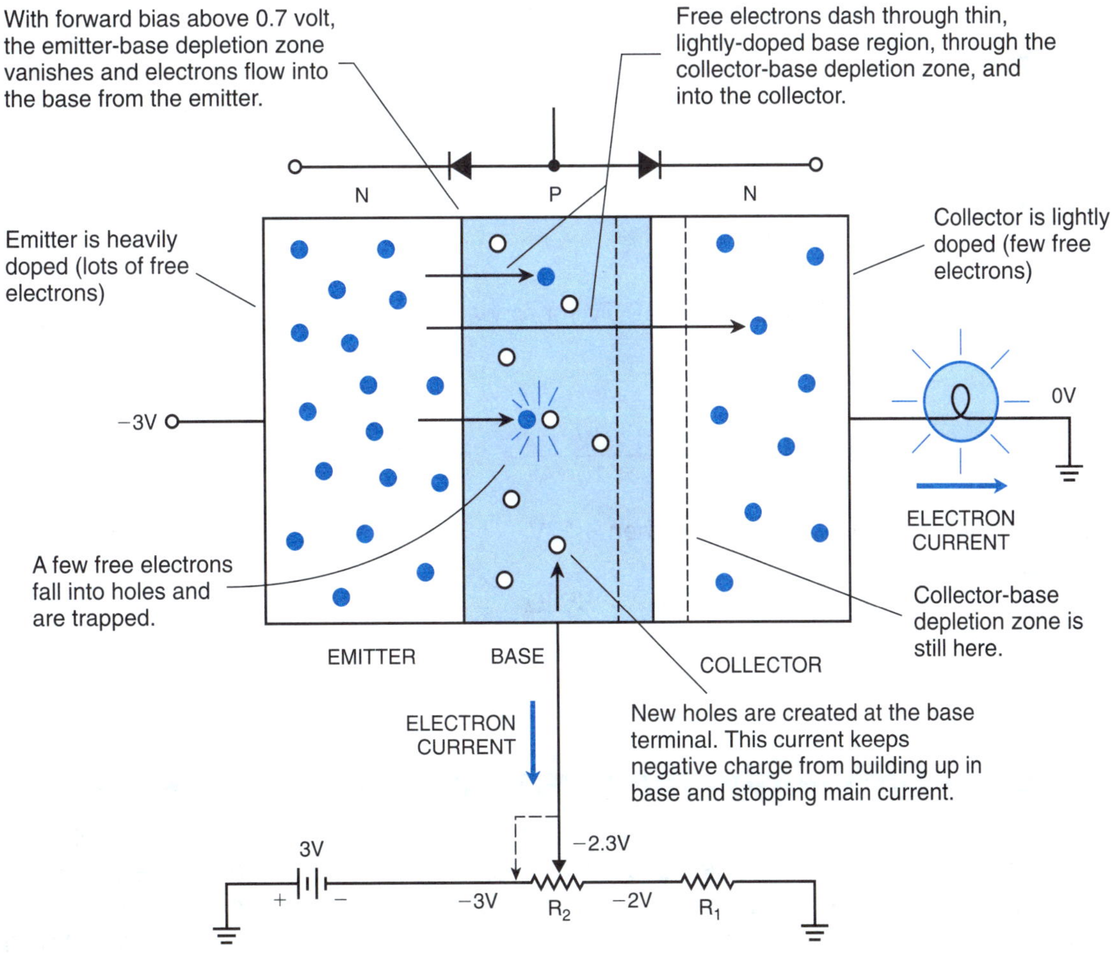

With forward bias above 0.7 volt, the emitter-base depletion zone vanishes and electrons flow into the base from the emitter.

Free electrons dash through thin, lightly-doped base region, through the collector-base depletion zone, and into the collector.

Emitter is heavily doped (lots of free electrons)

Collector is lightly doped (few free electrons)

A few free electrons fall into holes and are trapped.

Collector-base depletion zone is still here.

EMITTER BASE COLLECTOR

ELECTRON CURRENT

ELECTRON CURRENT

New holes are created at the base terminal. This current keeps negative charge from building up in base and stopping main current.

3V   −3V −2.3V −2V
      R_2 R_1

Figure 4-14. With an emitter-base bias of at least 0.7V, the emitter emits free electrons through the base, which pass quickly to the collector. Base current keeps negative charge from accumulating and stopping the flow.

Here is an important point about transistor action: *For every hole that is filled in the base region when a free electron falls into it, a tiny negative charge is trapped in the base region*. The only way to get rid of this charge is to draw electrons from the base terminal. Therefore, to maintain the collector current, we have to drain from the base all the electrons that fall into holes. This is why *the collector current is controlled by the base current*. An increase in base current is caused by a very tiny increase in the emitter-base bias. As in any other silicon diode, this bias never gets much greater than about 0.7 volt.

Current Gain

As a result, large collector currents are controlled with small base currents. In fact, an important transistor characteristic is h_{FE}, which is the ratio of the collector current to the base current. Both signals are referred to a common emitter. It is called the *common-emitter* current gain. When the base-emitter junction is forward biased, variations in the base current will cause the same variations in collector current, but of a much larger magnitude. As a result, the transistor is used as an amplifier. Common-emitter current gains of from 50 to 200 are common in modern transistors.

Example 4. Determining Resistor Values for an Amplifier Circuit

If an NPN transistor with an $h_{FE} = 100$ is used, what value should R_B have in order to produce 20 milliamperes (20mA) collector current? What value should R_C have so that V_{CE} is 5 volts? Current directions are conventional current.

$$h_{FE} = \frac{I_C}{I_B} \text{ and } I_B = \frac{I_C}{h_{FE}}$$

Given that $I_C = 20mA$ and $h_{FE} = 100$

$$\therefore I_B = \frac{20mA}{100} = 0.2mA$$

Since $I_B = 0.2mA$ and $V_{BE} = 0.7V$ for forward bias

$$\therefore +12V - I_B R_B - 0.7V = 0$$

$$\therefore R_B = \frac{12V - 0.7V}{0.2mA} = \frac{11.3V}{0.2 \times 10^{-3}A} = 56,500\Omega$$

If $I_C = 20mA$ and $V_{CE} = 5V$, then $+12V - I_C R_C - 5V = 0$

$$R_C = \frac{12V - 5V}{20mA} = 0.35 \times 10^3 = 350\Omega$$

$R_B = 56.5$ kilohms, $R_C = 350$ ohms

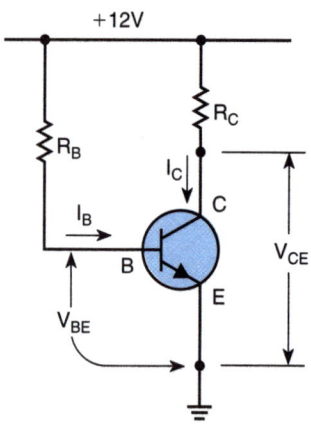

Common-Emitter Amplifier

Used Differently, the Same Transistor Acts As a Switch

When a transistor is used as a switch, the circuitry is similar, but different from the circuit for an amplifier. The transistor is held in two states, either all the way off or driven all the way on, without stopping in between as in an amplifier. Such a circuit is shown in *Figure 4-15a*. In the OFF state, the base-emitter junction is shorted so there is no current from emitter to collector. In the ON state, the base-emitter junction has a voltage across it to make sure it is forward biased. The base is flooded with electrons.

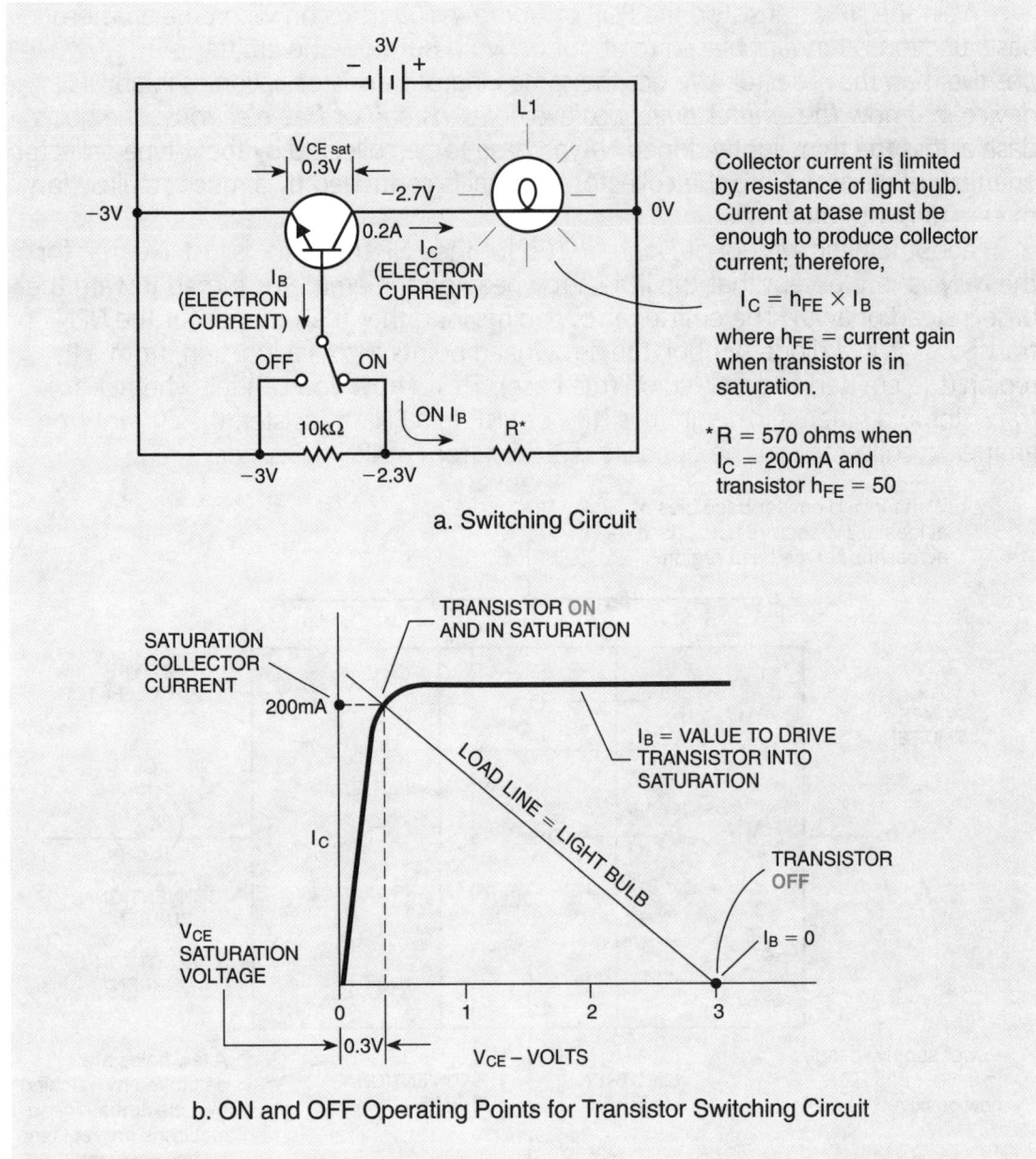

Figure 4-15. When a transistor is used as a switch, it is saturated when turned on all the way. The current gain may be much smaller than as an amplifier, but it is still enough for reliable, stable operation as a switch.

When a transistor is turned all the way on, it is in a condition called *saturation*. The effective resistance between the emitter and collector is very low and the emitter-collector bias (voltage difference) is only a fraction of a volt. In fact, in many cases, the base-collector junction becomes forward biased. *Figure 4-15b* shows the ON and OFF operating points for a switching transistor circuit.

In a PNP Transistor, Free Electrons and Holes Exchange Roles

There is another kind of transistor called the PNP transistor. It has a P-type emitter, an N-type base, and a P-type collector. This transistor works just like the NPN variety, except everything is in reverse. The two kinds of device are said to be *complementary* to each other. That means each does something the other cannot do—carry current in a particular direction.

As in the NPN transistor, the PNP of *Figure 4-16a* turns on when the emitter-base junction is forward biased to about 0.7 volt. But now the emitter is more positive than the collector. We get the same kind of transistor action as in an NPN device, but now *the emitter emits positive holes instead of free electrons*. The holes dash across the thin, lightly doped N-type base to be collected by the P-type collector. The flow of electrons into the collector terminal is controlled by a much smaller flow of electrons into the base.

The schematic symbol of *Figure 4-16b* for the PNP transistor is just like that for the NPN variety, except that the little arrowhead on the emitter is turned toward the base instead of away. The emitter arrow points in, rather than out, as for the NPN transistor. As in a diode symbol, the arrowhead points across a junction, from a P region (the emitter) to an N region (the base). This is how you can tell whether a transistor in a schematic diagram is NPN or PNP. In a PNP transistor, the current from emitter to collector is in the opposite direction from a NPN transistor.

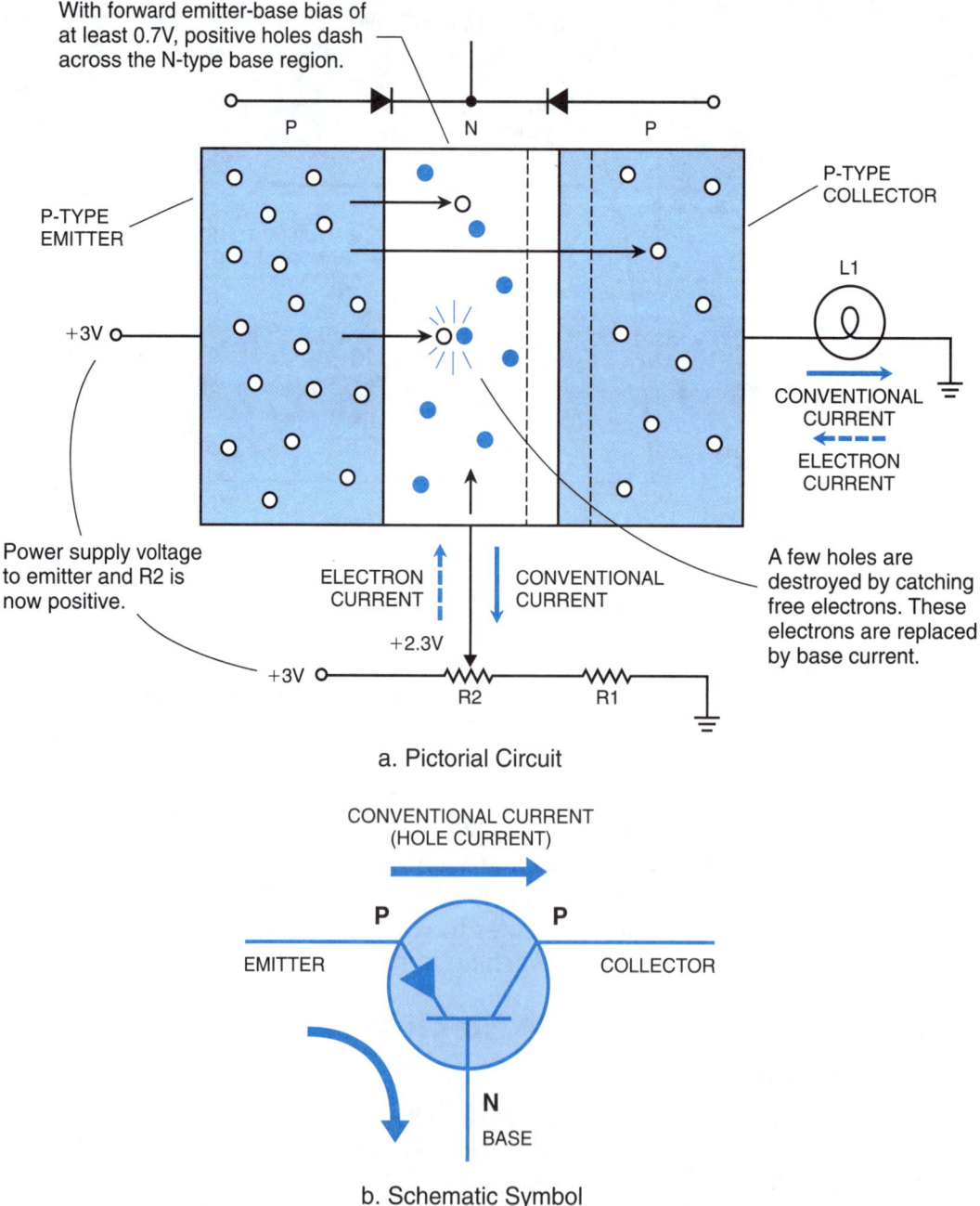

Figure 4-16. A PNP transistor is like an NPN transistor with everything reversed.

PNP Transistors Can Complement NPN Transistors in a Circuit

PNP transistors are used for the same purposes as NPN devices, but the circuitry is designed differently so that there is current the other way. Silicon NPN transistors are typically easier and less expensive to make than silicon PNP devices. So circuits are usually designed to use more NPN than PNP transistors.

But there are some situations when it is very efficient to use transistors of both kinds together. An example of an amplifier for ac signals, such as in a loudspeaker system, is shown in *Figure 4-17*.

The NPN transistor amplifies the *forward* half of each input wave from the microphone, while the PNP transistor is turned off. The PNP transistor, in turn, amplifies the *reverse* half of each input wave, while the NPN transistor is turned off. Thus, the two kinds of transistor complement each other in a *symmetrical* way— meaning the waves are treated the same on both sides, forward and reverse.

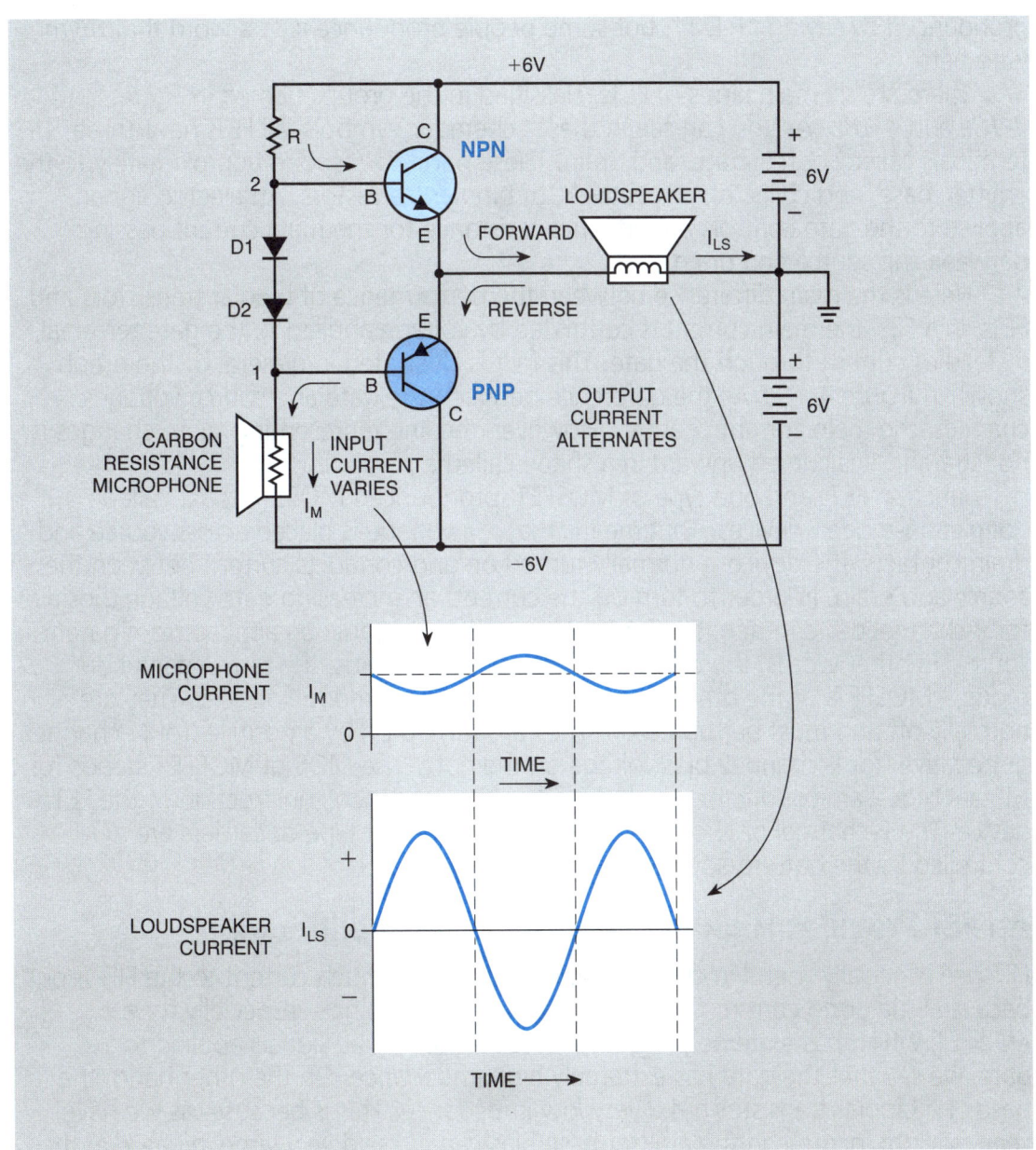

Figure 4-17. Simplified example of a PNP and an NPN transistor used together in a simple, but efficient, loudspeaker amplifier.

When the microphone resistance increases (current I_M decreases), the voltage at point 1 increases, tending to cut off the base current of the PNP transistor. The voltage at point 2 also increases and it increases the base current of the NPN transistor. Increasing base current of the NPN transistor increases the forward I_{LS} through the loudspeaker. A decrease in the microphone resistance (I_M increases) lowers the voltage at points 1 and 2. This causes the PNP base current to increase and the base current of the NPN transistor to decrease. Therefore, the reverse current increases. The NPN transistor amplifies the decrease in microphone current, and the PNP transistor amplifies the increase in microphone current. The diodes D1 and D2 maintain a constant bias voltage between points 1 and 2.

Field-Effect Transistors Are Controlled by Voltage Instead of Current

NPN and PNP transistors are called *bipolar* transistors. There is another important group of transistors called *field-effect transistors*. The abbreviation, FET, is usually pronounced by saying "F-E-T", but some people pronounce it as a word that rhymes with bet.

There are six main kinds of FETs, classified in the groups named in *Figures 4-18a* and *4-18b*. As you can see in these schematic symbols, all FETs have three terminals called source, gate, and drain. These parts correspond approximately to the emitter, base, and collector, respectively, of bipolar transistors. An electric signal applied to the gate controls the effective resistance for the main current passing between the source and drain.

Here is the main difference between the performance of bipolar transistors and FETs. In a FET, the main current is controlled by *voltage* applied to the gate terminal, instead of current through the gate. This fact is illustrated in general by the graph shown in *Figure 4-18c*. As the difference between the gate and source voltages changes, the main current changes. However, the line representing these changes is not straight. It is curved upward in a shape called a parabola.

Junction FETs and one type of MOSFET (pronounced MOSS-fet) are called "depletion-mode" devices. That means that if a voltage is placed across source and drain contacts, the device is normally turned on and conducts current between the source and drain. In order to turn off the current, an increasing gate voltage (positive for P-channel devices, negative for N-channel devices) must be applied from gate to source. In other words, the current must be depleted, thus, the term "depletion mode." MOSFETs of the other type are "enhancement-mode" devices. They are normally off and must be turned on. They are turned on by a positive (for N-channel) or negative (for P-channel) bias voltage on the gate. The MOS of MOSFET stands for Metal-Oxide-Semiconductor which describes the sandwich construction of the FET device. The P-channel or N-channel term identifies what type of carriers are controlled by the gate-to-source voltage.

An FET Provides High Impedance at an Amplifier Input

FETs are especially useful in circuits that cannot supply much current to the FET input because little or no current flows through the gate terminal—especially for a MOSFET, which has extremely high input resistance. For ac signals applied to the gate, we say that the gate has extremely high *impedance*. On the other hand, the base of a bipolar transistor has a very *low* impedance. This is because, as we have seen, a slight increase in the emitter-base bias can cause a very large increase in the base current.

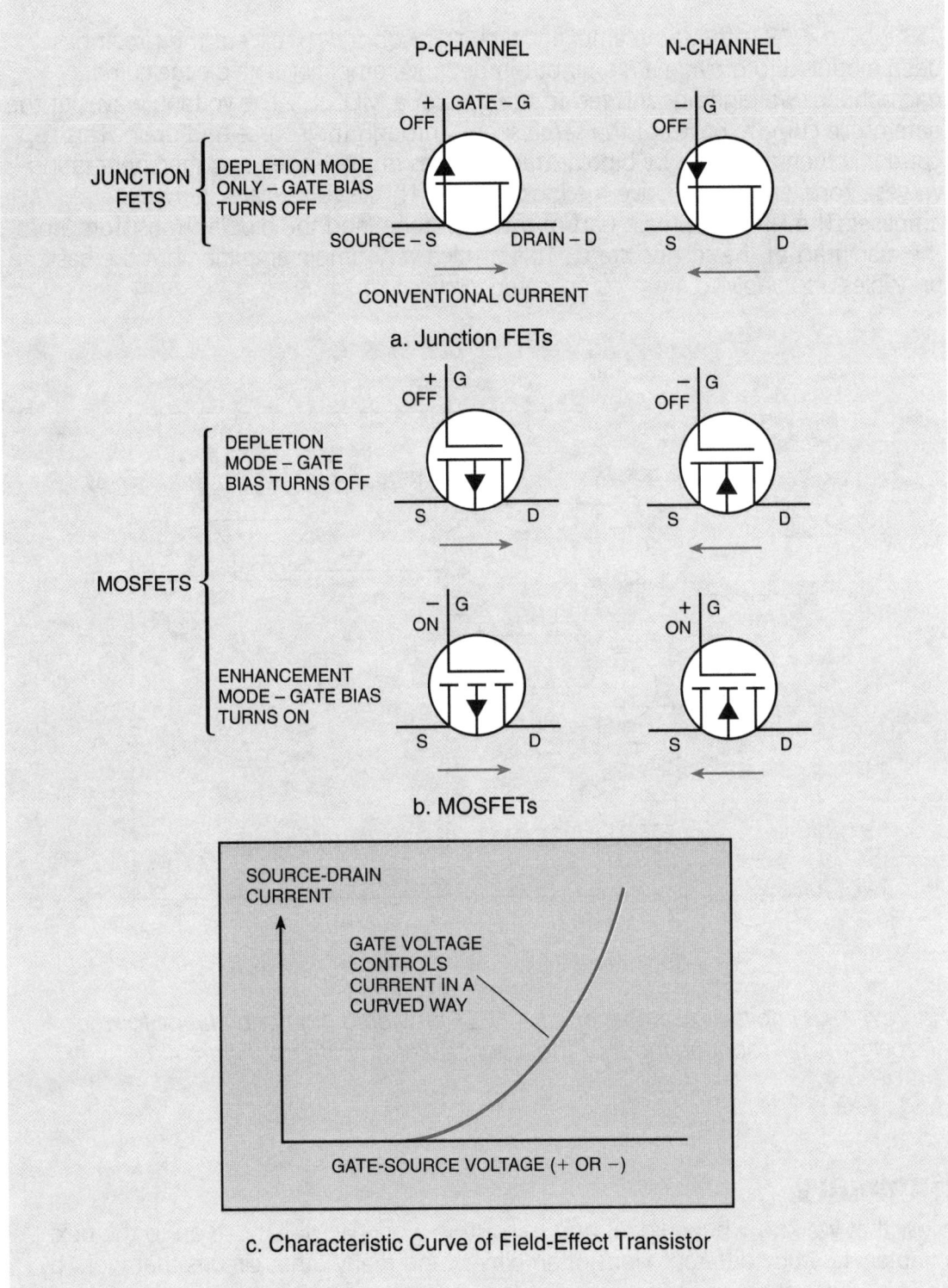

Figure 4-18. In field-effect transistors (FETs), current is turned on or off by positive or negative gate voltage (bias with respect to the source).

An important type of circuit application where an FET comes in handy is at the input to an amplifier. Many amplifiers need high input impedance, because whatever creates the input signal cannot provide very much current. An example is a crystal microphone. Sound waves vibrating the piezoelectric (PEE-zo-electric) crystal generate strong voltage waves across the crystal, but these waves weaken and disappear if current greater than a few microamperes is required. We need an amplifier with a high input impedance to amplify the signals from the piezoelectric microphone.

Figure 4-19 is the same amplifier circuit as *Figure 4-17* except the circuit has been modified to provide a high input-impedance amplifier. The piezoelectric microphone's voltage signal is fed to the gate of a MOSFET. The voltage waves at the gate cause current waves of the same shape through the source and drain. This current is then amplified by bipolar transistors as in the earlier amplifier, generating waves strong enough to drive the loudspeaker. This circuit is called a *two-stage* amplifier. The FET is the main part of the first stage, and the bipolar transistors are the main part of the second stage. This is still a very simple amplifier, but the basic principles also apply to more complicated circuits.

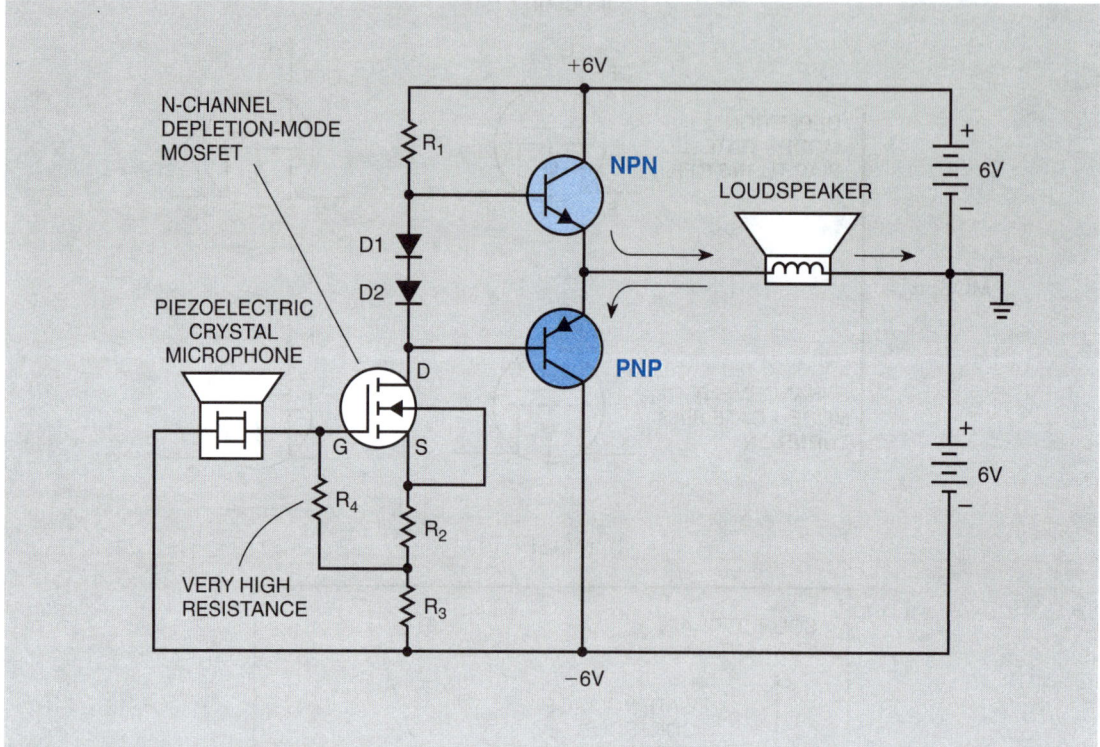

Figure 4-19. Simplified example of a MOSFET providing high impedance for the piezoelectric microphone input of the loudspeaker amplifier of Figure 4-17.

Summary

Now that we know how diodes and transistors work, we will move on to the next chapters to study different kinds of amplifiers and many other circuits that work by the switching and amplifying action of transistors.

Quiz for Chapter 4

1. A diode turns on when it is:
   a. doped with aluminum
   b. reverse biased
   c. forward biased
   d. negatively charged
2. The more phosphorous that is added to silicon:
   a. the lower its resistance
   b. the more holes produced
   c. the more inert it becomes
   d. the more depleted it becomes
3. Doping silicon with aluminum produces a free current carrier called a:
   a. proton
   b. electron
   c. neutron
   d. hole
4. The thin layer of silicon on both sides of the junction that has no carriers at all is called the:
   a. depletion zone
   b. twilight zone
   c. inert zone
   d. repulsion zone
5. The threshold voltage of a silicon P-N junction diode is approximately:
   a. 0.2 volt
   b. 0.4 volt
   c. 0.7 volt
   d. 1.4 volt
6. The P-type silicon of a diode is the:
   a. cathode
   b. anode
   c. base
   d. emitter
7. An atom uses the electrons of its outermost shell to:
   a. stabilize the nucleus
   b. cancel the mass of the proton
   c. form bonds
   d. emit light
8. The bond between charged atoms is called an _____ bond.
   a. ionic
   b. covalent
   c. chemical
   d. atomic
9. Pure silicon, or any other semiconductor material, is:
   a. an insulator
   b. a conductor
   c. a diode
   d. a transistor
10. In an NPN transistor, the P region is the:
    a. cathode
    b. emitter
    c. base
    d. collector
11. The _____ region is actually very thin and it is doped much more lightly than the other regions.
    a. emitter
    b. base
    c. anode
    d. collector
12. When a transistor is turned all the way on, it is in a condition called:
    a. complementary
    b. saturation
    c. forward
    d. inverted

Answers: 1c, 2a, 3d, 4a, 5c, 6b, 7c, 8a, 9a, 10c, 11b, 12b

Questions for Chapter 4

1. A semiconductor diode is basically an electrically-controlled _____ .

2. A diode turns off when it is _____ biased and turns on when it is _____ biased.

3. The two regions of a diode are called the _____ and the _____ .

4. The place where the two types of silicon in a diode meet inside the crystal is called the _____ .

5. The P-type silicon in a diode is called the _____ and the N-type silicon is called the _____ .

6. The extensions stretching between atoms of a silicon crystal are called _____ bonds.

7. The greater the height of the orbit of an electron above the nucleus, the more _____ the electron has.

8. What is the unfilled outer shell of an atom called?

9. The part of an atom consisting of filled shells and the nucleus is what we have called the _____ .

10. Charged atoms are called _____ .

11. What is the characteristic of the electron shells of inert gases?

12. Adding a tiny amount of either element is called _____ the silicon.

13. Silicon doped with phosphorus is called _____ silicon.

14. A missing bond, with one positive charge, is called a _____ in the crystal lattice.

15. The collector current of a transistor is controlled by the _____ current.

CHAPTER 5
How Amplifiers and Oscillators Work

In studying transistors, we have investigated several simple amplifier circuits. The general name, "amplifier," applies to several different types of circuits. Also, there are many circuits that usually are not called amplifiers, although they really are special amplifiers or are made out of one or more amplifiers.

Amplifiers are the most important, most basic type of electronic circuit. In fact, the circuitry of most electronic systems, such as television receivers, consists almost entirely of different kinds of amplifiers. They are connected in various ways, usually with resistors and capacitors or transformers between them.

In this chapter, we will discuss some general facts about amplifiers and some variations in their circuitry. This will help you understand what is going on when you study circuit diagrams of amplifiers. Then we will learn about the oscillator — a special kind of amplifier that is used in several types of electronic equipment.

A Basic Voltage Amplifier Stage

Let's begin with the two specialized amplifier concepts, one of which we used as an example in Chapter 4. We will modify these designs to make a more useful, general-purpose amplifier. It will be a *one-stage voltage amplifier,* meaning it amplifies voltage signals in one step. So we will call this a voltage amplifier stage. It can be connected before or after other stages to make an amplifier with several stages that amplifies the voltage to a large signal.

Figure 5-1a shows an earlier NPN amplifier circuit *(Figure 4-15)* that was used to demonstrate transistor operation, redrawn to fit generally accepted representations of amplifiers in circuit diagrams. *Figure 5-1b* shows a very simple circuit to amplify microphone signals. We don't show the battery as part of each amplifier circuit. Instead, we just show the points in the circuits where the battery (or a power-supply circuit) would be connected. Each NPN transistor amplifier circuit has a *positive power-supply connection* (+3V here) and a *ground connection*. The voltages are shifted to operate from +3V to ground.

Figure 5-1c shows a generalized version of these amplifier configurations drawn as one stage. In the place of the carbon microphone is a fixed resistor called the *bias resistor*. In the place of the loudspeaker or the lamp is another fixed resistor called the *load resistor*. Since this load is built into the amplifier stage, it is called an *internal load*.

The input signal to be amplified doesn't come from varying the bias resistance as it did in the circuits we analyzed earlier. Instead, it is a varying voltage from another amplifier stage or some other device. The varying voltage causes variations in the base current of the transistor. The transistor amplifies the base current variation and produces larger current variations in the internal load resistor connected to the

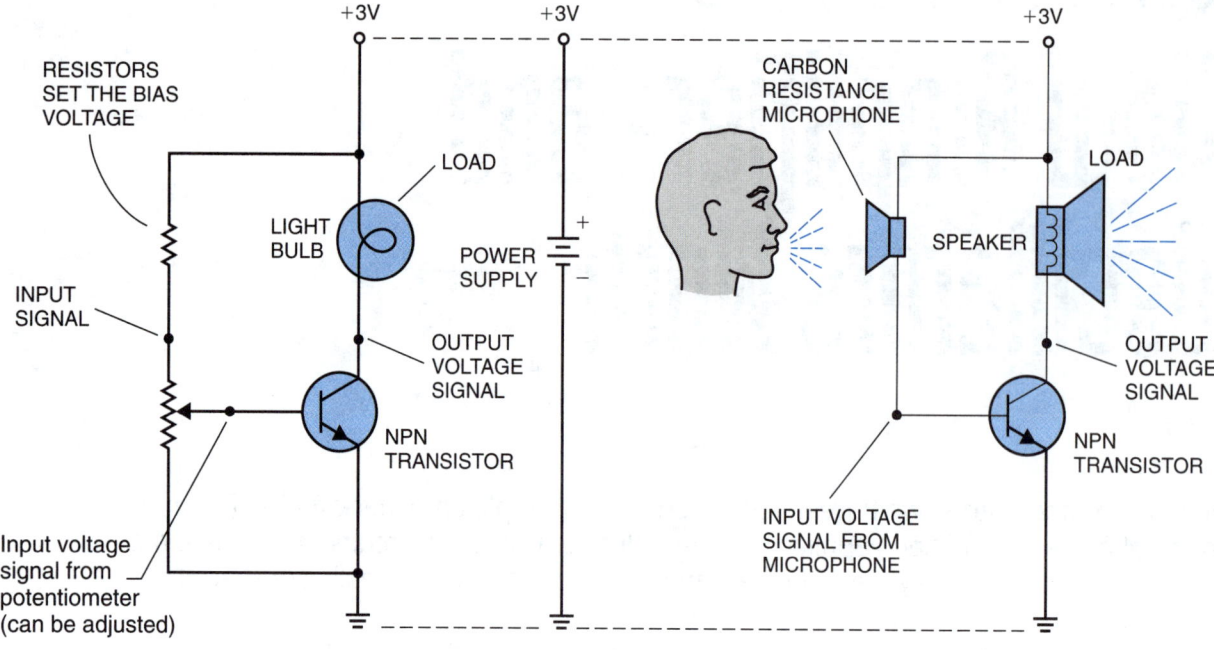

a. Chapter 4 Demonstration Circuit

b. Loudspeaker Amplifier

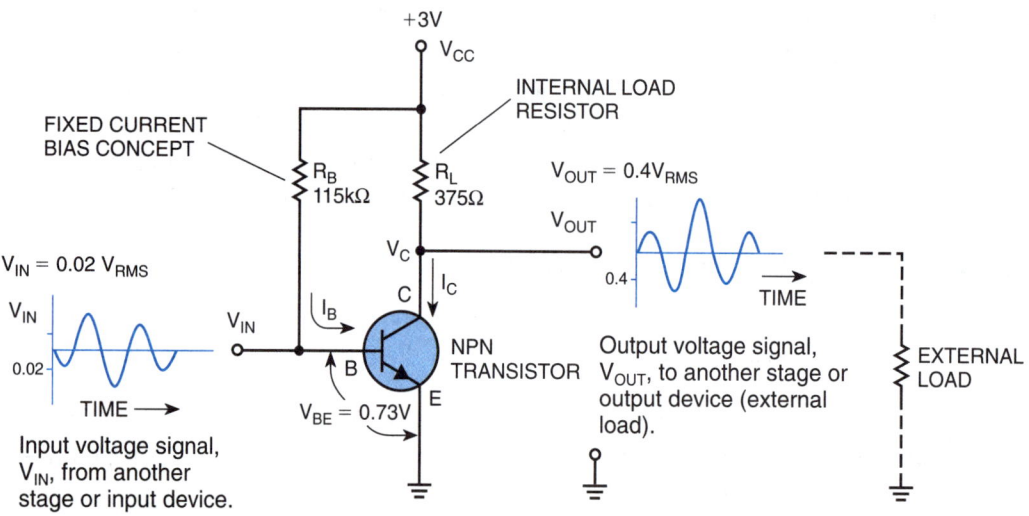

c. Fixed-Current Biasing – NPN Common-Emitter Voltage Amplifier Stage

Figure 5-1. Voltage Amplifiers – A practical voltage amplifier stage is very similar to the amplifier example in Chapter 4.

transistor collector. The amplified current variations produce the output signal voltage variations across the internal load resistor. Thus, we take the output voltage from the point where the internal load resistor connects to the collector. We attach a wire to this point and connect the other end to another stage or device, which is called the *external load*.

As a result, we have a common circuit used as a *voltage amplifier stage*. Small voltage variations in the input are turned into larger voltage variations in the output. The multiplication factor is called the *voltage gain*. It depends on the common-emitter current gain h_{FE}, sometimes mistakenly called Beta (β), of the transistor and the value of the load resistance. (Remember Example 4 in Chapter 4.) This way of setting the emitter-base bias is called *fixed-current bias*. You will learn about other bias methods later.

This general type of amplifier stage is called the *common-emitter configuration*, mentioned in Chapter 4. The emitter is at ground, which is a common signal point for input and output. Two other frequently used basic configurations are called *common-base* and *common-collector* (these are not shown).

Voltage Gain

Very small input signal voltages increase and decrease the base voltage which causes a change in base current. The changing base current changes the collector current through the load which produces an enlarged (amplified) output voltage. The voltage gain of the amplifier stage is a number obtained by dividing the output voltage by the input voltage. Let's look at an example of how to calculate the voltage gain.

Example 1. Calculating Voltage Gain

If the input voltage of the circuit of *Figure 5-1c* has an ac rms value of 0.02 volt and it produces an ac rms output voltage of 0.40 volt, then the voltage gain of the amplifier stage is:

$$\text{GAIN} = \frac{\text{output voltage}}{\text{input voltage}} = \frac{0.40 \, V_{RMS}}{0.02 \, V_{RMS}} = 20$$

Thus, an amplifier with a voltage gain of 20 will have an output voltage that is 20 times the input voltage.

The various currents and voltage drops can be found by circuit analysis methods like you learned in Chapter 2.

Example 2. Calculating Fixed-Bias Base Current and Collector Current and Voltage

To calculate the fixed-bias base current I_B in *Figure 5-1c* according to Ohm's law, divide the voltage across the base bias resistor R_B by the value of the resistor, 115 kΩ. The voltage drop across the base resistor is the difference between the 3V supply and the base-emitter voltage (3V − 0.73V = 2.27V). So, the base current I_B is:

$$I_B = \frac{2.27 \, V}{115 \, k\Omega} = 0.02 \, mA$$

If the transistor has a common-emitter current gain (h_{FE}) of 200, since $h_{FE} = I_C/I_B$, then the collector current I_C can be found by:

$$\begin{aligned} I_C &= I_B \times h_{FE} \\ &= 0.02 \, mA \times 200 \\ &= 4 \, mA \end{aligned}$$

And the fixed-bias voltage V_C at the collector without any signal can be found by subtracting the voltage across the load resistor R_L from the power supply voltage:

$$\begin{aligned} V_C &= V_{CC} - I_C R_L \\ &= +3 \, V - (4 \, mA)(375 \, \Omega) \\ &= +3 \, V - 1.5 \, V = +1.5 \, V \end{aligned}$$

So, one-half of the power supply voltage is dropped across the load resistor and one-half across the transistor from collector to emitter. Remember, mA means 10^{-3} amperes, or 0.001 A.

A Transistor Can Have Different Operating Points

Not all common-emitter voltage amplifier stages have the same circuit design. One difference is how far the transistor is turned on when no signal is applied at the input. The transistor's condition at that time is called its *operating point*. A more complete name is *quiescent operating point*. (Quiescent is pronounced kwy-ESS-ent. It means quiet or inactive. For this discussion, it means the no-signal condition.) For a given transistor, power-supply voltage, and load resistance, the operating point depends on the bias resistances in the circuit. *Figure 5-2* shows two possibilities for the quiescent operating point. *Figure 5-2a* is the same circuit as *Figure 5-1c*.

Class A Operation

First, if we want the output voltage wave to have very nearly the same shape as the input wave shown in *Figure 5-2b,* then we pick a bias resistance (Figure 5-2a) that makes the transistor turn about *halfway on* when there is no input signal. This is called *Class A operation*. The output waveform is shown in Figure 5-2c.

When the input voltage goes *above* the quiescent operating point, the transistor turns on more. That pulls the output voltage *down*. Conversely, when the input voltage goes *below* the quiescent operating point, the transistor is turned on less. That pulls the output voltage *up*. Therefore, the output signal is an amplified, but upside-down, copy of the complete input signal as shown in *Figure 5-2c*. The waveform for Example 1 is shown dotted in *Figure 5-2c*. This is what we would want for good sound reproduction by the loudspeaker system amplifier in *Figure 5-1b*.

You have noticed that the output signal is upside-down or *inverted;* that is, the output voltage goes less positive (or more negative) when the input voltage goes more positive. The output voltage is said to be *out of phase* with the input voltage. We usually don't care about that. Also, the output's quiescent (zero or neutral) voltage may be shifted so that the top part of the maximum wave is larger than the bottom portion. If the quiescent point is shifted until the transistor is just barely conducting, we get into Class B operation.

Class B Operation

Figure 5-2d shows what happens when we pick a higher bias resistance that allows only an extremely small base current. It operates the transistor just *barely turned on* when there is no signal. For all practical purposes, the transistor is turned off. This is called *Class B operation*. With no input signal, the turned-off transistor lets the load resistor pull the output voltage up to the power supply voltage (3 volts here). When the input voltage goes above the quiescent level, the transistor turns on, pulling the output voltage down. When the input voltage goes low (below $V_{BE} = 0.7V$), the transistor just stays off, leaving the output voltage high.

So in Class B operation, *only one-half of the input wave is amplified*. Look back at the complementary amplifiers in Chapter 4 for a moment. Remember, we said each transistor amplified a different half of each wave. Now we can see that both the NPN and PNP transistors are biased to give Class B operation.

Different Biasing Methods Can Be Used

There are several ways to design the bias circuit. *Figure 5-3* shows two ways to bias a common-emitter voltage amplifier stage. Remember, the bias resistors determine the quiescent (no-signal) bias voltage between the base and the emitter. Look at the general circuits in *Figures 5-1c and 5-2a*. Both use a concept called *fixed-current bias*. They have one bias resistor connected to the positive voltage, V_{CC}.

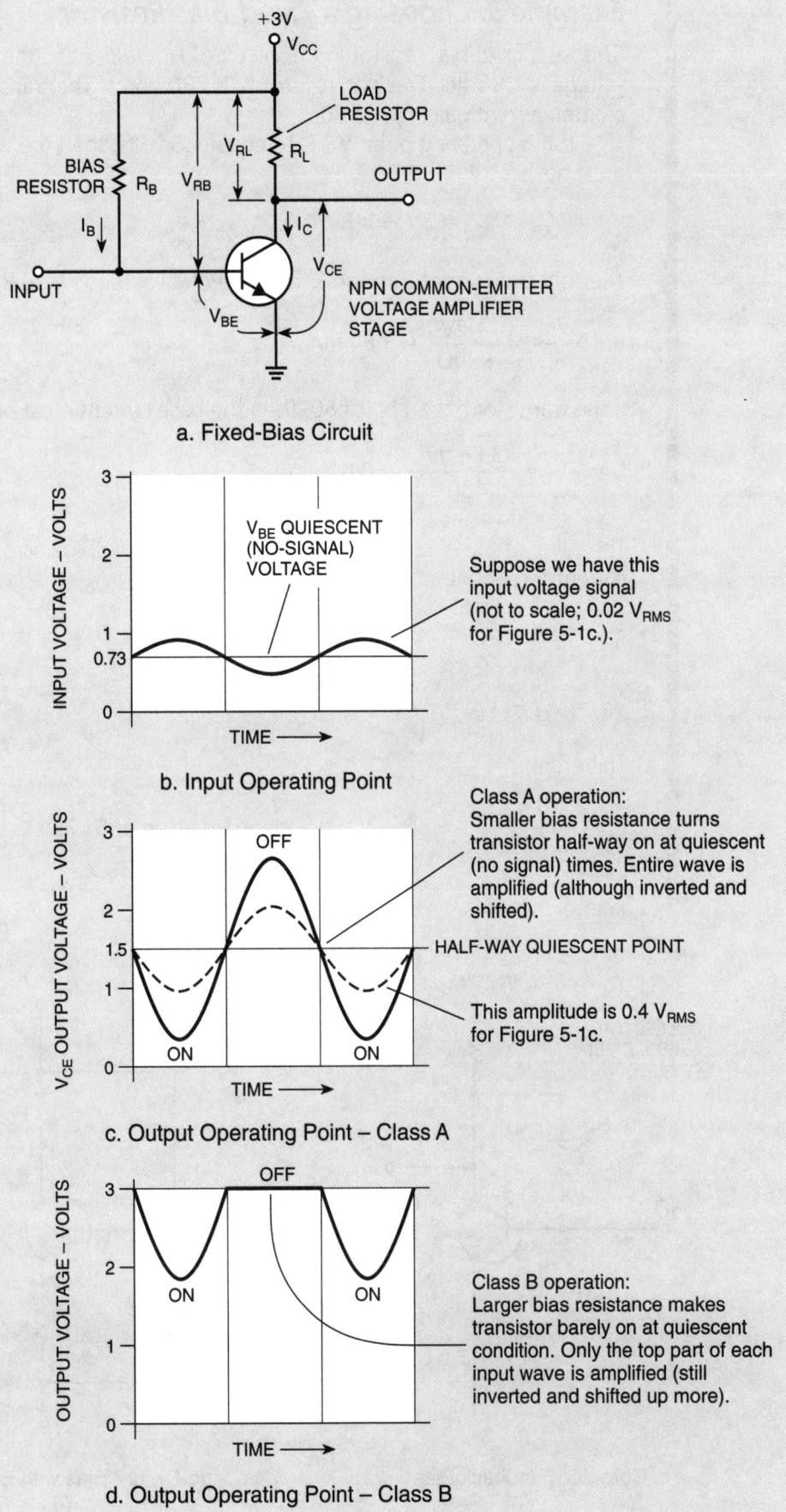

Figure 5-2. Amplifier Operating Points – Setting the quiescent operating point of the transistor determines how much of an input wave is amplified.

Example 3. Choosing a Fixed-Bias Resistor

Choose a fixed-bias resistor in the circuit of *Figure 5-2a* so the V_{CE} operating point voltage is 1.8 volts. The load resistor R_L is 330 ohms. The transistor has a common-emitter current gain, h_{FE}, of 80.

If the operating point V_{CE} is to equal 1.8 volts, then the voltage drop across R_L must be:

$$V_{RL} = V_{CC} - V_{CE} = 3V - 1.8V = 1.2V$$

The collector current through R_L must be of a value to cause R_L to have 1.2V across it:

$$I_C = \frac{V_{RL}}{R_L} = \frac{1.2V}{330\Omega} = 3.64 \text{ mA}$$

If the transistor has an h_{FE} of 80, then the base current must be:

$$I_B = \frac{I_C}{h_{FE}} = \frac{3.64 \text{ mA}}{80} = 0.045 \text{ mA}$$

The value of the biasing resistor R_B is equal to the voltage across it divided by the current through it:

$$R_B = \frac{V_{RB}}{I_B}$$

$$V_{RB} = V_{CC} - V_{BE} = 3V - 0.73V = 2.27V$$

Therefore,

$$R_B = \frac{2.27V}{0.045 \text{ mA}} = 50.4 \text{ k}\Omega. \text{ Use standard value of 51 k}\Omega.$$

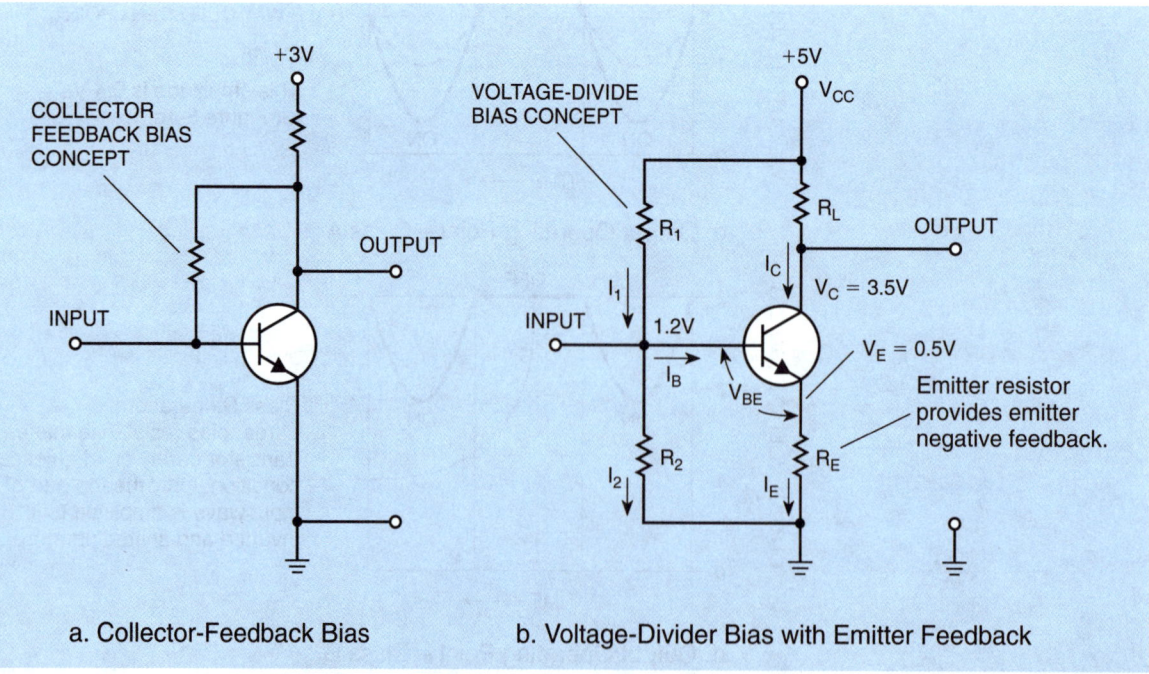

a. Collector-Feedback Bias b. Voltage-Divider Bias with Emitter Feedback

Figure 5-3. Other Amplifier Biasing – Two other forms of bias circuitry that help keep a stable operating point even though the transistor's electrical characteristics change.

Collector Feedback Bias

A slight variation from the fixed-bias circuit is shown in *Figure 5-3a*. It is called the *collector-feedback* bias concept. The resistance value usually is smaller in this case; however, the big difference is that the bias resistor is connected to the variable output voltage instead of the fairly constant voltage from the power supply. As we noted before, this amplifier stage is an *inverting amplifier,* but there is an added feature. When the input voltage goes high and the transistor pulls the output voltage down, the input gets pulled down a little bit too. Conversely, when the input voltage goes low, the increasing output voltage tries to hold up the input voltage. We say that a portion of the output signal is fed back to the input. Because this *feedback* works against the amplification process, we call it *negative feedback* or *degenerative feedback*. The disadvantage is that the voltage gain is less than before. The advantage is that the operation is more stable; for example, the operating point doesn't vary as much when the transistor's characteristics change as the temperature goes up and down. In addition, the amplifier has less distortion and more bandwidth.

Voltage-Divider Bias

The other bias method shown in *Figure 5-3b* is called the *voltage-divider bias* concept. The quiescent base voltage is established by a nearly constant current through the two resistors connected between the power supply and ground. The resistors form a voltage divider with their common connection tied to the transistor base. If the upper resistance is smaller, the base bias voltage is pulled higher and the quiescent base current is greater. Conversely, if the lower resistance is smaller, the bias voltage is pulled lower and the quiescent base current is reduced.

Notice that a resistor has been added between the emitter and ground. The emitter resistor provides some negative feedback. When the input signal voltage goes high and turns on the transistor more, the increased current through the emitter resistor raises the emitter voltage a little bit. This decreases the emitter-base bias and opposes the effect of the input signal. Here again, we get less voltage gain, but more stable operation.

Coupling Signals to Different Voltage Levels

Usually only one amplifier stage will not provide the required signal gain; therefore, two or more stages are connected in series to increase gain. If the proper coupling method is not used, we may have the problem mentioned earlier — that of different quiescent (average or neutral) voltages at inputs and outputs.

We can't simply connect the output of the first stage of *Figure 5-4a* directly to the input of the second stage. Here's why. Suppose both amplifiers are designed for Class A operation, so that the transistors are halfway on when there is no signal. The first output is about 1.5 volts, but it needs to feed an input at about 0.7 volt. If these two points are connected, the second transistor would be on all the time, with far too much forward bias voltage applied to the base. This would throw off the operating point of the second stage, and, in addition, the transistor would probably "burn out," meaning it would get too hot and fail.

Capacitive Coupling

One simple and efficient solution is to put a capacitor in series between the two stages, as shown in *Figure 5-4a*. This is called *capacitive coupling*. It is the most common form of ac coupling between amplifier stages. The left plate of the capacitor is at the first stage's output voltage, and the right plate is at the second stage's input voltage. *A capacitor blocks direct current (dc), but allows alternating current (ac) signals to pass through it.* This characteristic prevents the dc output

Example 4. Choosing Voltage-Divider Bias

Look at the common-emitter amplifier with voltage-divider bias shown in *Figure 5-3b*. Choose the values for R_L, R_E, R_1, and R_2 when the operating point of the transistor has a collector voltage of 3.5V, an emitter voltage of 0.5V, and a collector current of 0.83 mA. Solve for R_L first.

$$V_{CC} - I_C R_L = 5V - 0.83 \text{ mA} \times R_L = 3.5V$$

$$\therefore 0.83 \text{ mA} \times R_L = 1.5V$$

$$R_L = \frac{1.5V}{0.83 \text{ mA}} = 1.8 \text{ k}\Omega$$

Since in any transistor, $I_E \cong I_C$, use this fact to solve for R_E ($\cong$ means approximately equal).

$$R_E = \frac{0.5V}{0.83 \text{ mA}} = 602\Omega \text{ (make 620}\Omega\text{, a standard value)}$$

If the *minimum* common-emitter current gain, h_{FE}, of the transistor is 40, then the maximum base current is:

$$I_B = \frac{0.83 \text{ mA}}{40} = 0.02 \text{ mA}$$

The current through the voltage divider should be at least 10 times the maximum base current, but the larger the value, the more stable the bias. We chose it to be 0.36 mA (so resistor values can be common values), which is almost half of the collector current. Now solve for R_2. Since $V_{BE} = 0.7V$:

$$V_{R2} = V_{BE} + V_{RE} = 0.7V + 0.5V = 1.2V$$

$$\therefore R_2 = \frac{V_{R2}}{I_2} = \frac{1.2V}{0.36 \text{ mA}} = 3.3 \text{ k}\Omega$$

R_1 will have $I_1 = I_2 + I_B$ through it.

$$\therefore I_1 = 0.36 \text{ mA} + 0.02 \text{ mA} = 0.38 \text{ mA}$$

Now solve for R_1:

$$R_1 = \frac{5V - 1.2V}{0.38 \text{ mA}} = \frac{3.8V}{0.38 \text{ mA}} = 10 \text{ k}\Omega$$

voltage of the first stage from affecting the dc bias voltage on the second transistor, while coupling the ac signals between the stages.

An important point to remember about capacitive coupling is that *the signal must vary with time* to couple it through the capacitor, as a result, the coupling capacitor affects the amplifier's frequency response. Frequency response is an indication of how the amplifier's gain changes with the frequency of the input signal. Frequency, you remember, is a measure of how many times a signal changes in a second. The effective ac resistance of the capacitor, called *capacitive reactance*, increases as the signal frequency decreases, and attenuates the input signal to the amplifier at low frequencies compared to little attenuation at high frequencies.

Capacitive coupling works as long as we only want to amplify ac signals, such as the audio (sound) frequencies in a loudspeaker circuit. We couldn't use capacitive coupling with the lamp control amplifier we saw earlier because it must amplify direct current.

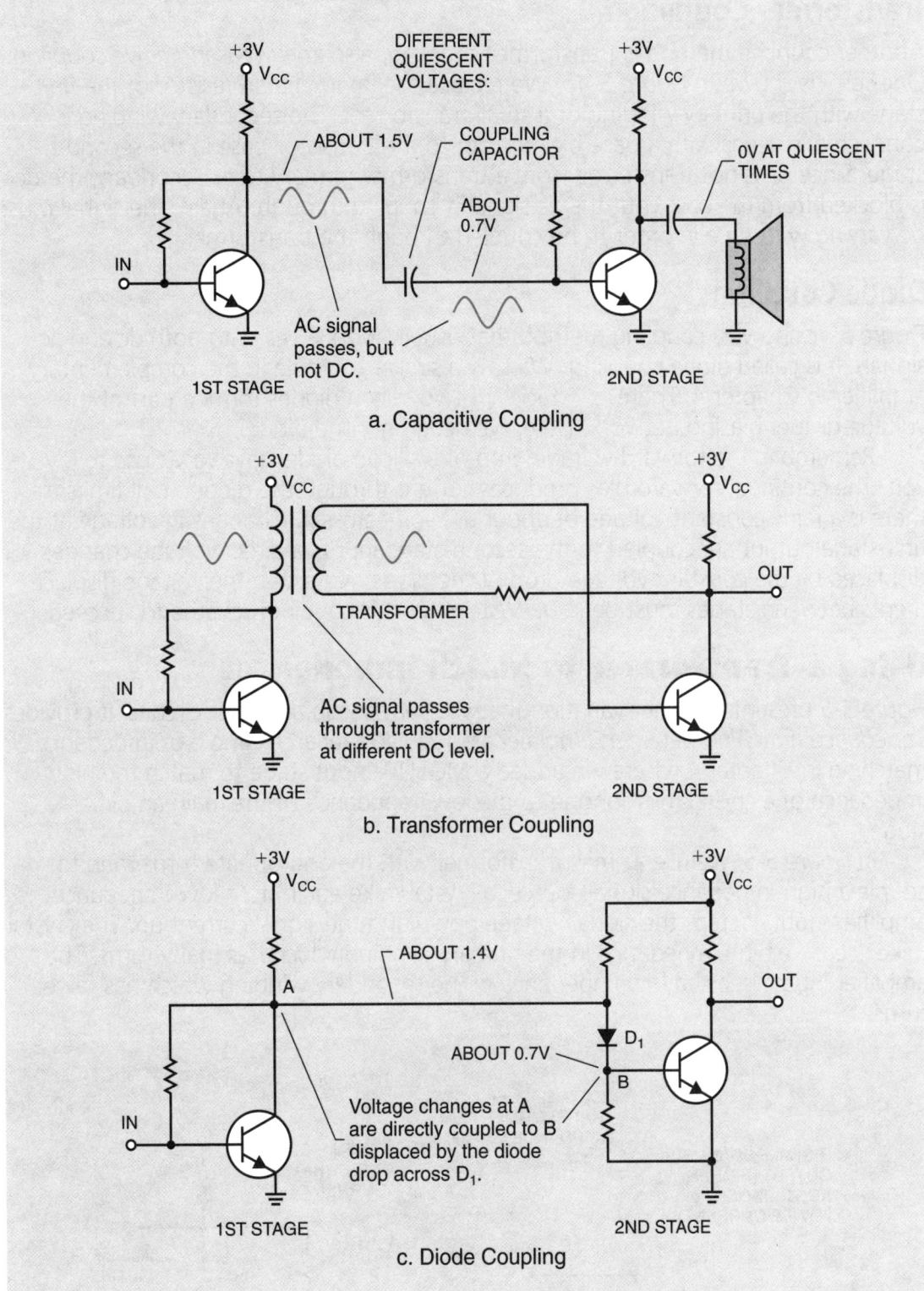

Figure 5-4. Amplifier Coupling – Three ways to couple an output to an input at a different quiescent voltage level.

The circuit in *Figure 5-4a* also includes a capacitor to couple the second stage's output to a loudspeaker. The other loudspeaker lead is connected to ground, which pulls the right capacitor plate to ground voltage at quiescent condition. When an ac input signal is applied to the amplifier, the voltage to the loudspeaker varies up and down around zero volts. Without the capacitor, the amplifier output would cause fluctuating direct current in the loudspeaker coil, as we saw in Chapter 4.

Transformer Coupling

Another coupling method is transformer coupling, also known as *inductive coupling*. One form is shown in *Figure 5-4b*. We replace the internal load resistor of the first stage with the primary winding of a small transformer. The secondary winding is connected in series with a base bias resistor to the transistor base in the second stage. Since dc is not transferred from a transformer primary to its secondary, the dc is blocked from the base, but the ac signal is coupled through. Again, the signal *must be varying with time* in order to be coupled through the transformer.

Diode Coupling

Figure 5-4c shows a coupling method that usually works well with both dc and ac signals. It is called diode coupling. We saw a similar example in the complementary amplifier in Chapter 4. There, a "stack" of two silicon diodes formed part of the voltage-divider biasing network. Here, we need only one diode.

Remember, the forward voltage drop of a silicon diode stays very close to 0.7 volt. The continual forward bias produces current through the diode at all times, so there is a fairly constant voltage of about 0.7 volt across it. Changes in voltage at the first-stage output are coupled to the second-stage input as direct voltage changes displaced by the constant voltage drop of the diode. Note that the voltage displacements between stages must be in 0.7V steps for this coupling scheme to succeed.

Using a Transformer to Match Impedances

Figure 5-5 presents another variation that you may see in amplifier circuits. It provides *impedance matching* between amplifier stages. We saw an example of impedance matching in Chapter 4 where we added a MOSFET input stage to match the high impedance of a crystal microphone to the low impedance of the main amplifier stage.

In *Figure 5-5*, we use a small transformer with the appropriate turns ratio to couple a high-impedance signal source (a crystal mike again) to a low-impedance amplifier input. It steps the signal voltage down and the signal current up. The crystal mike "sees" a high impedance in the primary winding which has many turns. The amplifier input "sees" a low impedance in the secondary winding which has fewer turns.

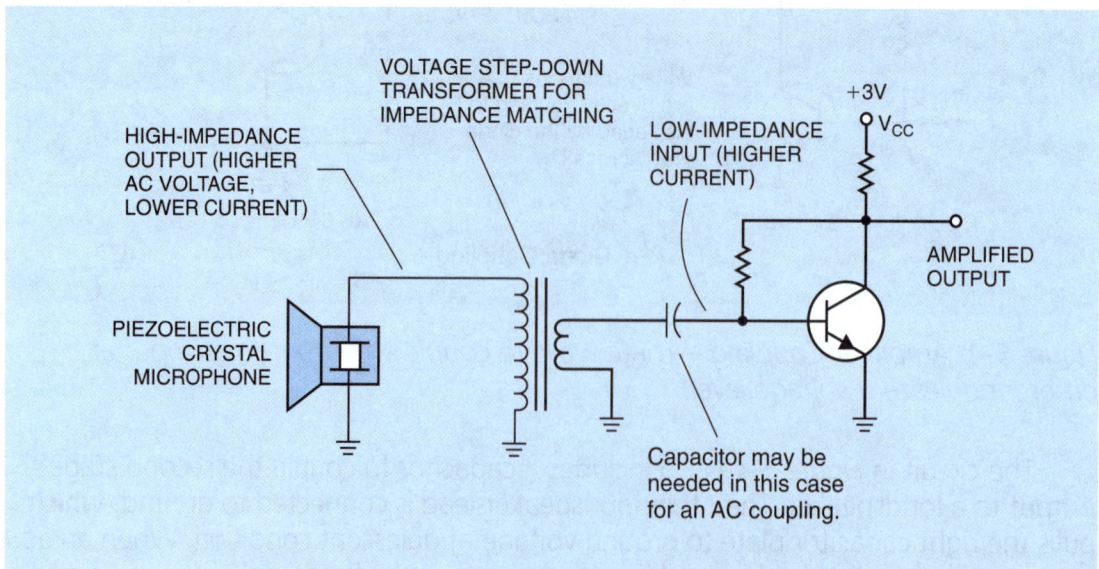

Figure 5-5. Impedance Matching – Using a small transformer to match the impedance of an amplifier input to the impedance of the signal source.

Similarly, we could use a transformer to match a low-impedance dynamic microphone to a high-impedance amplifier input. In that case, we would turn the transformer around and step the voltage up instead of down.

Notice carefully that *a transformer is not an amplifier*. Stepping voltage or current up with a transformer is not the same thing as amplifying it, because *no power is added;* in fact, some power may be lost. Power output will be equal to the power input times the transformer efficiency (usually less than 1.0).

A Triangle Symbol Simplifies Amplifiers

The preceding material should give you a start in understanding some variations you are likely to see in amplifier designs. With the advent of integrated circuits, many amplifiers used in circuits and systems are self-contained in an IC package and are not built from discrete components. As a result, schematic diagrams are simplified by using symbols for the amplifiers.

In many circuits and systems that use amplifiers, or that are made out of amplifiers, *we often don't need to know details of the amplifier circuits themselves.* It's enough to know that amplifiers are used in certain places, and maybe some details like the voltage gain, frequency response, and input and output impedances.

For those situations, a symbol that means any kind of amplifier is often used on schematic diagrams; this is especially true for systems using ICs. The symbol is a triangle, pointing like an arrowhead in the direction of signal flow. As shown in *Figure 5-6b,* the input goes into the wide end of the triangle and the output comes out of the point. Sometimes power-supply connections are shown going to the top and bottom (angled sides) of the symbol, but in most cases are left out.

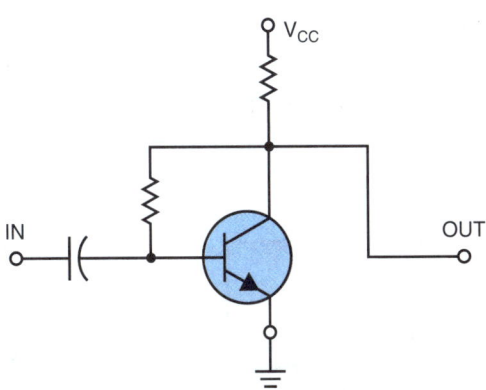

a. Simple Single-Input Amplifier

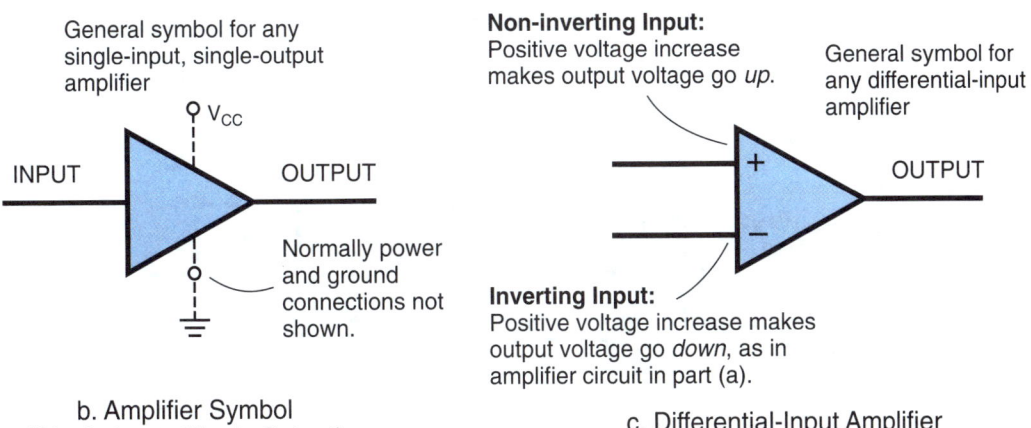

b. Amplifier Symbol
(Single Input, Single Output)

c. Differential-Input Amplifier

Figure 5-6. Amplifier Symbols – To simplify schematic diagrams, amplifiers are often shown as triangles.

Sometimes you will see two inputs on an amplifier symbol as shown in *Figure 5-6c*. One input is labeled with a plus sign and the other input is labeled with a minus sign. This is a *differential-input amplifier,* usually just called a differential amplifier.

We won't look into how differential amplifiers work. Just remember that the input with the plus sign is called the *non-inverting input*. A positive-going voltage on this input makes the output voltage go positive (up). The input with the minus sign is called the *inverting input*. A positive-going voltage on this input makes the output voltage go less positive (down, or negative, as in the simple common-emitter amplifier we discussed earlier). A differential amplifier amplifies the *difference between the voltages* at the two inputs. The output moves in the direction determined by which input is most positive with respect to the other input. This can be a very useful thing to do in many circuits and systems.

Positive Feedback Makes Any Amplifier an Oscillator

Finally, let's talk about a common amplifier-type circuit that is not usually called an amplifier. This is the oscillator (pronounced OSS-ilator, sort of rhymes with elevator). It is a circuit that generates a wave signal all by itself. The output voltage oscillates, meaning the voltage swings up and down in a regular pattern. So the output signal is a simple wave of some kind. *Figure 5-7b* shows the most common wave types: sine wave, square wave, and sawtooth wave.

In principle, nearly every oscillator is *an amplifier with positive feedback*. Positive feedback is also called *regenerative feedback*. It's the opposite of the negative or degenerative feedback that we discussed in connection with biasing. Positive feedback is a signal from the output of an amplifier that is fed back to its input in a manner that *reinforces* the input.

For the simplest situation, suppose we have a differential-input amplifier, represented by a triangle symbol. It has both an inverting and a non-inverting input. The signal fed back from the output to the non-inverting input reinforces the input

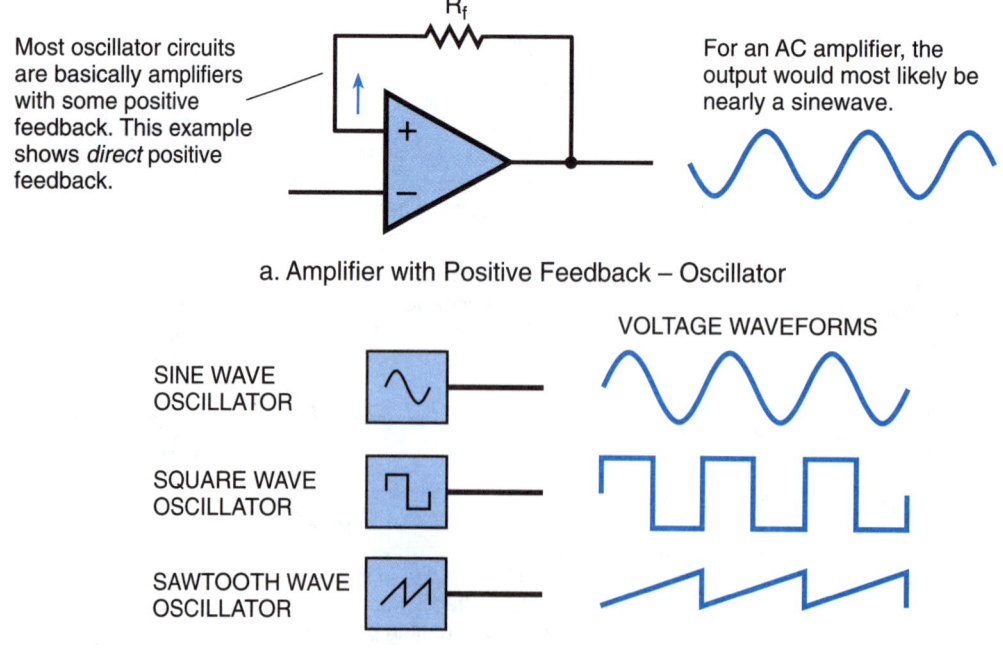

a. Amplifier with Positive Feedback – Oscillator

b. Oscillator Waveforms

Figure 5-7. Oscillators – An oscillator is basically an amplifier with positive feedback from output to input.

signal. In *Figure 5-7a*, the output is connected right back to the input through a resistor. We assume we don't have to do any coupling to match the quiescent voltage levels. This gives us the strongest, most direct positive feedback that we can get. A resistor is needed in the feedback connection, as shown in *Figure 5-7a*, to keep from burning out a transistor in the amplifier. The circuit will oscillate at what is called its natural resonant frequency.

Propagation Delay and Phase Shift

You might think the output voltage (and the input voltage) would immediately push each other all the way up or down and stay locked in place. That might happen if we could make the voltage change very slowly. But a phenomenon inside the amplifier called *phase shift* keeps that from happening.

You see, the output voltage from an amplifier never changes at *exactly the same time* as the input voltage. Instead, there is a certain time delay, caused mainly by stray capacitance and inductance in the wires and circuit devices. This is called the *propagation delay*. The output always changes a little bit later than the input. An output wave is *shifted* slightly later than the input wave that causes it. This is called a *phase shift*. It is measured in degrees, with 360 degrees for one full wave or cycle.

As an example, *Figure 5-8* shows how a sine wave is generated. A vector with a constant amplitude of A rotates from 0° through 90°, 180°, 270° and 360° (which is back to 0° for a new cycle). If the sine of the rotated angle θ (represented by the vector X), is plotted against degrees along the horizontal axis, and the points are connected, a sine wave results. This is shown as curve (1) A Sin θ. If we have an amplifier with an inverted output, the vector for the output will have an 180° phase shift. It is said to be 180° out of phase.

Curve (2) A Sin (θ +180°) represents the 180° out-of-phase signal. Note that it is exactly opposite from the A Sin θ curve. In other words, if this signal is fed back to the input, it would oppose the input signal. Now look at curve (3), it has a phase shift greater than 180°. In this case, note that portions of the cycle will reinforce the input and cause the circuit to oscillate. Thus, an oscillator is formed from an amplifier circuit by feeding back to the input from the output a signal with a phase shift greater than 180° and *with a gain greater than one*.

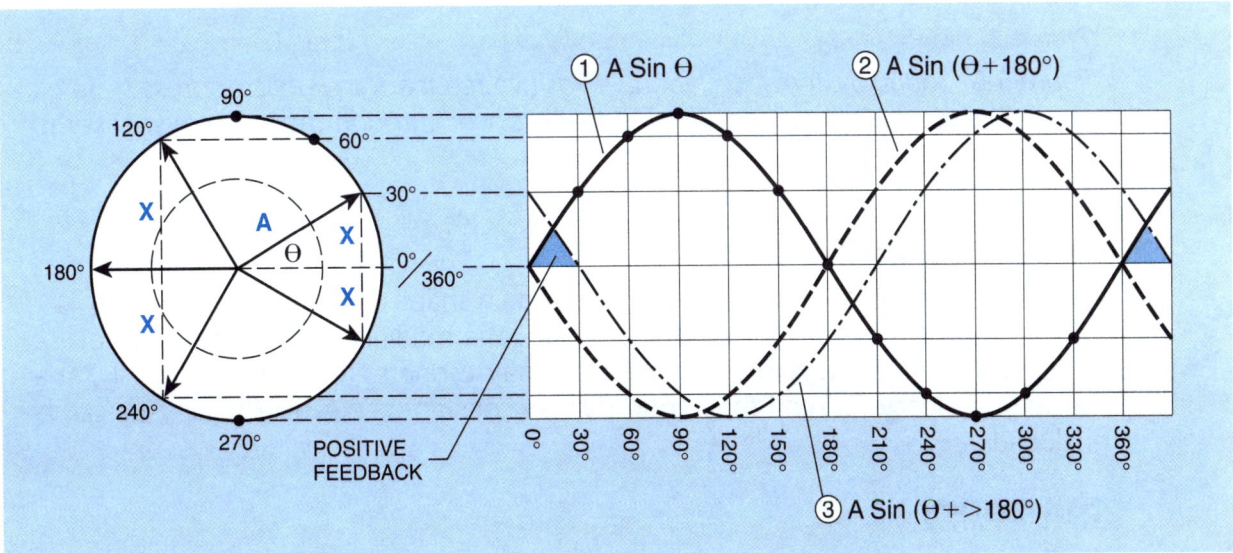

Figure 5-8. Plots of Sine Waves – The amplitude of the vector X represents the value of the sine of the angle θ as the vector A rotates through an angle θ from 0° to 360°. Plotting the value results in a curve A Sin θ. Other curves that are "out-of phase" can be plotted to show their time relationship.

Natural Oscillation Frequency

With strong positive feedback, an amplifier just naturally starts oscillating at a frequency which is high enough for the *phase shift to be greater than 180° and the gain around the loop is one or greater*. That is, the propagation delay is greater than half of one wave period. Whatever polarity the input signal has, the output reinforces it. So the output is forever supporting the input. The oscillations occur at the frequency at which the phase shift and gain rules are satisfied.

You sometimes hear feedback in a public-address system consisting of a microphone, amplifier and loudspeaker, as shown in *Figure 5-9*. If the microphone is placed in a position so that it "hears" the loudspeaker, positive feedback occurs and the system oscillates. The system oscillates at a frequency determined by the time delay for getting the audio signal from the speaker through the air feedback loop back to the microphone with a phase shift greater than 180° and a gain greater than one. The result is a very loud squeal.

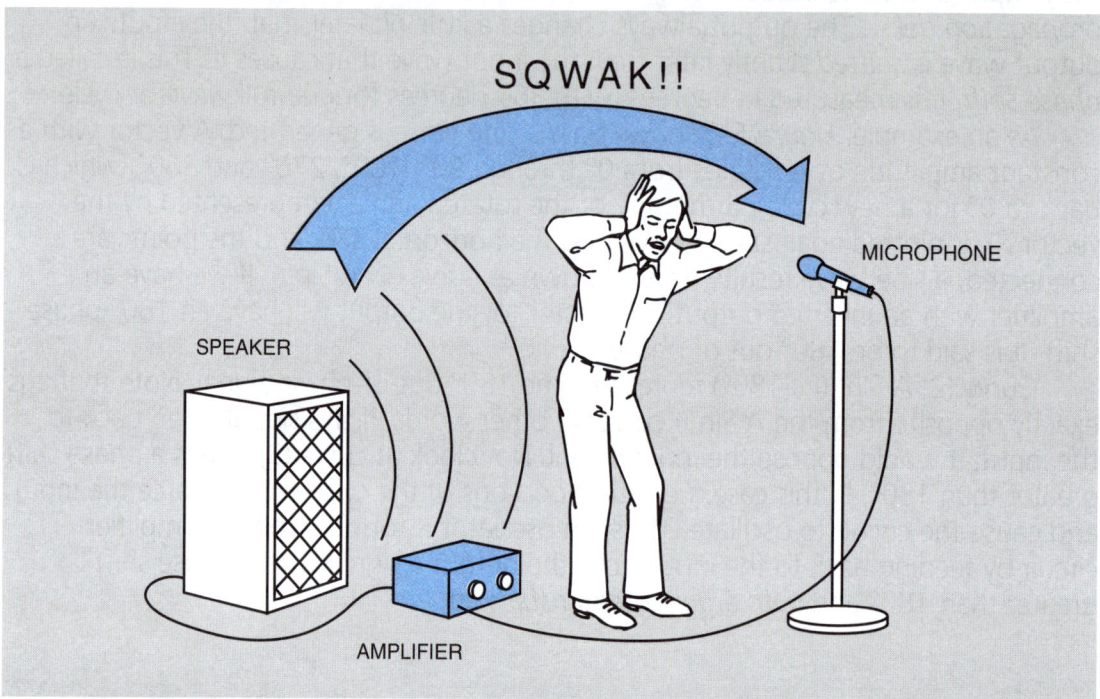

Figure 5-9. Sound System Oscillations – Positive feedback in Public Address system causes the system to oscillate. The feedback loop is through the air from speaker to microphone.

Controlled Oscillation Frequency

To make an oscillator produce waves of a certain shape and frequency, various modifications are made in the feedback path. The modifications usually involve inductors and capacitors in a configuration that resonates (bounces electrically) at a certain frequency. To keep the frequency very constant, a piezoelectric crystal can be included.

Summary

In this chapter we have found out about various types of transistor circuits used as amplifiers and oscillators. We have found out how to bias the circuits so they will operate properly. In the next chapter we will learn how such amplifiers are used in radio transmitters and receivers.

Quiz for Chapter 5

1. The circuitry of most electronic systems, such as television receivers, consists almost entirely of
   a. modulators
   b. oscillators
   c. amplifiers
   d. rectifiers

2. A circuit that amplifies voltage signals in one step is called a voltage amplifier _____ .
   a. bridge
   b. stage
   c. path
   d. network

3. In a transistor amplifier, the resistor that is connected from the base to the V_{CC} power supply is called a _____ resistor.
   a. bias
   b. load
   c. feedback
   d. stabilization

4. If the input voltage to an amplifier is 0.05 volt and it produces an output voltage of 2.0 volts, then the voltage gain is
   a. 40
   b. 120
   c. 200
   d. 400

5. For a given transistor, power-supply voltage, and load resistance, the operating point depends on the
   a. signal voltage
   b. bias resistance
   c. coupling capacitor
   d. bypass capacitor

6. If a transistor is biased about HALFWAY ON when there is no signal it is said to be class
   a. A
   b. B
   c. C
   d. D

7. If the input signal of an amplifier is 0.04 volt and it has a voltage gain of 30 then the amplifier stage will produce an output voltage of
   a. 0.3 volts
   b. 1.2 volts
   c. 2.4 volts
   d. 3.0 volts

8. Only one-half of the cycle of input waves are amplified in class _____ operation.
   a. A
   b. B
   c. C
   d. D

9. When the base bias resistor is tied to the collector output instead of the power supply it results in
   a. increased gain
   b. increased distortion
   c. positive feedback
   d. negative feedback

10. The emitter resistor causes _____ feedback.
    a. harmonic
    b. inductive
    c. positive
    d. negative

11. An amplifier that has two inputs, called the inverting and noninverting inputs, is called a _____ amplifier.
    a. para-phase
    b. duo-voltage
    c. differential
    d. push-pull

12. An amplifier will oscillate if it has sufficient
    a. supply voltage
    b. temperature
    c. positive feedback
    d. degenerative feedback

Answers:
1 c, 2 b, 3 a, 4 a, 5 b, 6 a, 7 b, 8 b, 9 d, 10 d, 11 c, 12 c

Questions for Chapter 5

1. What is the most basic kind of electronic circuit?

2. What is the fixed resistor connected from the base of a transistor to the power supply called?

3. What is the fixed resistor connected from the collector of a transistor to the power supply called?

4. What is a transistor's condition called at those times when there is no signal being applied at the input?

5. In what class is an amplifier operating if its bias resistance is low enough to make the transistor turn about *HALFWAY ON* when there is no signal applied to its input?

6. If a transistor is biased so only one-half of the cycle of input waves is amplified, what class is it?

7. If the bias resistor is tied to the output instead of the power supply, it is called _____ bias and results in _____ feedback.

8. Adding an emitter resistor to a common-emitter transistor amplifier stage causes a form of _____ feedback.

9. What are the advantages and disadvantages of negative feedback?

10. What is the most common type of ac coupling between amplifier stages?

11. When would a transformer be used for coupling instead of a capacitor?

12. An amplifier that has positive feedback is called an _____ .

13. An amplifier that has a voltage gain of 25 and has an input voltage of 0.040 volts (40 millivolts) would have what output voltage?

14. If a transistor has an h_{FE} of 120 and a base current of 0.050 mA (50 µA), what is its collector current?

15. If a common-emitter transistor amplifier has a power supply voltage V_{CC} of 9 volts and a collector current of 2 mA through a load resistor of 2 kΩ, what is the collector voltage?

CHAPTER 6
Radio Transmitters and Receivers

Radio is one of the most fascinating subjects that you can learn about and work with in electronics. It involves sending invisible energy through the earth's atmosphere and space without using wires. The energy consists of *radio waves*, which travel at the speed of light, which is about 186,000 miles per second, or 300,000 kilometers per second, or roughly 1,000,000,000 feet per second.

Special electronic circuits called *radio transmitters* send out radio waves which carry information for various purposes. The information usually consists of voice, music, pictures, or digital data, or combinations of these. Other electronic circuits called *radio receivers* catch radio waves passing by. The receiver circuits select particular radio waves and retrieve the information from them so it can be used.

Electronic circuits were developed in the early 1900s to make radio communication practical. Earlier electromechanical methods didn't work very well with the high-frequency electrical signals that are involved in radio. High frequencies and radio waves go hand in hand — you can't understand one without understanding the other.

High-Frequency Signals Make Electric Waves In a Long Wire

To understand how electronic circuits transmit and receive radio waves, we first have to think about how *electric* waves travel along a wire. Take a look at *Figure 6-1a*. Imagine that an oscillator in the box on the left is sending an alternating current signal through the wire to some receiving circuitry in the box on the right. The wire carrying the signal is called the *transmission line* and we first imagine it is about six inches long. The frequency of the ac signal is 100,000,000 cycles per second (usually written as 100 megahertz or 100 MHz).

Like any other alternating current, this signal carries energy from the oscillator to the receiving circuitry. This transfer of energy is as if you were working an old-fashioned hand pump very rapidly. You are the oscillator and the pump is the receiving circuitry. The transmission line is like a short piece of heavy rope between your hand and the pump handle. The up and down motion of your hand is like the alternating electric current. With each quick stroke up or down, energy flows from your arm to the pump. The rope moves in a sort of wavy fashion, but we can't see any waves in the rope because it is so short.

In *Figure 6-1b,* we see the same situation, but with a much longer transmission line and a longer rope tied to the pump handle. You whip your end of the heavy rope rapidly up and down. Now we can see *waves traveling along the rope*. At the other end, those waves flip the pump handle up and down as though you actually had your hand on it.

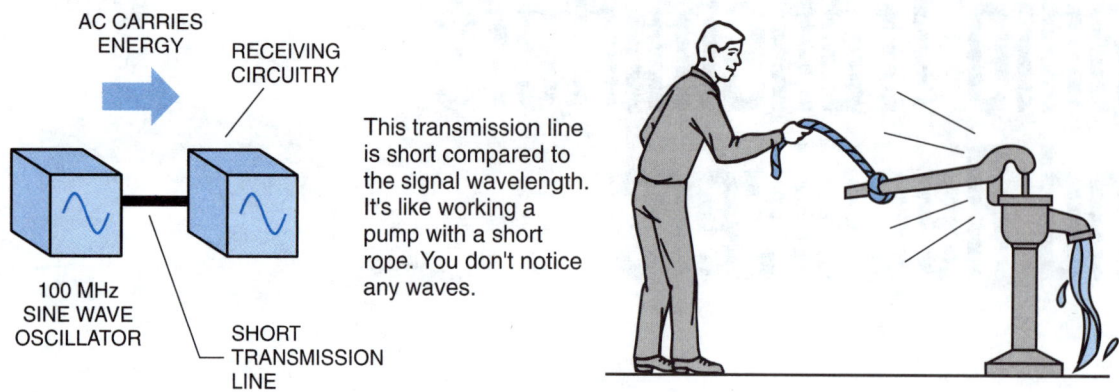

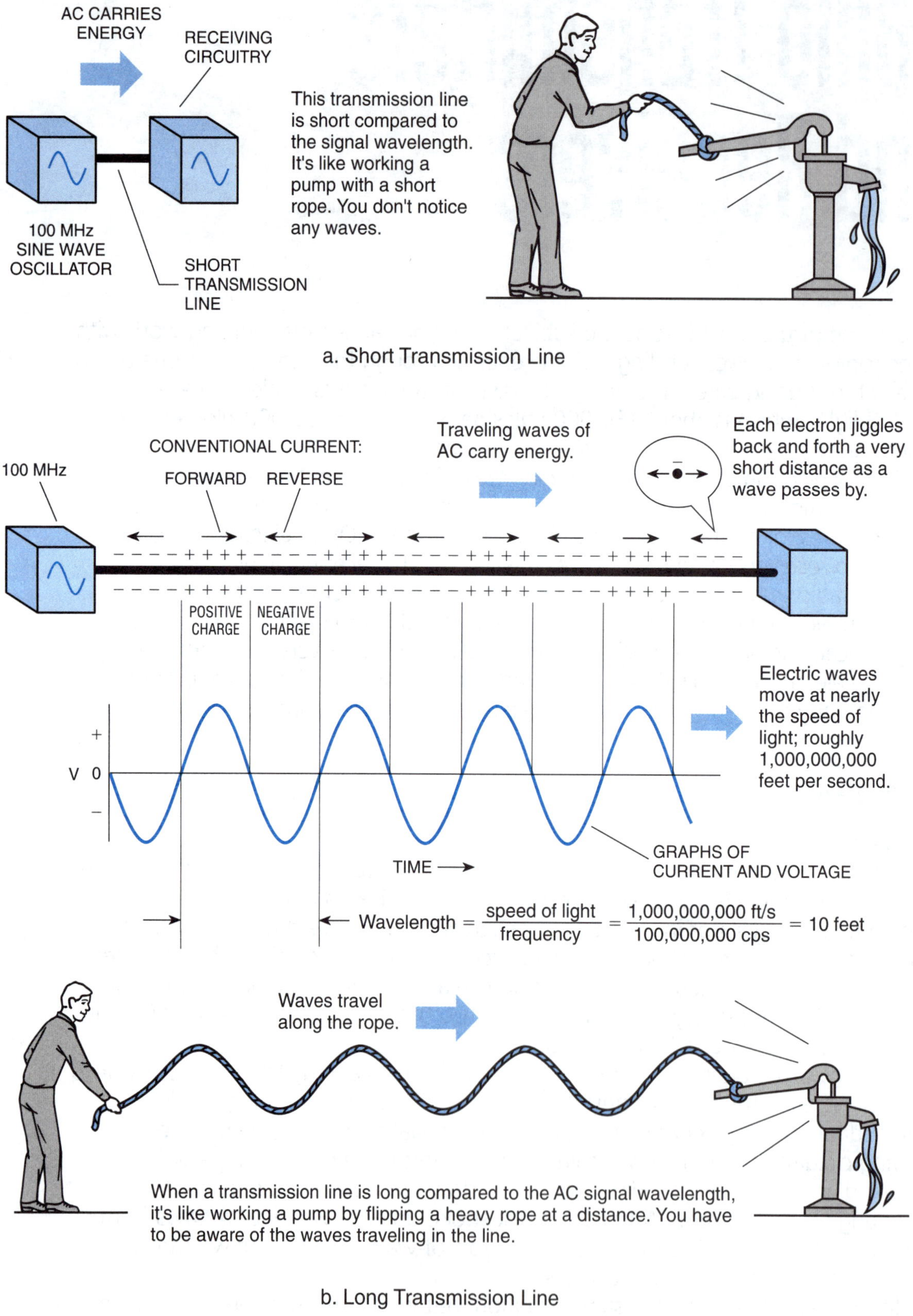

Figure 6-1. Alternating Current in Wires – To understand how radio waves are generated, you must first think of how ac waves travel in a wire.

We get very similar *waves of alternating current* in the transmission line. Notice the graph of voltage and current along the line. They are shown as if a picture was taken at some instant in time. The voltage and current form sine waves, and these waves move together through the wire. Along the peak of each wave, the wire has a slight positive charge, and there is conventional current in the forward direction (to the right). In the valley of each wave, the wire has a slight negative charge, and current is directed back toward the oscillator. As each wave passes any given point in the wire, electrons at that point surge forward and then backward. Waves in the rope travel about 30 miles per hour, depending on how tightly the rope is stretched and how heavy it is, but *electrical waves travel at nearly the speed of light*.

The *wavelength* of the waves is very important. Wavelength is the distance between any two adjacent peaks (or any two valleys) of the signal along the wire. The higher the frequency, the shorter the wavelength. To be exact, *the wavelength equals the speed of light divided by the frequency*. In this case, the wavelength of 100 MHz waves is roughly ten feet.

$$\text{Wavelength} = \frac{1{,}000{,}000{,}000 \text{ feet/second}}{100{,}000{,}000 \text{ cycles/second}} = 10 \text{ feet}$$

AC Generates Oscillating Electric and Magnetic Fields

Whenever a single wire is long enough to hold at least one-fourth of a wavelength, then it becomes an efficient radiator of radio waves. That is, the wire becomes a fairly efficient transmitting antenna. It happens because of the way the electric and magnetic fields around the wire change together.

Figure 6-2 shows how these fields work together. There are three snapshots of the transmission line at the same instant. *Figure 6-2a* shows only the electric fields. *Figure 6-2b* shows only the magnetic fields. *Figure 6-2c* shows a side view and an end view of the electric fields and the magnetic fields together in the same space around the wire. The field lines cannot be seen, but in order to describe them, they are considered as imaginary lines in space and drawn as shown to be able to talk about their effect.

The electric field is also called an electrostatic field. In *Figure 6-2a*, arrowheads on the lines indicate the direction of the electrostatic force on any positive charges in that space. For example, if there were some positively charged dust particles near the wire at the instant shown in the figure, they would be trying to move along the lines. *Each arrow runs from a positive charge to a negative charge*. In this case, the electric field lines run from segments of the wire that are positively charged to neighboring segments that are negatively charged.

Figure 6-2b shows the magnetic field lines. Arrowheads on the lines show the direction in which the north-seeking end of a tiny compass needle would point at various positions around the wire. For example, if there were tiny magnetized particles of iron dust in the air at the instant represented in the figure, their north-seeking poles would be trying to point along these lines.

The magnetic fields are created by current in the wire, and the lines form circles around the current. In *Figure 6-2b,* where conventional current is forward (to the right), the arrows circle clockwise as we look from the left end toward the right. Where conventional current is back to the left, the arrows circle counterclockwise as we look from the left end.

Now think of one particular point along the transmission line and consider the fields in space around that point, as indicated in *Figure 6-2c*. As electric waves pass by, *the electric and magnetic fields increase and decrease together*; that is, they get stronger in one direction, then weaken and disappear, then get stronger in the other direction.

Also notice that *the electric and magnetic field lines always cross each other at a 90° angle*. The electric field lines run along the wire, expanding out from the wire and then collapsing back to the wire. The magnetic lines are circles around the wire, again expanding and collapsing as the current increases and decreases. The combination moves in the direction the wave travels.

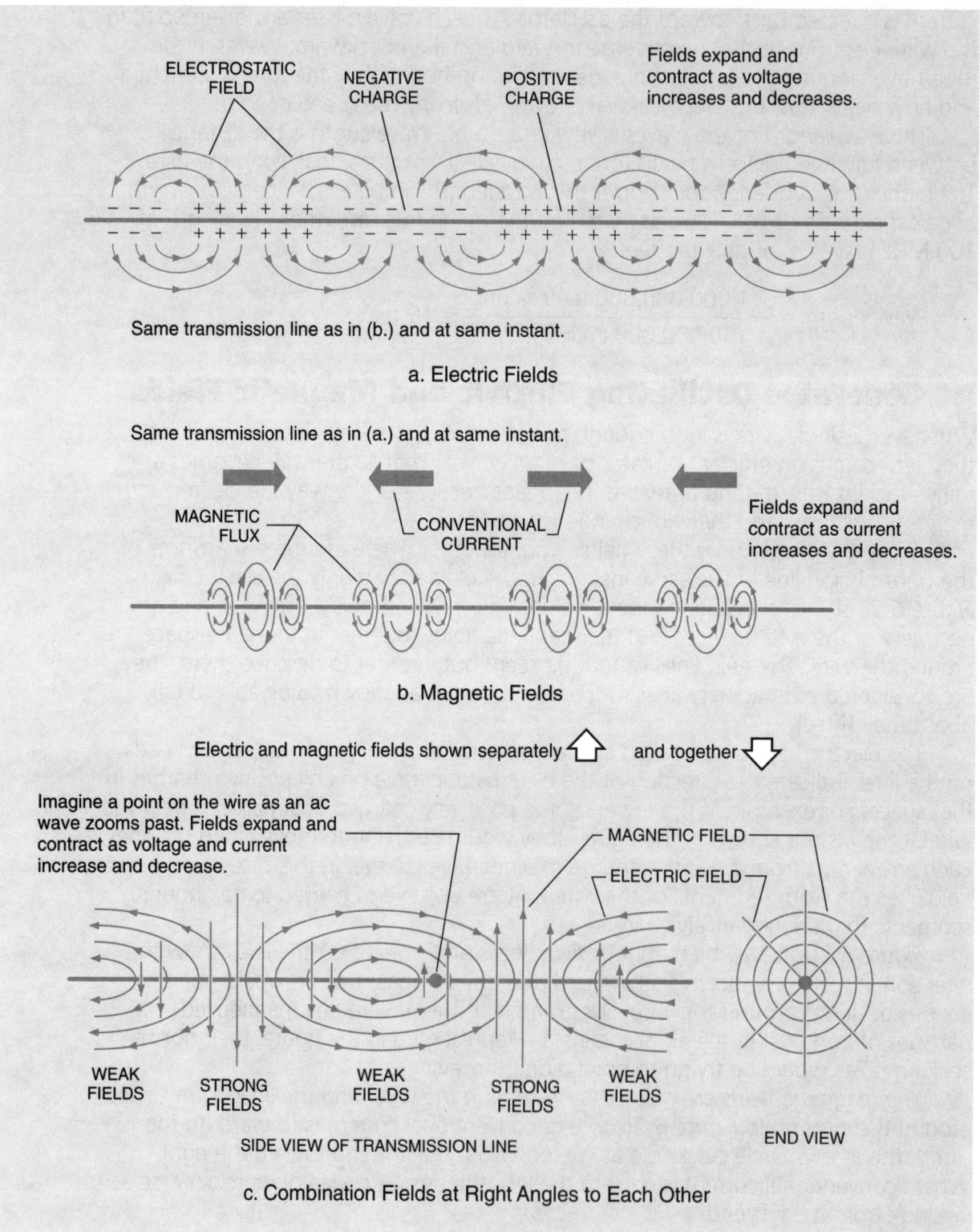

Figure 6-2. Electric and Magnetic Fields – At a point in a transmission line, the electric and magnetic fields around the point increase and decrease as the ac wave passes, but stay perpendicular to each other.

Radio Waves Are Electric and Magnetic Fields Oscillating Crosswise to Each Other

Now the amazing fact is, *this is just what radio waves are*. They are electric and magnetic fields moving in space — not in wires — oscillating crosswise to each other at 90° angles. Since they are a combination of electric and magnetic fields, another name for radio waves is *electromagnetic waves*.

Radio waves are the same electric and magnetic fields caused by alternating current in a wire, but they have been released, or sent, into space. It's as though the fields break off and fly away at the speed of light. This is shown in *Figure 6-3a*.

The reason that these oscillating, crosswise fields travel out into space is that they keep each other going. An increasing or decreasing magnetic field creates an electric field perpendicular to itself. An increasing or decreasing electric field, in turn, creates a magnetic field perpendicular to itself. This happens even though there are no charged particles traveling with the fields. Both fields contain energy, which they hand off to each other as they oscillate.

Here's an important point to remember: *The frequency of the radio waves is always the same as the antenna electric waves* that generate them. In our example, the oscillator feeding the antenna is creating 100 MHz ac waves, so we get 100 MHz radio waves. (This frequency is near the middle of the FM band on a broadcast radio receiver.) The speed of the radio waves in space is nearly the same as that of electric waves in wires, so both wavelengths are approximately 10 feet.

Also notice in *Figure 6-3a* that the direction in which the oscillating crosswise fields travel is *perpendicular to both of them*. For this reason, we say that radio waves are *transverse* waves, meaning crosswise. Waves in a stretched rope or strings in a musical instrument are also transverse. In contrast, sound waves and alternating current waves move back and forth along the path of the waves. Waves of that type are called longitudinal or pressure waves.

Because of the shapes of the electric and magnetic fields around the wire, the transmitted radio waves are strongest in all directions perpendicular to the wire. This is indicated in *Figure 6-3b* in the view looking at the end of the wire. The radio waves *radiate* outward from the wire like ripples in a bathtub when you vibrate one spot on the water surface with a fingertip.

This comparison with water ripples also helps us understand why an antenna isn't very effective for generating radio waves that are very much longer than the antenna. The ripples that you make with a finger are never much longer than the width of your finger, but if you use your fist, corresponding to a longer antenna, the waves are longer.

Transmission Lines Prevent Radio Waves from Radiating

As shown, electric and magnetic waves in wires radiate into space when they are delivered to a radiating antenna. The antenna length is cut to a particular wavelength to deliver the energy into radio waves traveling through space.
To keep from radiating energy from normal wires, transmission lines are used. A transmission line is used to move a signal along the line from one place to another to *prevent* radio waves from radiating away from the line.

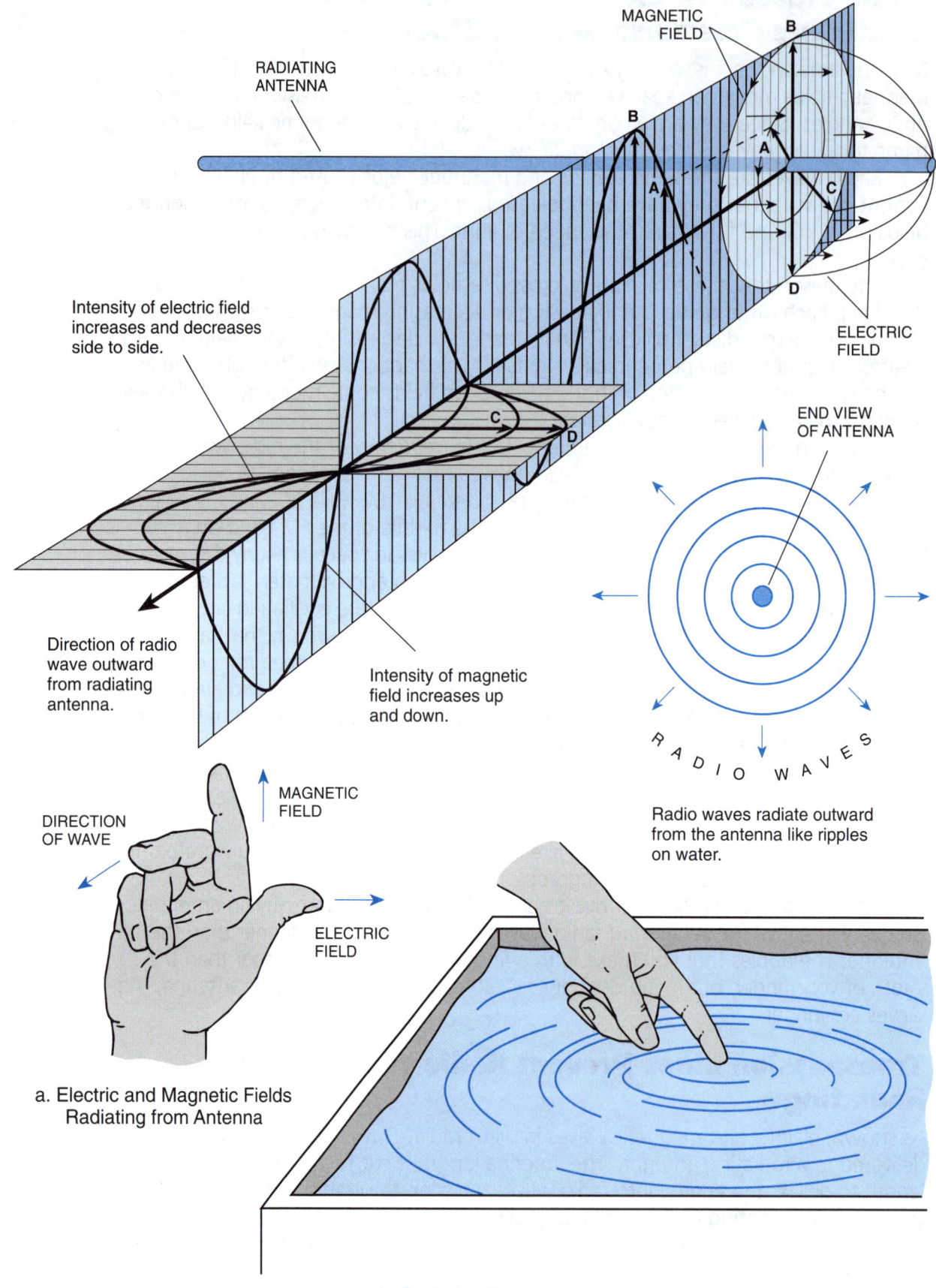

Figure 6-3. Radio Waves – Oscillating, perpendicular electric and magnetic fields radiate out from antenna, forming radio waves.

Example 1. Wavelength of a Radio Wave of Given Frequency

What is the wavelength in feet of a radio wave with a frequency of 540 kHz?

Wavelength is the distance a cycle of the radio wave occupies in space as the wave travels. If we stand in space and watch a wave go by, the distance from a peak in the wave to the next peak that passes is the wavelength. The distance between the peaks occurs in one cycle.

Distance is a velocity multiplied by a time. The time of one cycle of a wave is 1 divided by its frequency, or 1/f. Since the velocity of radio waves is 300,000,000 meters/second, wavelength (identified by the Lambda symbol λ) is:

$$\lambda = 300{,}000{,}000 \text{ m/s} \times \frac{1}{f} \text{ s} = \frac{300{,}000{,}000}{f} \text{ m}$$

λ for a 540 kHz radio wave is:

$$\lambda = \frac{300{,}000{,}000}{540{,}000 \text{ Hz}} = 555.55 \text{ meters}$$

Since one meter = 3.28 feet, the velocity of radio waves in feet per second is:

$$\lambda = 300{,}000{,}000 \times 3.28 \text{ ft/s} = 984{,}000{,}000 \text{ ft/s}$$

∴ λ of a 540 kHz radio wave in feet is:

$$\lambda = \frac{984{,}000{,}000}{540{,}000} = 1822.2 \text{ feet}$$

Example 2. Convert to Frequency from Wavelength

What is the frequency of a radio wave that has a wavelength of 328 feet?

$$f = \frac{984{,}000{,}000 \text{ ft/s}}{328 \text{ ft}} = 3{,}000{,}000 \text{ Hz or 3 MHz}$$

Coaxial Lines

One of the most common transmission lines is a *coaxial cable*. As shown in *Figure 6-4a,* a coaxial cable has an insulated wire (inner conductor) running along the centerline of a hollow tube (outer conductor) made of aluminum foil, braided copper wire, or solid copper. The outer conductor acts as a shield that keeps energy from radiating from the center conductor.

In some cases, the outer conductor is grounded at both ends of the line. In other cases, the inner and outer conductors at each end are connected to a transformer winding as shown in *Figure 6-4a*. One transformer transmits equal and opposite signals in the two conductors, and the other transformer receives the signals. In either case, the electric and magnetic fields around the center conductor stay bottled up inside the cylinder formed by the outer conductor.

Dual Lines

Another way to keep radio waves from radiating very much is to use a *dual transmission line* or *twin lead* as shown in *Figure 6-4b*. In this type, two parallel wires are kept a short, constant distance apart by insulation. Equal and opposite ac signals are carried in each line. The signals are usually transmitted and received by transformers. In effect, any radio waves that one wire might create are canceled out by the other wire. As a result, most of the energy flows along the dual line instead of

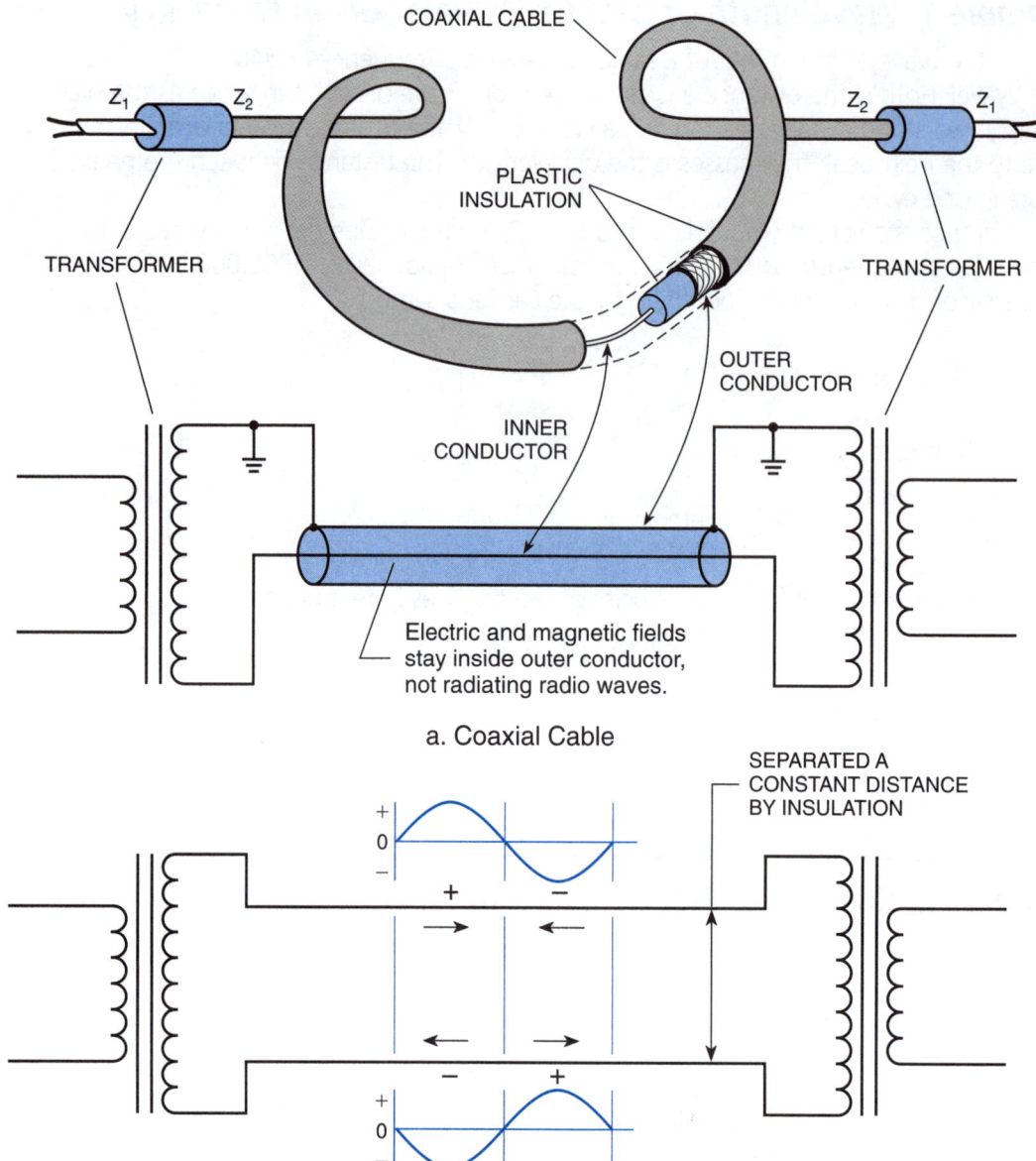

Figure 6-4. Transmission Lines – Coaxial cable and dual transmission line keep radio waves from radiating, so little energy is lost.

radiating as radio waves. The fields do extend a short distance out from the wires, so to keep from losing energy, the dual line is kept away from conductive or magnetic surfaces.

Efficient Transmitting Antennas Use Standing AC Waves

The purpose of a transmitting antenna is the opposite of a transmission line. We want to send out as much energy as possible as radio waves. To do this, we set up *standing electric waves* in the antenna. The waves are much like those in a transmission line, but the waves don't travel. They just bounce back and forth in one place and radiate radio waves.

Figure 6-5 shows the idea of standing waves in a rope such as we used to illustrate traveling waves with the pump. This time, we tie the far end of the rope to a tree or a post so that no energy comes out there.

First, suppose you flip the rope up and down in a way that makes the middle of the rope bounce up and down as shown in *Figure 6-5a*. This is called a *standing half wave*. At any instant, the rope forms either one peak or one valley, but not both. Assuming your elbow is stationary, the distance from your elbow to the tree is half of one wavelength.

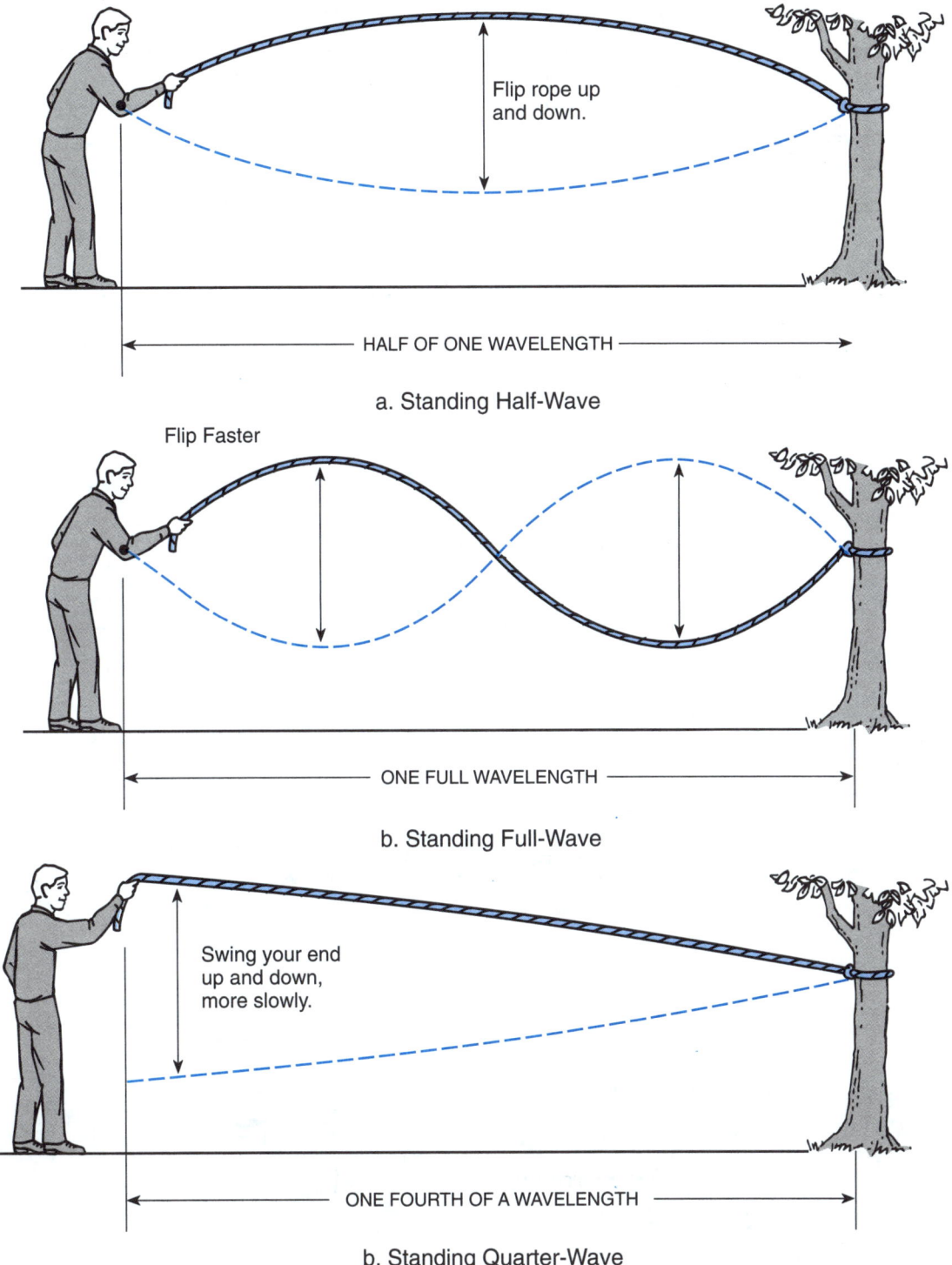

a. Standing Half-Wave

b. Standing Full-Wave

b. Standing Quarter-Wave

Figure 6-5. Standing Waves – Radio transmitting antennas contain ac waves that are standing, not traveling as in a transmission line. You can demonstrate the most common kinds of standing waves by flipping a rope in certain ways.

Next, if you flip the rope twice as fast, you can get a *standing full wave* as shown in *Figure 6-5b*. The middle of the rope is stationary, but the halves on each side of the middle bounce up and down. At any instant, the rope forms one peak and one valley. There is one full wave between your elbow and the tree.

Third, if you move the free end far up and down instead of flipping it from your elbow, you can get a *standing quarter wave* as shown in *Figure 6-5c*. It's as though you had hold of the middle of a rope with a standing half wave. The distance from your hand to the tree is one-fourth of one wavelength.

There are many different ways to create standing electric waves in transmitting antennas. *Figure 6-6* shows two common concepts called the *quarter-wave* or Marconi antenna and the *half-wave dipole* or Hertz antenna. In both cases, the antenna is usually at the end of a coaxial or dual transmission line.

A quarter-wave antenna is just a wire whose length is about one-fourth of the wavelength of the ac signal that is to be radiated as radio waves. An example is shown in *Figure 6-6a*. For our 100 MHz oscillator, the antenna would be about 2½ feet long. A ground plane is a very important part of this antenna.

For a half-wave dipole antenna, we take the two ends of a dual transmission line and bend them out at right angles, in line with each other, as shown in *Figure 6-6b*. Each free end is a quarter wavelength long, so the two tips are a half wavelength apart. The result is like two quarter-wave antennas pointing in opposite directions, with opposite signals in them.

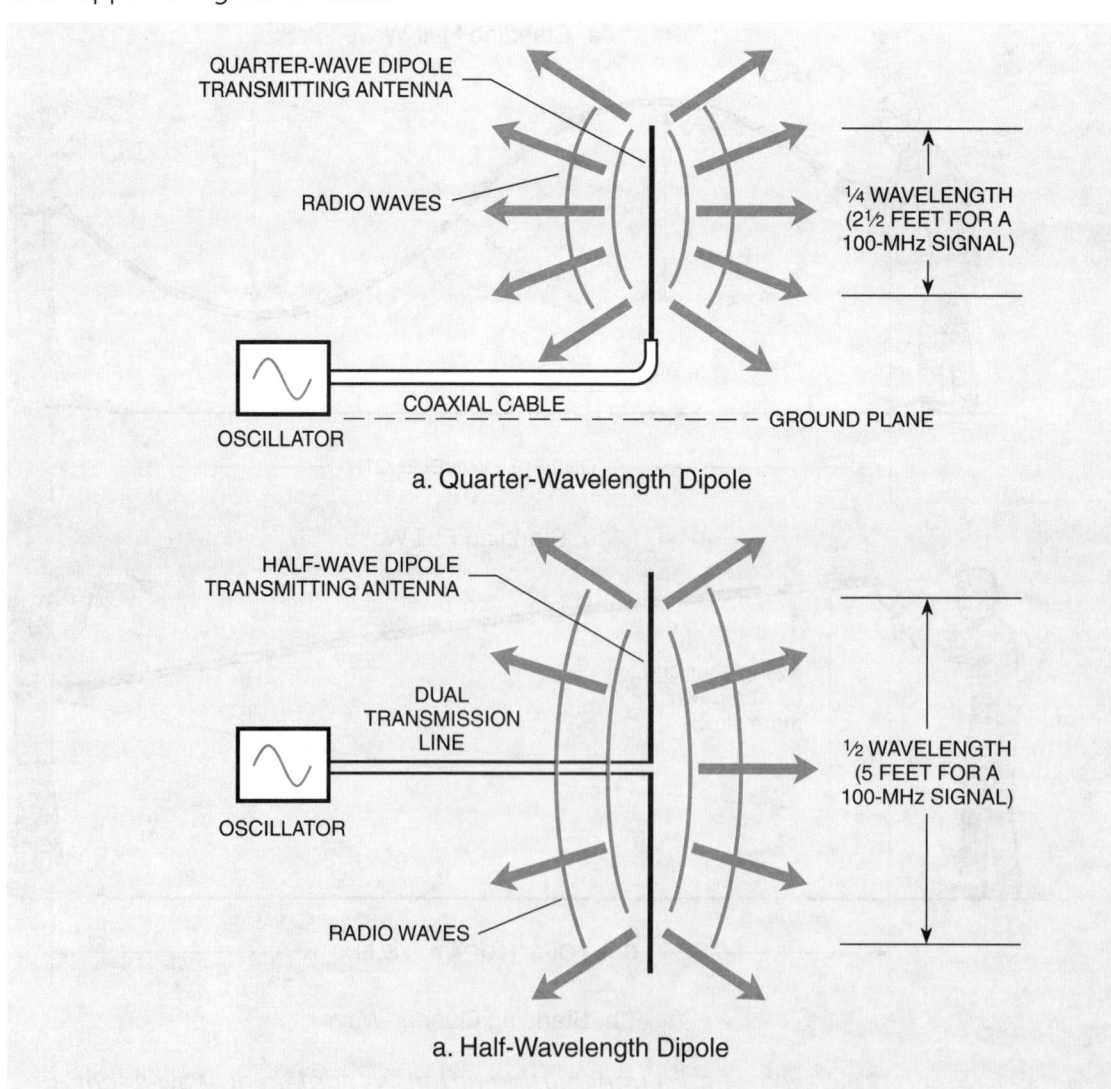

Figure 6-6. Radio Antennas – Two common styles of radio transmitting antennas create standing ac waves.

In both cases, but especially in the half-wave dipole, the alternating current in the antenna *resonates*. That means the antenna is tuned to the frequency of the signal to be radiated. It's a special case where the ac bounces back and forth by itself. The same thing happens with sound in a musical instrument, whether it is a violin string, a kettle drum head, or an organ pipe. The oscillator signal keeps the resonance going, radiating the incoming energy with high efficiency so little energy is lost.

For lower frequencies, transmitting antennas can get quite long. The lowest radio frequency in the AM broadcast band is about 540 kilohertz (kHz). Waves at that frequency are about 1800 feet long, so a half-wave dipole antenna would be about 900 feet from end to end.

Example 3. Half-Wavelength Antennas

What is the length in feet of a half-wavelength antenna tuned to a frequency of 150 MHz radio wave?

$$\lambda = \frac{984,000,000}{150,000,000} \text{ ft} = 6.56 \text{ ft}$$

∴ a half-wavelength antenna is 6.56 ft/2 = 3.28 feet long.

Example 4. Quarter-Wavelength Antennas

What is the length in meters of a quarter-wavelength antenna tuned to a frequency of 900 MHz?

$$\lambda = \frac{300,000,000}{900,000,000} = 0.333 \text{ meters}$$

∴ a quarter-wavelength antenna is 0.333/4 = 0.083 meters long.

Since 1m = 3.28 ft, 0.083m × 3.28 ft = 0.273 ft or about 3 inches.

Radio Waves Induce AC Signals In a Receiving Antenna

We now know how radio waves are created and transmitted. At a radio receiver, the process is just the reverse.

Whenever a radio wave passes by any piece of conductive material, a voltage is *induced* in the material by the electromagnetic radio wave. The conductive material may be a metal building, a wire fence, or the antenna of a radio receiver. The electromagnetic field induces a voltage in the receiving material (antenna). This voltage produces a current in the input receiving circuit that is a mirror image of the current that generated the radio waves. The total action from transmitter to receiver is summarized in *Figure 6-7*.

The magnetic field around the transmitting antenna radiates out from the antenna and induces a voltage in any receiving antenna it crosses. The first receiving antenna is very close so the action is more like a transformer, but the second receiving antenna is at a great distance and shows the radio waves traveling through space before they cut across the receiving antenna. The electric field is also present at 90° to the magnetic field, but it is not shown to keep *Figure 6-7* simpler.

The induced ac signal is a very weak copy of the original signal in the transmitting antenna. Electronic circuitry in the radio receiver amplifies these weak signals and reproduces the same information that was fed into the transmitter. We'll talk about that a bit later.

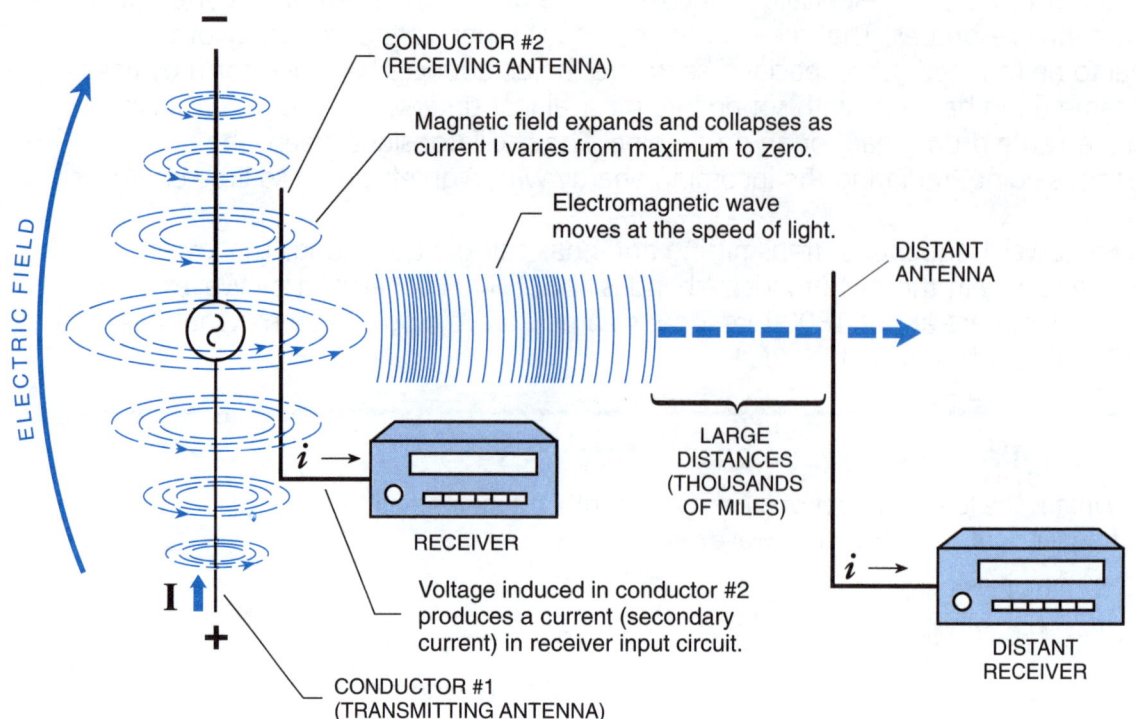

Figure 6-7. Transmitter to Receiver – Radio waves from transmitting antennas induce signals in receiving antennas as they pass by.
Source: *Listening to Shortwave,* Ken Winters, ©1993 Master Publishing, Inc., Richardson, Texas

Receiving Antennas

A wire works best as a receiving antenna if it is pointing parallel to the electric field of the radio waves as indicated in *Figure 6-7*. We say the antenna is *vertically polarized*. Since the electric field at the transmitting antenna is vertical (goes from top to bottom or vice versa), then the receiving antenna should be vertical to catch the maximum signal. If the antenna is parallel to the magnetic field (in this case, horizontal), or parallel to the direction of travel, you will get very little, if any, induced ac. If the transmitting antenna were horizontal, the electric field would be horizontal and the radio waves would be *horizontally polarized*.

The ac waves will build up to much greater strength if the current in the receiving antenna is allowed to *resonate* as in a transmitting antenna. By making the antenna length a half wavelength or a full wavelength, the antenna is said to be tuned to a particular frequency. However, if the receiver must pick up more than one frequency, you don't usually change the antenna length whenever you switch the station. One antenna length will work reasonably well for all frequencies over a range of about three to one; for instance, one length will work for the AM broadcast band which ranges from about 540 kHz to about 1600 kHz.

You may see another kind of antenna, like the one shown in *Figure 6-8,* in some AM radios. It's called a ferrite bar antenna or loopstick. It looks like a round stick with some wire coiled around it. This is actually a sort of transformer with a straight core made of magnetic iron oxide called ferrite. Like any other magnetic material, the ferrite draws in and concentrates any magnetic flux in the immediate neighborhood. The magnetic field of a passing radio wave causes vibrations in the flux through the core, inducing ac in the coils. This kind of antenna works best when the bar is pointed parallel to the magnetic field of the radio waves, which, of course, would be 90° to the way the electric field is polarized.

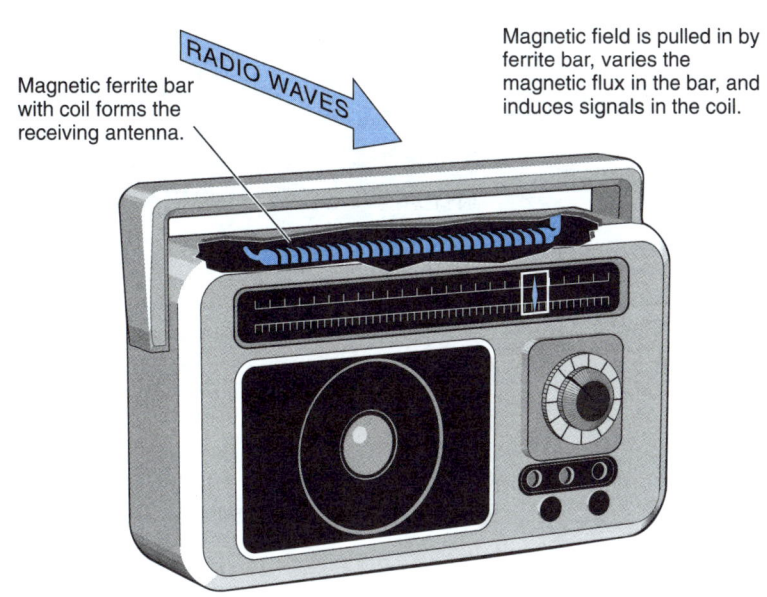

Figure 6-8. Ferrite Receiving Antenna – The electromagnetic radio waves are drawn in by the ferrite magnetic material, vary the bar flux, and induce signals in the antenna coil.

A Modulator Puts Information Into Radio Waves

We have seen how plain radio waves of a particular frequency are transmitted and received. These waves are called *carrier waves* because they can carry useful information; usually sound, digital data, or pictures. To do that, the transmitter has to change the carrier waves in some way that corresponds to the information, and the receiver has to detect the changes to recover the information. Changing the carrier to put information into it is called *modulation*. Getting the information out at the receiver is called *detection* or *demodulation*. Three ways of modulating are shown in *Figure 6-9*.

CW or On-Off Modulation

The simplest and oldest form of modulation is called *continuous wave modulation*, or CW modulation. It only works for on-off information, like that created by telegraph signals and computer data.

In an old-fashioned CW radiotelegraph transmitter, the telegraph operator moves the telegraph key down and up to close and open a circuit that turns an oscillator on and off. As indicated in *Figure 6-9a,* this turns the transmitter output on and off so the carrier signal is broken into short and long pulses. These pulses correspond to the dots and dashes of the International Morse Code. The radiotelegraph receiver generates a tone in a loudspeaker or earphones when the carrier is on and is quiet when the carrier is off. The receiving operator hears short and long beeps and writes down the letters they stand for.

Amplitude Modulation

The sound on a broadcast AM radio and the picture on a television receiver are carried by a method called *amplitude modulation*, abbreviated AM. The audio information from a microphone and the video signal from a TV camera consist of very complex voltage waves. A graphic trace of this information signal would typically look like random squiggles.

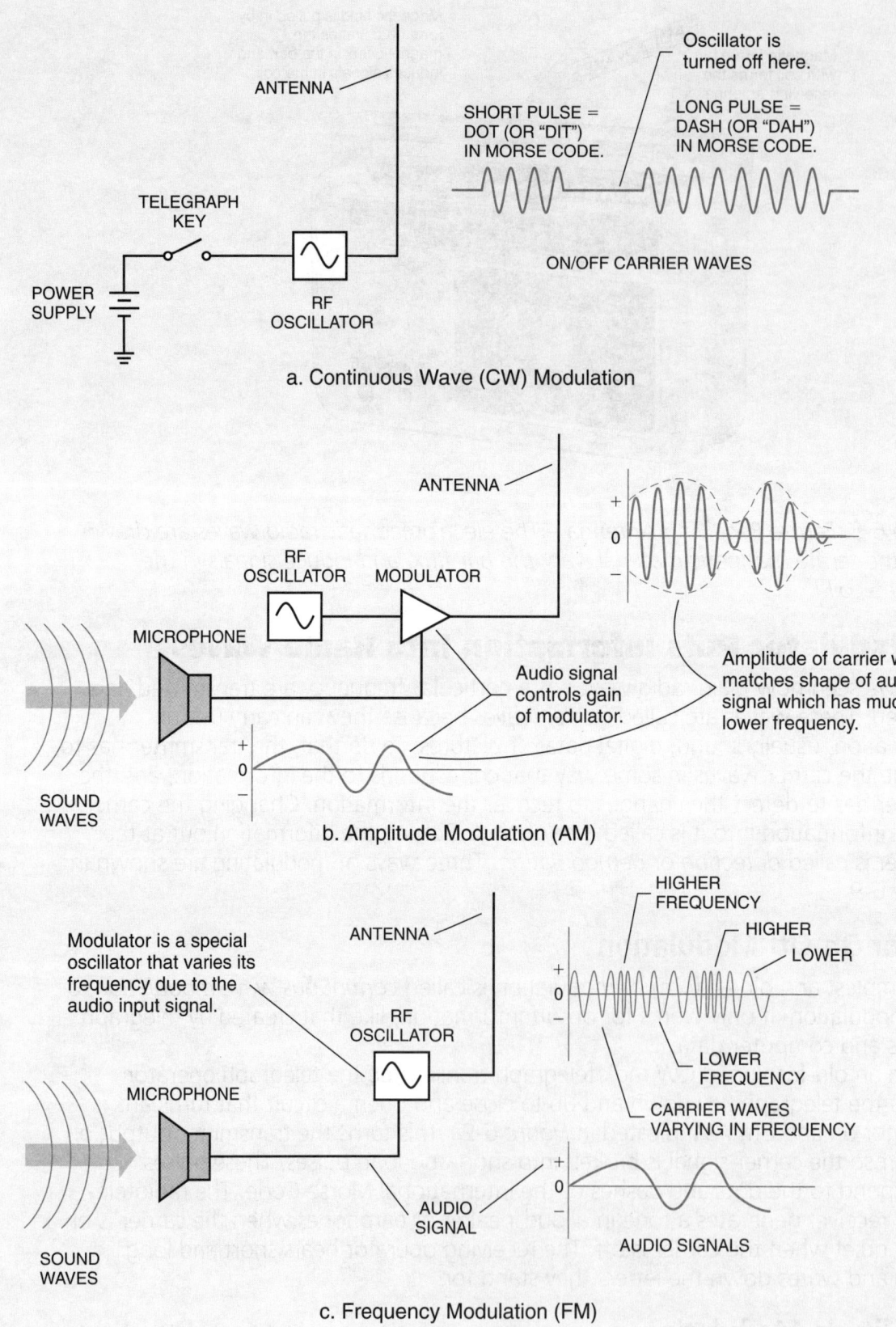

Figure 6-9. Modulation – Three ways to modulate carrier radio waves to send information.

The amplitude modulator circuit is a special type of amplifier. Its gain is controlled by the information signal that is applied to one of its inputs as shown in *Figure 6-9b*. The other input to the amplifier comes from an oscillator called the radio frequency (RF) oscillator or carrier oscillator. The amplitude (height or strength) of the carrier wave is varied to match the voltage of the audio or video signal. A more positive voltage causes taller waves and a less positive or negative voltage causes lower waves. The output is amplitude-modulated carrier waves, which are usually amplified to produce high-powered copies that go to the transmitting antenna.

Frequency Modulation

The sound that is broadcast for FM radios and television receivers is input to the carrier by a method called *frequency modulation*, abbreviated FM. As shown in *Figure 6-9c*, the modulator is a special oscillator whose frequency is varied by the information signal. When the audio signal voltage goes up, the carrier frequency is increased a little bit, so the carrier wavelength gets a little bit shorter. When the audio signal voltage goes down, the carrier frequency is decreased a little bit, so the wavelength gets a little bit longer.

A Diode Makes a Simple AM Radio Detector

To give you an idea of how radio receivers work, let's consider the very simplest and oldest type of electronic receiver — the diode detector. It works for AM sound signals if the radio waves are strong enough.

There are several different ways to use a diode detector. At the very least, all you have to do is connect one lead of a dynamic earphone to a long wire strung in the air and connect the other lead to earth ground through a water pipe, as shown in *Figure 6-10a*. Connect a diode — almost any type will work — across the earphone leads. Polarity of the diode doesn't matter, so you can connect either end to ground. When this type of receiver was invented about 1900, the diode was made of a crystal of lead sulfide, so the receiver was called a "crystal radio set." Obviously, this is a very crude type of crystal set. More commercial sets can still be purchased today in toy or science stores.

At the antenna connection, we mainly have a combination of AM signals from all the strongest radio broadcasting stations. The rf current can easily pass through the earphone coil, but the earphone mechanical diaphragm can only respond to the audio sound signals so the carrier frequency is not heard. Only the amplitude variations of the carrier come through — which, of course, is the sound we want.

The diode *rectifies* the signal. As shown in *Figure 6-10b*, only the radio frequency (RF) signals that are positive above ground cause current through the earphone coil. Any RF signals that are negative to ground are shorted out by the diode as it conducts and holds any voltage variations to just the forward diode voltage (0.5V-0.7V). The power for the signals comes entirely from the radio waves, so a battery is not needed.

The detector works because *the current driving the diaphragm varies according to the amplitude of the ac signal*. The greater the amplitude, the more force to move the diaphragm. Thus, we have a copy of the original AM audio signals from the earphone. You won't hear any FM stations, because changes of a carrier frequency don't affect the rectified current.

Since this receiver doesn't have any frequency selective circuits (a tuner), you are likely to hear several stations; the stronger ones will be loudest with some weaker stations in the background. Remember, when this type of receiver was used many years ago, only a few stations were on the air, so interference usually was not a problem.

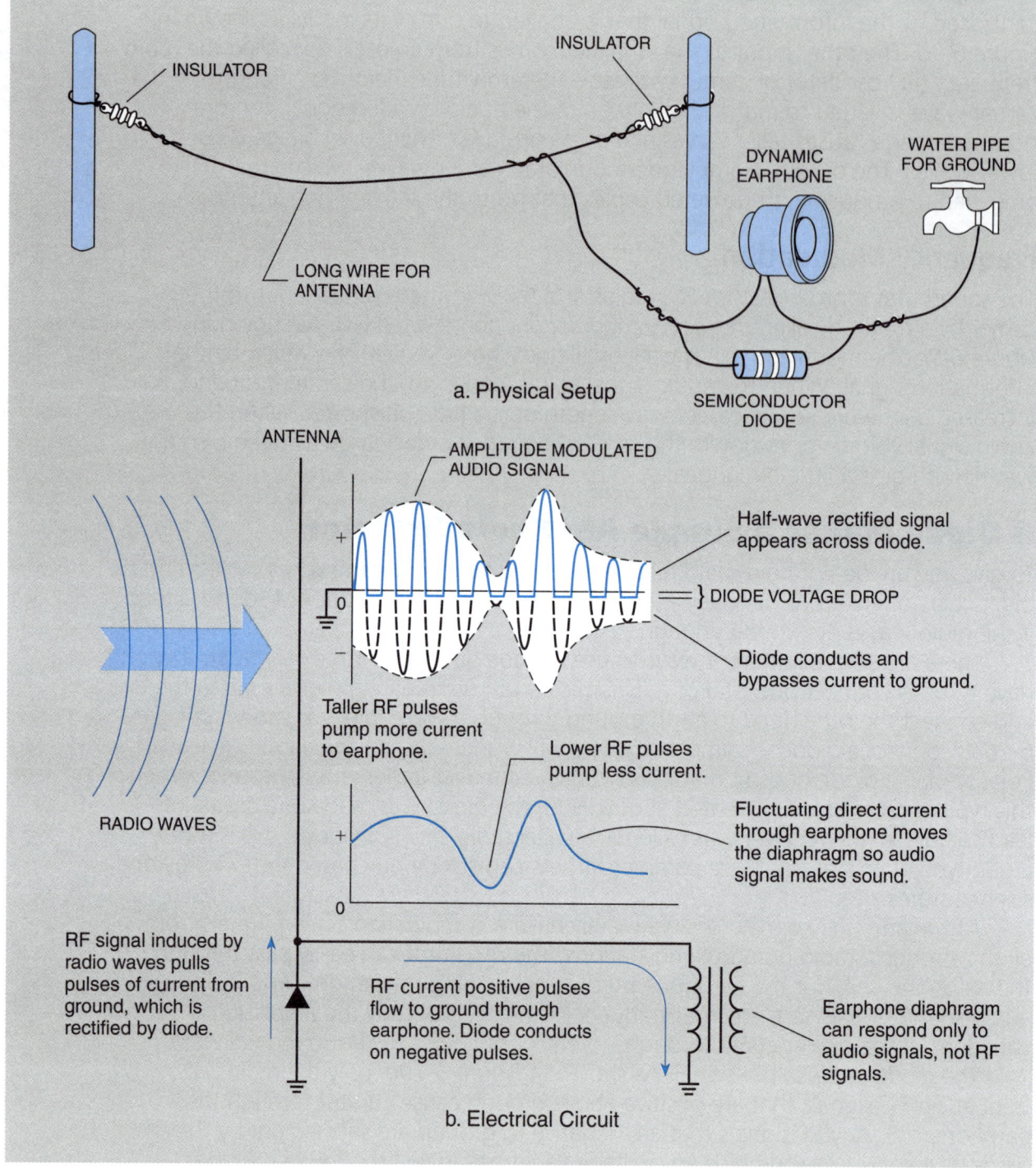

Figure 6-10. Detection – A very crude "crystal" radio receiver – using a diode to detect the amplitude modulation.

Summary

Many voltage amplifier circuits, as well as amplifiers designed with frequency selective circuits, are built into modern radio receivers to provide the detection of weak signals at selected frequencies so you can tune your receiver to a particular station at a particular frequency. AM and FM signals are available using modern receivers. All of these signals are analog signals. Next we will learn about digital signals.

Quiz for Chapter 6

1. Radio waves travel at the same speed as
   a. a fast race car
   b. a jet plane
   c. sound
   d. light

2. The distance between any two adjacent peaks of a signal along a wire is called the _____ of the signal.
   a. bandwidth
   b. wavelength
   c. space-wave
   d. valley

3. The _____ field lines only run along the antenna.
   a. electric
   b. magnetic
   c. gravity
   d. force

4. The _____ field lines only run in circles around the antenna.
   a. electric
   b. magnetic
   c. gravity
   d. force

5. The _____ of the radio waves is always the same as the electric waves that generate them.
   a. amplitude
   b. phase
   c. frequency
   d. volume

6. The higher the frequency of a radio wave the _____ its wavelength.
   a. shorter
   b. longer
   c. stronger
   d. weaker

7. The transmitted radio waves are strongest in all directions _____ to the wire.
   a. parallel
   b. perpendicular
   c. adjacent
   d. close

8. A _____ is a transmission line made from an insulated wire running along the centerline of a hollow tube made of aluminum foil, braided copper wire, or solid copper.
   a. waveguide
   b. dipole
   c. coaxial cable
   d. twin lead

9. Waves that just bounce back and forth in one place, while they radiate radio waves are called
   a. travelling waves
   b. transverse waves
   c. longitudinal waves
   d. standing waves

10. A half-wave dipole antenna for use in the middle of the AM broadcast band would be from end to end about
    a. 20 feet
    b. 500 feet
    c. 2500 feet
    d. 4 miles

11. A wire works best as a receiving antenna if it is pointing _____ of the radio waves.
    a. parallel to the electric field
    b. parallel to the magnetic field
    c. opposite to the electric field
    d. in the same direction as the magnetic field

12. Getting the information out at the receiver is called
    a. modulation
    b. oscillation
    c. detection
    d. carrier suppression

Answers: 1 d, 2 b, 3 a, 4 b, 5 c, 6 a, 7 b, 8 c, 9 d, 10 b, 11 a, 12 c

Questions for Chapter 6

1. What is the wire carrying the radio signal called?

2. What is the distance between two adjacent corresponding points on a wave (peaks or valleys) called?

3. What is the speed of electromagnetic (radio) waves?

4. What are radio waves?

5. Radio waves are _____ waves (meaning crossways).

6. To prevent radio waves from going out from the line that connects the transmitter to the antenna, a _____ cable would probably be used.

7. We set up _____ waves in the antenna to send out as much energy as possible as radio waves.

8. When the current alternates back and forth freely by itself, it is said to _____ .

9. A wire works best as a receiving antenna if it is pointing _____ to the electric field of the radio waves.

10. What is used for the core of an AM radio receiver loopstick antenna?

11. Plain radio waves of a particular frequency that are transmitted and received without information on them are called _____ waves.

12. Changing the carrier to put information in is called _____ .

13. What type of modulation is used to carry the picture portion of a television signal?

14. What type of modulation is used to carry the sound portion of a television signal?

15. What was the earliest and simplest form of AM receiver?

CHAPTER 7
What Digital Circuits Do

One general way to classify electric or electronic circuits is according to whether they are alternating-current or direct-current circuits. Another very important way to classify circuits is according to whether they are *analog* or *digital*. Most of the circuits we have been studying so far in this book, both ac and dc, have been analog. That includes amplifiers and similar circuits, such as oscillators and modulators.

Now it's time to start talking about digital circuits; that is, computers and similar equipment. There is such a thing as *analog* computers, but what we call computers nowadays are *digital* systems. The basic principles of digital electronics involve dc, not ac. That helps make these principles very easy to learn.

Actually, analog and digital refer to different ways of transmitting and storing information. In this chapter, we will see how digital circuits transmit *information*.

Analog Signals Carry Information by Varying Smoothly and Continuously Over a Range

To begin to understand the difference between analog and digital forms of information, let's consider some example *analog* circuits.

First, take a look back at the simple loudspeaker circuit in *Figure 5-1b*. Think of how information is carried in this circuit. The microphone converts sound waves into voltage and current electric waves. The electric waves are turned back into sound waves by the loudspeaker.

Sound waves, of course, are waves of air pressure. The wires between the microphone and loudspeaker carry *information* about the rapidly fluctuating air pressure at the microphone. A higher voltage and current means a higher pressure, and a lower voltage and current means a lower pressure. We say that the voltage or current signal is an *analog* of the air pressure. An analog of something means a copy in another form; that is, analogous. It also says that the signal is varying in a continuous way. It is not broken into disconnected parts.

For another example of analog information, think about a frequency modulated (FM) signal such as we discussed in *Figure 6-9*. How is information carried? The FM modulator converts voltage changes coming from a microphone into *frequency changes* in the carrier waves. The frequency of the carrier waves, not their voltage or current, carries the information. So the frequency is an analog of air pressure in sound waves.

Figure 7-1 shows still another example of analog information. This is an old-fashioned kind of electric fuel gauge in a car. A float fastened to a pivoted arm inside the fuel tank moves a potentiometer contact. A wire from this contact leads to a moving-coil voltmeter on the dashboard. The needle of the voltmeter points to a scale that is marked in fractions of a full tank instead of in volts.

The voltage in the wire is an analog of the fuel level, carrying information from the level sensor to the meter. The position of the needle is an analog of the voltage and the fuel level. Zero volts means empty, and some higher voltage (maybe 12 volts) means full. The voltage can have any value in between, according to the fuel level.

So, in an electric analog system, some characteristic of electricity that can be measured, such as current, voltage, power or frequency, varies *continuously over a certain range*. That measurable electric signal stands for the transmitted information, which is also something measurable.

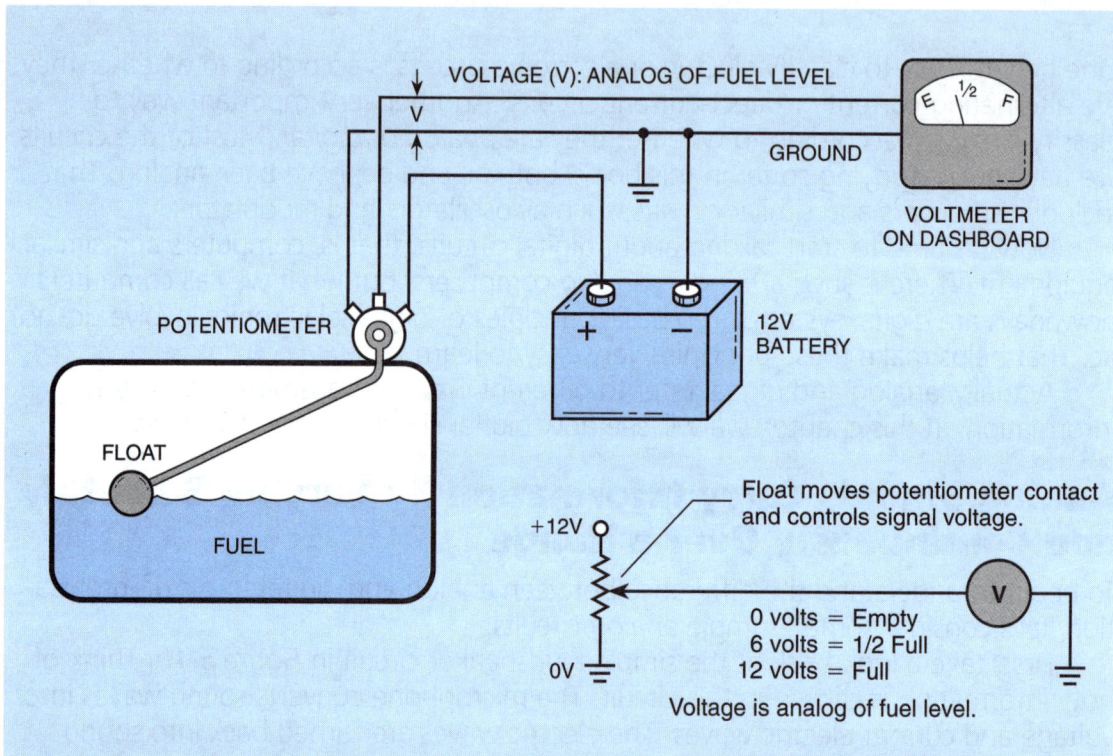

Figure 7-1. Fuel Gauge – An automobile fuel gauge shows how a measurable electrical quantity carries information about something else. In this case, voltage is an analog of fuel level.

Digital Signals Carry Information as Patterns of States

Now let's see how information is handled in a *digital* circuit.

As an example, consider a homemade telegraph circuit as shown in *Figure 7-2*. It works like the old-fashioned clicking kind that was common. To simplify our discussion, the receiving instrument is a light bulb instead of a clicking sounder. It could just as well be a buzzer, or a tone generator feeding earphones or a loudspeaker.

Using International Morse code, one operator is sending a message to the other operator. Suppose the message happens to be about the level of fuel in a tank. It is the first letters of the message spelling out "THREE FOURTHS OF A TANK". It could have been the numbers "0.75".

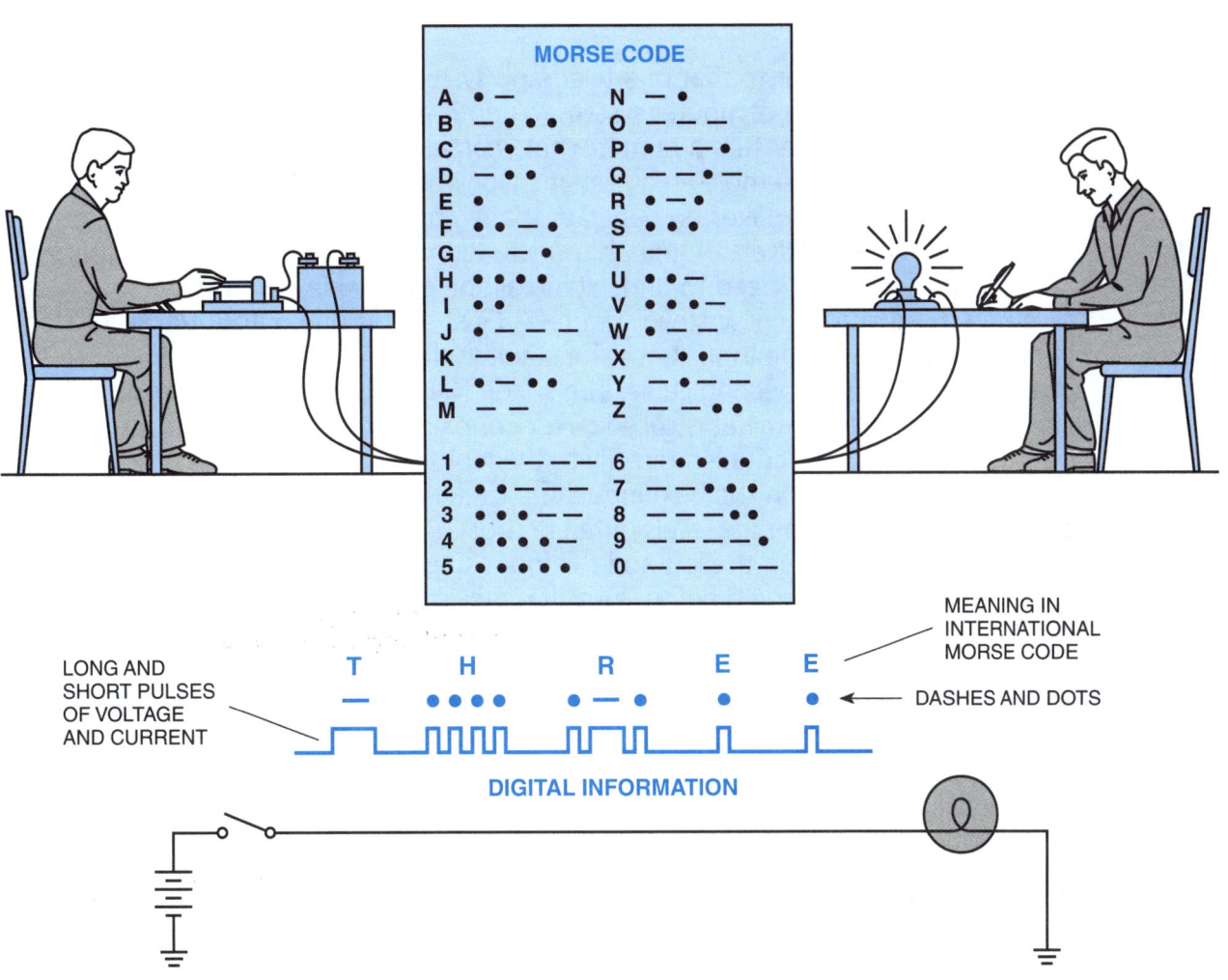

Figure 7-2. Telegraph Circuit – Digital systems send information as patterns of switched signals, similar to a telegraph circuit.

Telegraph code uses short and long pulses of electric current called dots and dashes. The letter T, for instance, is sent as a single dash (long pulse). The letter H is sent as four dots (short pulses). The numeral 7 is two dashes followed by three dots. The sending operator keeps all his dots at one length and all his dashes at another length. And he pauses for about the same time between each letter. This is so the receiving operator can tell a dot from a dash and can tell when a new letter is beginning.

As in the loudspeaker system and fuel-gauge system, electric signals are carrying information. But *nothing about the electric signals is an analog of the information*. At any instant, the transmitter key is in one of two different states or conditions: closed (on) or open (off). And the receiving light is in either one of two states, bright or dark. **The information is carried as coded patterns of changes from one state to another as time passes. This is what makes the system digital instead of analog. Only two levels are used to identify the signal at any instant in time.**

Transistor Switching Circuits Handle Two Electrical States

A digital concept similar to that of a telegraph system is used when information is sent from one part of a computer system to another. As an example, let's see how a personal microcomputer tells a printer what characters to print. A printer is a machine which may resemble an automatic typewriter. Characters are the marks that a printer or typewriter makes on paper, or that a computer shows on its video display screen. That includes letters, numerals, punctuation marks, and even empty spaces.

For now, we don't need to understand all the circuitry involved in sending and receiving this digital signal. Instead, *Figure 7-3* just shows one output transistor stage in the computer and one input transistor stage in the printer. In between, we see two of the wires in the connecting cable. The silicon NPN transistors in the *Figure 7-3* circuit are connected in what is called *direct-coupled transistor* (DCT) circuitry. The output of one circuit is coupled directly to the input of the next circuit. A dc voltage on an output appears on the next input. The circuitry used today may be a bit more complicated and the voltage levels different, but the basic concepts are the same.

The output stage on the left and the input stage on the right look like amplifier stages that we have studied before. But *the transistors are used as switches* instead of variable resistors. We talked about that idea in *Figure 4-15,* using a light bulb circuit as an example. Each transistor is either all the way on (called "saturated"), or all the way off (called "cut-off"). In switching back and forth between these two states, a transistor zips rapidly through the part-way-on condition as driven by the input signals. It is made to do that by the particular voltages on the input and output and the values of the resistances used in the circuitry.

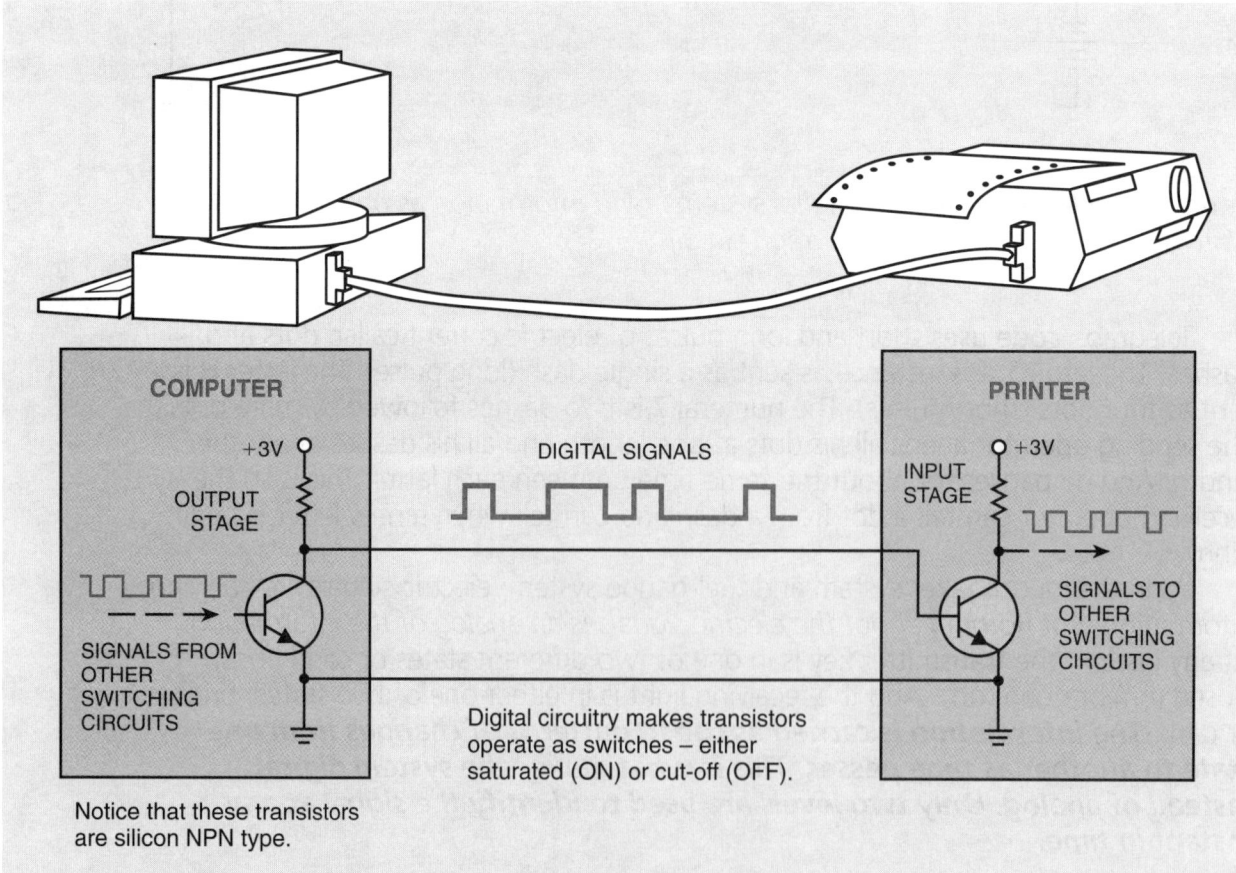

Figure 7-3. Computer to Printer – An example of real digital signals: sending characters from a computer to a printer.

Switching Circuits Drive Other Switching Circuits

To send digital signals representing characters, the computer turns the left NPN transistor on and off in a coded pattern. When this left transistor is on (saturated), it pulls the output voltage nearly all the way to zero (ground). This voltage signal will be about 0.2 volt. Therefore, the right transistor is turned off, because its base voltage must be at least about 0.6 volt to start conduction (0.7V for full conduction). These conditions are shown in *Figure 7-4a*.

When the left transistor is cut off as shown in *Figure 7-4b*, it allows the base voltage of the right transistor to rise above the cut-off level of about 0.6 volt and enough base current is supplied to saturate the right transistor. The current depends on the resistance and voltage supply back at the cut-off transistor. But the right transistor's base-emitter junction keeps the signal voltage from getting much higher than about 0.7 volt. We will assume the signal is about 0.8 volt.

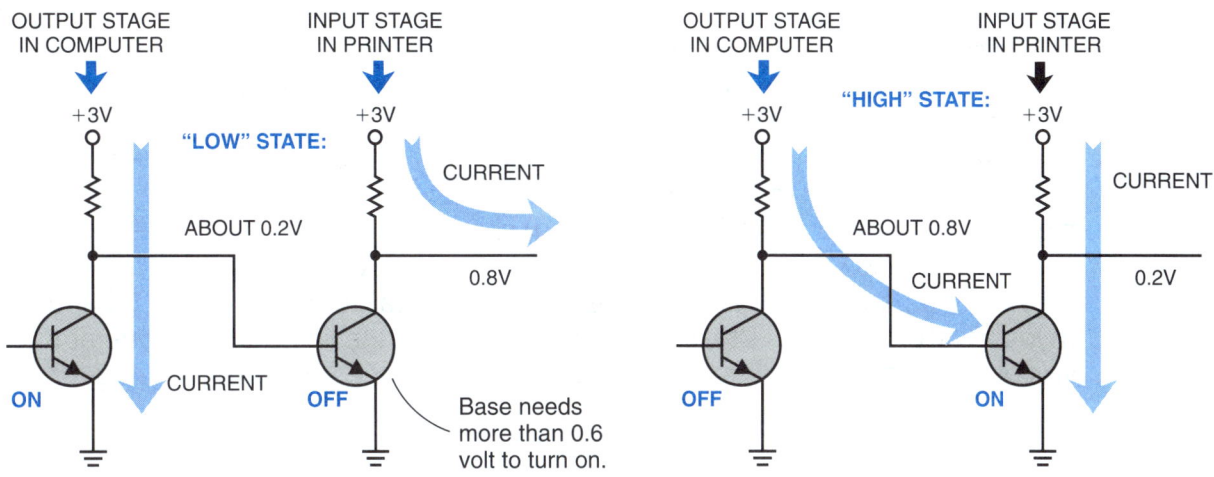

a. ON State Holding Next Stage OFF   b. OFF State Driving Next Stage ON

Figure 7-4. How NPN direct-coupled transistor switching circuits turn each other on and off.

We also assume that the right transistor is connected to other switching circuitry in the printer in the same way. Those circuits will be explained later. So the output from the right transistor also switches between about 0.2 volt and about 0.8 volt. In this way, switching signals are passed from one circuit to another in the printer.

This example illustrates several very fundamental things about typical digital systems.
1. *Switching-type circuits are driving other switching-type circuits*. Throughout a digital system, thousands of little transistor switching circuits send signals to one another like this.
2. As in the telegraph example, a conductor carrying information can be in either of two states. In this case, the states are defined in terms of voltage. The two electrical states are called *high* and *low*.
3. Notice that *the voltages or currents don't have to be exact* to work okay as high or low. In this particular example, the low voltage only has to be below about 0.6 volt, and the high voltage only has to be above about 0.7 volt. So there is a safety margin built into the circuit design. This means circuits, voltages, and transistors can vary, but the circuits still perform without error.

Timed Changes Are Used Instead of Dots and Dashes

How do digital systems transmit information by using these two electrical states? To find out, let's continue with the example of a computer sending characters to a printer through a single wire plus a ground wire. Look at *Figure 7-5*. When no information is being sent, the signal wire is in the low state. To begin sending a character, the computer switches the signal high for a certain short period of time. Let's say the period is exactly one millisecond (one thousandth of a second). This first one-millisecond signal that is in the high state is called the start bit because it tells the printer to get ready to receive a character.

Then, to send the character, the computer switches the signal high or low several times in a certain coded pattern. To allow the printer to recognize each change of voltage, the changes are only permitted at certain times. In our example, *changes can occur only at the end of each millisecond* after the beginning of the start bit. Seven milliseconds are allowed after the end of the start bit to send one character. In each of those seven milliseconds, the signal voltage can be either high or low. At the end of that time, the computer waits at least a couple of milliseconds before sending the next start bit and character.

These seven voltage signals serve much the same purpose as dots and dashes in telegraph code. Different patterns or combinations of the signals stand for different characters in a special code. For example, as shown in *Figure 7-5*, "high-low-low-low-low-low-high" stands for the capital letter "A", and "low-high-low-low-high-high-low" stands for the "&" (ampersand) character.

There is a special name for the fixed frequency at which changes of state are allowed in a particular digital transmission line. It is called the *baud rate* ("baud" rhymes with "Claude"). One baud would mean that the line changes state at intervals of one second. In our example, the baud rate is one *kilobaud,* meaning a

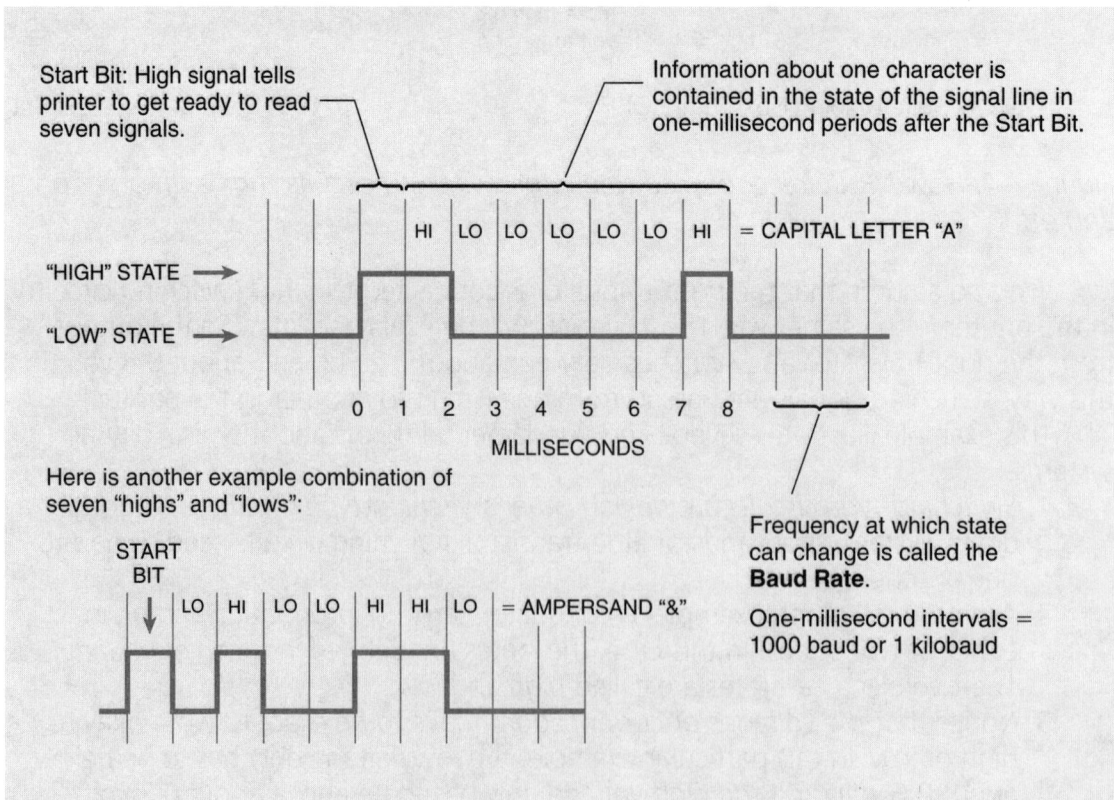

Figure 7-5. Serial Transmission – Digital information is sent in a single wire by switching states at fixed intervals. Seven intervals are enough for one character.

thousand possible changes in each second. There is also a special name for the transmission. It is called *serial transmission* because each bit of information needed to determine the character is sent in series, one bit after another in a timed sequence.

Two Electrical States Stand for Bits in Binary Numbers

Now here is a very important point for your understanding of digital electronics: To write and keep track of the many possible combinations of high or low signals in digital information, the combinations are treated as binary numbers. Binary numbers are also called base two numbers. An example is shown in *Figure 7-6*.

These days, most children learn about decimal numbers in elementary school. When written in ordinary decimal form, integers (whole numbers) are expressed as so many ones, so many tens, so many hundreds, and so forth. As shown in *Figure 7-6a*, each place in the number has a value ten times the place to the right. This requires using *ten* (for decimal) different symbols called decimal numerals or digits: 0, 1, 2, 3, 4, 5, 6, 7, 8, and 9. But in binary form, whole numbers are expressed as so many ones, so many twos, so many fours, and so forth. Each place in the number has twice the value of the next place to the right. Binary numbers use only two digits instead of ten: just 0 and 1.

So, in binary form, as shown in *Figure 7-6b*, a number is written as a string of ones and zeroes. For instance, the integer one is written as 1. Two would be 10, which we read as "one-zero". It means a two and no one. Three would be 11 ("one-one", not eleven), meaning a two and a one. Four would be 100, called "one-zero-zero", meaning a four, no two, and no one. Five would be 101, meaning a four, no two, and a one. One hundred would be 1100100. That means a sixty-four, a thirty-two, no sixteen, no eight, a four, no two, and no one.

*Each zero or one in a binary number is called a **bi**nary digi**t**, or bit for short.* Bit also means a little piece of information. In fact, a bit is the smallest possible piece of information in a digital system. It expresses a choice between only two possibilities. In a binary number, for instance, a particular bit can say either, "Yes, there is a 64 in this number," or, "No, there is not a 64 in this number."

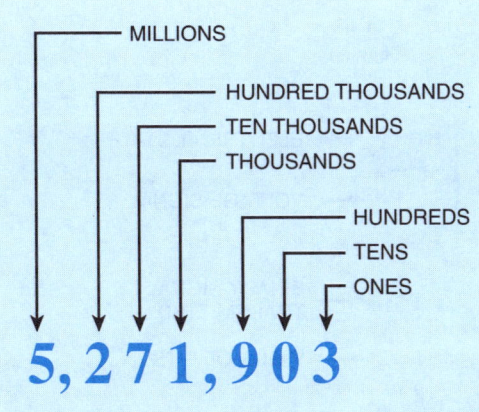

Decimal Number:

Each place has a value **ten times greater** than the next place to the right. **Ten numerals or digits** are used.

a. Decimal System

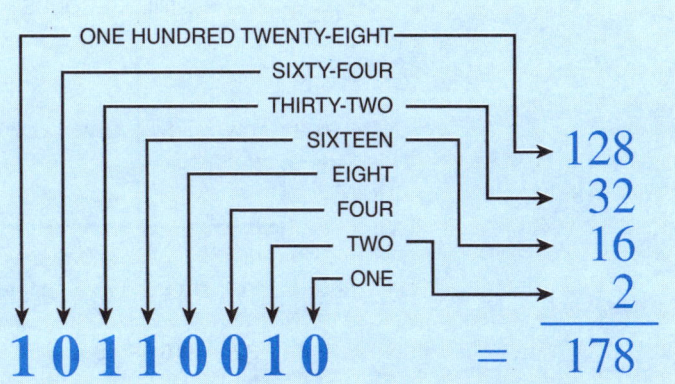

Binary Number:

Each place has a value **twice as great** as the next place to the right. **Two numerals or digits** are used:

0 (Zero = "No")
1 (One = "Yes")

These digits are called bits.

b. Binary System

Figure 7-6. Binary Numbering System – In digital systems, binary numbers are used to write and keep track of the many possible combinations of the two electrical states.

Example 1. Decimal Value of Binary Numbers

What is the decimal value of the following binary codes?
Step 1: Assign binary value to each bit.
Step 2: Add binary values for all bits in each binary code.

128	64	32	16	8	4	2	1	
								◀ Step 1: Set Bit Value for Each Bit
								▼ Step 2: Add Bit Value for Decimal Value
1	0	0	1	1	1	0	0	128+16+8+4=156
1	1	1	0	1	1	1	0	128+64+32+8+4+2=238
0	1	1	0	0	0	1	1	64+32+2+1=99
0	0	0	1	0	0	0	1	16+1=17

In a digital system, the two electrical states stand for bits. As indicated in *Figure 7-7*, usually the higher (or more positive) voltage in a particular conductor stands for 1 and the lower (or less positive) voltage stands for 0. This convention is called *positive logic,* and is the convention we will use in this book. (The other convention, called negative logic, let's the lower voltage stand for 1. We will not use it in this book.)

Using positive logic, the code for the capital letter "A" that we mentioned before (high-low-low-low-low-low-high) is the binary number 1000001 or the decimal number 65.

Each character that the computer sends to the printer is coded as a seven-bit binary number like this, plus the start bit at the beginning. So you could say that what the computer really sends to the printer is a series of binary numbers. The printer, in turn, interprets each number and prints the character that it stands for.

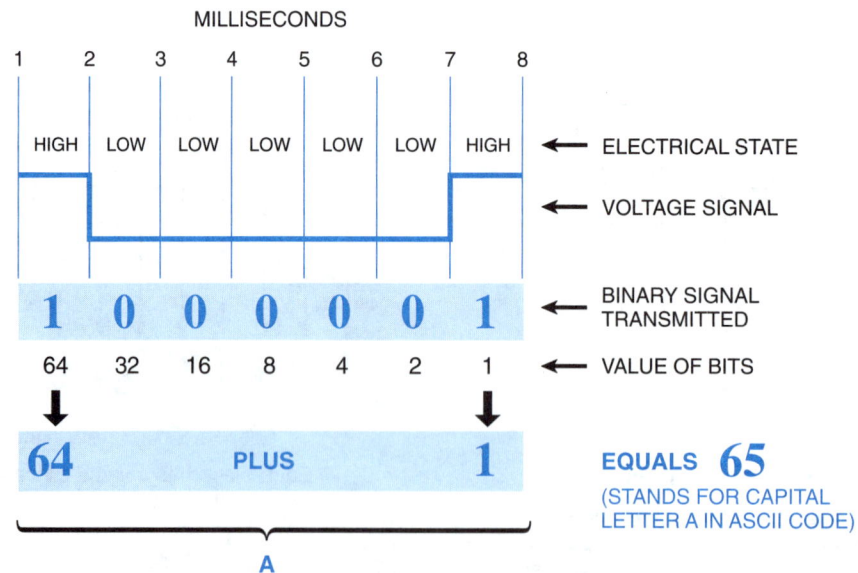

Figure 7-7. Digital Code for "A" – Using "positive logic", the high state is called 1 and the low state is called 0. This turns a confusing set of states into a binary number.

Table 7-1 lists the American Standard Code for Information Interchange that is used for transmitting characters and control codes. (The abbreviation, ASCII, is often used in conversations as a word by pronouncing it as ASK-ee.) It was developed before computers were invented for automatically sending typewritten telegraph messages. The machines used were called teletype machines. Today, we would call a teletype machine a crude computer printer.

Table 7-1. American Standard Code for Information Interchange

Bit Position

							0	1	0	1	1	0	0	1
							0	0	1	1	1	1	0	0
1	2	3	4	5	6	7	1	1	1	1	0	0	0	0
0	0	0	0				@	P	`	p	0	sp	NUL	DLE
1	0	0	0				A	Q	a	q	1	!	SOH	DC1
0	1	0	0				B	R	b	r	2	"	STX	DC2
1	1	0	0				C	S	c	s	3	#	ETX	DC3
0	0	1	0				D	T	d	t	4	$	EOT	DC4
1	0	1	0				E	U	e	u	5	%	ENQ	NAK
0	1	1	0				F	Z	f	v	6	&	ACK	SYN
1	1	1	0				G	W	g	w	7	'	BEL	ETB
0	0	0	1				H	X	h	x	8	(	BS	CAN
1	0	0	1				I	Y	i	y	9	)	HT	EM
0	1	0	1				J	Z	j	z	:	*	LF	SUB
1	1	0	1				K	[	k	{	;	+	VT	ESC
0	0	1	1				L	\	l	|	<	,	FF	FS
1	0	1	1				M	]	m	}	=	-	CR	GS
0	1	1	1				N	^	n	~	>	.	SO	RS
1	1	1	1				O	_	o	DEL	?	/	SI	US

Example 2. Determining Binary Codes

Using *Table 7-1*, what are the ASCII codes for M, T, Z, b, g, p, y

Bit Position ▶ Letter ▼	1	2	3	4	5	6	7
M	1	0	1	1	0	0	1
T	0	0	1	0	1	0	1
Z	0	1	0	1	1	0	1
b	0	1	0	0	0	1	1
g	1	1	1	0	0	1	1
p	0	0	0	0	1	1	1
y	1	0	0	1	1	1	1

Example 3. Determining Decimal Value of ASCII Codes

Assign the decimal value to each of the ASCII codes in Example 2.

Bit Position ▶	1	2	3	4	5	6	7	
Bit Value ▶	1	2	4	8	16	32	64	
Letter ▼								▼ Decimal Value
M	1	0	1	1	0	0	1	1+4+8+64=77
T	0	0	1	0	1	0	1	4+16+64=84
Z	0	1	0	1	1	0	1	2+8+16+64=90
b	0	1	0	0	0	1	1	2+32+64=98
g	1	1	1	0	0	1	1	1+2+4+32+64=103
p	0	0	0	0	1	1	1	16+32+64=112
y	1	0	0	1	1	1	1	1+8+16+32+64=121

Parallel Transmission Is Faster Than Serial

We have seen that digital information consists of strings of bits (ones and zeroes) that are usually interpreted as binary numbers. In the example we have been discussing, the bits are sent one at a time through a single wire plus a ground wire. This method is called *serial transmission*. But in various parts of a digital system, groups of several bits are often sent from one point to another at the same time. This method is called *parallel transmission*. For example, some computers transmit characters in parallel to printers.

A group of bits transmitted and stored together in a digital system is usually called one *byte* (pronounced like "bite"). *A byte customarily consists of eight bits*. Since ASCII needs only seven bits, the eighth bit can be used for some other purpose when that code is used.

In parallel transmission, all the bits of a byte have to appear at the same time. And the transmitting unit has to send a separate timing signal to tell the receiving unit when each byte is ready to be read. Timing signals are very important in digital systems. They are digital signals called *clock signals* or clock pulses. On the other hand, the actual information being sent is called *data signals*.

In the example in *Figure 7-8*, a computer is sending characters in parallel to a printer. There are eight data lines, one clock line, and a ground line. Each data line has a certain value in a binary number: 1, 2, 4, 8, 16, 32, 64, or 128. The eight lines are usually numbered from 0 through 7 instead of 1 through 8. Line 7, for the "128" bit, is not needed for seven-bit characters.

Whenever the clock line goes high, the computer changes all the data lines simultaneously to the next character, then holds that character until the clock line goes high again. The printer probably would read this data when the clock goes low. This would make sure the data is "stable"; that is, the printer wouldn't try to read while the data is changing to the next character.

Comparing *Figure 7-5* and *Figure 7-8,* we can see that parallel transmission is much faster than serial transmission. In *Figure 7-5,* it takes eight milliseconds to transmit the bit information for a character. In *Figure 7-8,* it only takes one millisecond to transmit the bit information for a character. It takes seven milliseconds to transmit the bit information for the message "Hello!". If the bit information were transmitted serially, it would take at least 7 x 8 milliseconds, not counting the couple of milliseconds between characters usually provided by serial transmission systems.

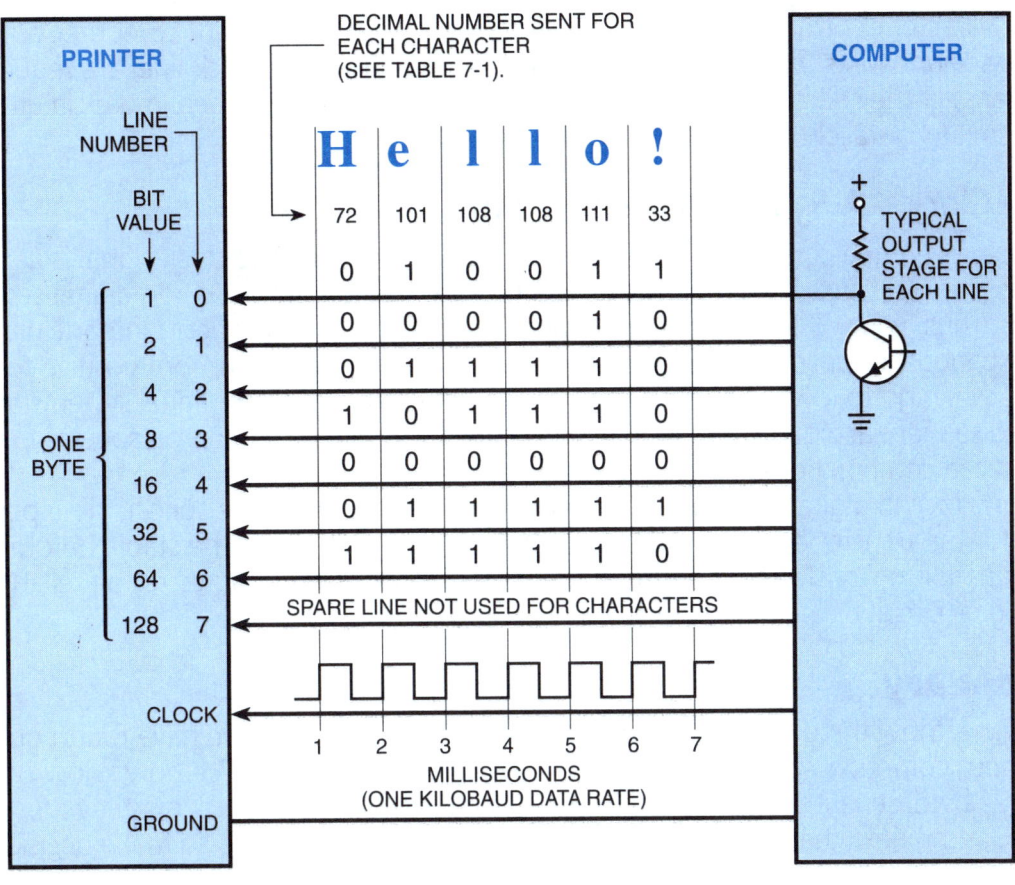

Figure 7-8. Parallel Transmission — Digital data is often sent in parallel (usually one 8-bit byte at a time) instead of serially (one bit at a time).

Example 4. Determining ASCII Codes Sent to Parallel Printer

A parallel printer receives the message, Dear Sir,. Using *Table 7-1*, what ASCII codes (in binary) were transmitted. Make sure you put each code in its proper time slot.

Bit No.	Time Slots							
	D	e	a	r	sp	S	i	r	,
1	0	1	1	0	0	1	1	0	0
2	0	0	0	1	0	1	0	1	0
3	1	1	0	0	0	0	0	0	1
4	0	0	0	0	0	0	1	0	1
5	0	0	0	1	0	1	0	1	0
6	0	1	1	1	1	0	1	1	1
7	1	1	1	1	0	1	1	1	0

The speed advantage is not without a price. As shown in *Figure 7-3*, only one computer output stage, one wire, and one input printer stage is required to transmit the character information serially. But as shown in *Figure 7-8*, eight computer output stages, eight wires, and eight printer input stages, beside the clock line, are required to transmit the character information in parallel. It takes a great deal more circuitry to transmit by parallel transmission.

The Clock – A System Time Standard

In *Figure 7-8,* the concept of a clock signal was introduced. The clock signal is a master timing signal that determines when certain actions can occur in digital systems. Thus, you will learn that digital system switching can only occur in particular time periods, or only when signals are going through a transition from positive to negative (high to low) or negative to positive (low to high). Such digital systems are called sequential digital systems. Digital circuits that do not depend on specific timing are called combinational digital systems.

A clock in digital systems is also very important to determine specific time periods for the transfer of data so that circuits can change state and be stable before the information is transferred. So timing circuits with a clock contributes to the reliability of digital circuits.

Summary

Now we know the difference between analog and digital, and we have found out how our computer systems are based on binary digital systems. We now know what a bit, a byte, serial transmission, and parallel transmission are, and how they are used. In the next chapter, we will see how digital circuits are used to make decisions.

Quiz for Chapter 7

1. An example of digital is
   a. volume control
   b. on-off light switch
   c. light dimmer
   d. playground swing

2. An example of analog is a
   a. sound
   b. telegraph
   c. calculator
   d. flashlight

3. The marks that a printer or typewriter makes on paper, or that a computer shows on its video display screen are called
   a. bits
   b. baud
   c. characters
   d. bytes

4. In digital circuits, transistors are used as
   a. variable resistors
   b. oscillators
   c. lamps
   d. switches

5. The voltages or currents in a circuit don't have to be exact to work okay as
   a. digital
   b. analog
   c. video
   d. audio

6. A high signal that tells the printer to get ready to receive a character is called
   a. a GO byte
   b. an ASCII character
   c. a baud
   d. a start bit

7. The special name for the fixed frequency at which changes of state are allowed in a particular digital transmission line is called
   a. bit rate
   b. byte rate
   c. baud rate
   d. character rate

8. The binary number that represents the decimal 7 is
   a. 0010
   b. 0111
   c. 1100
   d. 1010

9. Each zero or one in a binary number is called a
   a. bit
   b. byte
   c. baud
   d. nibble

10. The ASCII code for the capital letter G is
    a. 1111000
    b. 1010100
    c. 1000111
    d. 1001100

11. The method of transmission where whole groups of bits are sent from one point to another at the same time is called
    a. analog
    b. packet
    c. serial
    d. parallel

12. Besides the data signals sent by parallel transmission, a digital system must also send _____ signals.
    a. carrier
    b. clock
    c. analog
    d. baud

Answers:
1 b, 2 a, 3 c, 4 d, 5 a, 6 d, 7 c, 8 d, 9 a, 10 c, 11 d, 12 b

Questions for Chapter 7

1. What is meant by an analog signal? a digital signal?

2. Is a fuel gauge in an old-fashioned automobile digital or analog?

3. Is a telegraph system an example of an analog or digital system?

4. Does a digital circuit or an analog circuit require the most accurate or exact values of voltage?

5. A one-millisecond, high signal called the _____ tells the printer to get ready to receive a character.

6. Certain patterns or combinations of computer signals that stand for different characters is a special code called _____ .

7. What is the name for the fixed-frequency at which changes of state are allowed in a particular digital transmission line?

8. Binary numbers are also called base _____ numbers.

9. In binary form, a number is written as a string of _____ .

10. What is a binary bit?

11. What does it mean if the bits are sent one at a time through a single wire?

12. What is meant by parallel transmission?

13. All other things equal, which is faster, serial or parallel transmission?

14. What makes sure that the printer doesn't try to read while the data is changing to the next character?

15. If the higher voltage in a particular conductor of a digital system stands for 1 and the lower voltage stands for 0 means that _____ logic is being used.

CHAPTER 8
Digital Logic Circuits – How They Work

To get us into the subject for this chapter, look back at *Figure 7-3*. It shows a switching circuit in a computer sending digital signals to another switching circuit in a printer. We learned that the information is sent as HIGH and LOW voltage states, and that these states are treated as bits which form binary numbers. We learned that a digital system consists of thousands of switching circuits. They turn each other ON and OFF as the two circuits in *Figure 7-3* do.

Now, what about those other circuits that come before and after the ones shown in *Figure 7-3*? How can all the complicated jobs of a digital system be done by simple switching circuits?

The answer is that the complicated jobs are broken down into many very simple tasks that can be handled by special switching circuits. There are only *three basic kinds* of these simple circuits. They are called *gates*, like a gate in a fence. A more complete name is a *logic gate*. Gate circuits are connected together like the little pieces of a child's building set, to make digital systems of any size and variety. In this chapter, we will learn about the simplest building-block digital circuits and how they can be used together.

Digital Systems Manipulate and Store Information

First, it will help to consider what goes on in digital systems in general. Then we can more easily understand what the gate circuits do. Refer to *Figure 8-1*.

Generally, the purpose of any digital system is to work on *information*. The information can be about numbers and letters in a computer, images on a screen in a video game, sound waves in a compact disk player, control of valves and motors in a washing machine, or almost any type of task. Inside the system, all this information is in *digital* form. That is, it consists of bits, meaning ones and zeroes.

More specifically, a digital system first *takes in information* from outside (external to itself). This function is called input, or inputting information. For instance, a person puts information into a computer or calculator by pressing keys on a keyboard. For a video game, information usually comes in through a joystick, buttons or switches. A washing machine's digital control system gets information from control knobs or pushbuttons and from a device that measures the water level in the tub.

After translating all incoming information into digital form, the system uses it to create some *new* digital information. This is called translating data, processing data, or manipulating data. It is basically a matter of *making decisions*. For instance, a washing-machine controller decides when to open and close water valves, when to turn the drain pump on and off, when to spin the motor, and so forth.

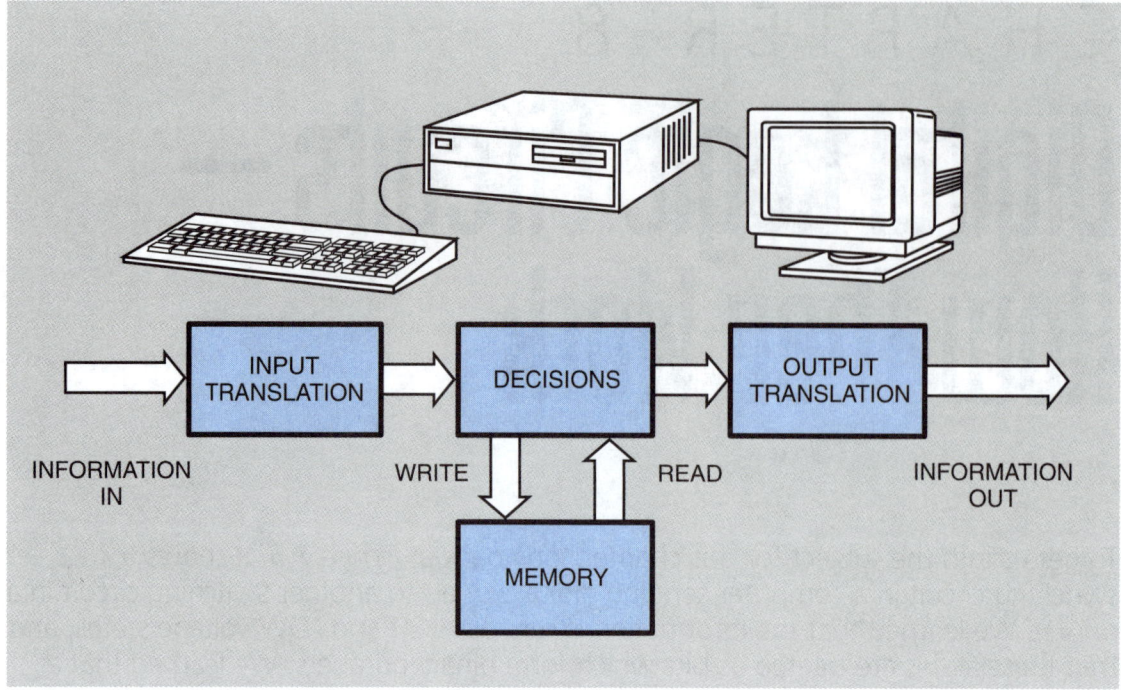

Figure 8-1. A computer or nearly anything else called a digital system handles information in four general functions: input, decisions, memory, and output. Decisions in various parts of a digital system are handled by circuits called logic gates.

To help make the necessary decisions, most digital systems store some information and then use it later. This function is called *memory*. It is a matter of recording ones and zeroes as though they were written on paper, placing them in specific locations, and then reading them later.

Finally, the system uses the results of its decisions to do something outside the system. This function is known as *output*, or outputting information. It is a matter of translating the new information inside the system into another form for use by an output device. For instance, the main output of a personal computer is characters and images on a screen. The output of a washing-machine controller is the control of contactors that switch power to the pump, the agitator motor, and the water valves at the correct times.

Actually, just about anything you could call a "system" involves these same four functions of input, decisions, memory, and output. Even analog systems—such as radio receivers or the human body—use these four functions. Your eyes, ears, and other senses handle the information input. Your brain takes care of decisions and memory. Your muscles are output devices.

The Basic Gate Circuits Make Three Logical Decisions

The purpose of the three kinds of gate circuits that we mentioned earlier is to use bits of digital information to *make very simple decisions*. However, when many of these gates are connected together in various ways, they can work very fast and make very complex decisions. Wherever decisions are made throughout a digital system, the job is done by logic gates.

Each kind of gate makes a different kind of simple decision, called a *logic function*. Logic means using rules for reasoning correctly; for example, "If Joe is taller than Beth, and Beth is taller than Tony, then Joe is taller than Tony." The three basic digital decisions or logic functions, shown in *Figure 8-2,* are called the *AND* function, the *OR* function, and the *NOT* function.

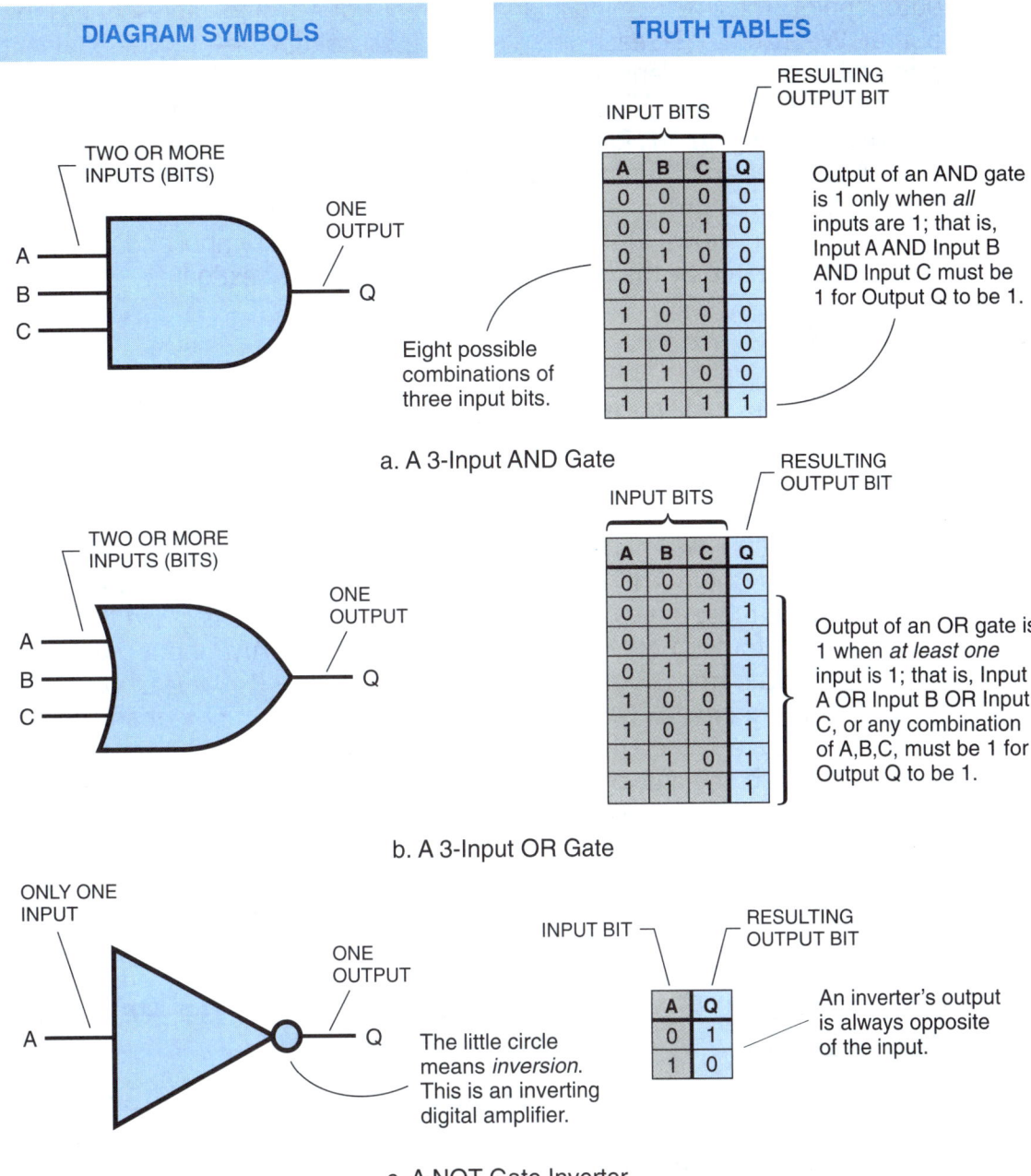

Figure 8-2. In a digital system, all decisions are broken down into the simplest three decisions—AND, OR and NOT—which are handled by circuits called logic gates.

Figure 8-2 shows the symbols for the AND gate, the OR gate, and the NOT gate, which perform the respective logic functions. (Notice that the names AND, OR, and NOT are customarily written in all capital letters.) To be strictly accurate, a circuit that performs the NOT function isn't really a gate. It is usually called an *inverter* instead. We classify the inverter with the AND and OR gates because its NOT function is just as basic and important.

A diagram made with interconnected logic symbols to describe a system or subsystem is called a *logic diagram*, and we will show some examples later. A logic diagram is similar to a schematic diagram of an electrical circuit, and the gate symbols are used sort of like amplifier symbols. The symbol for an AND gate resembles a stretched-out capital letter D. The symbol for an OR gate looks like a spearhead. The symbol for an inverter is an amplifier triangle with a little circle at the output point. The lines coming in and going out of a symbol represent its input and output. We assume that each line is an electrical conductor—a wire—that carries electrical information identifying a 1 bit or a 0 bit.

The AND gate and the OR gate have *at least two inputs* (never just one) to receive bits. The gates in *Figure 8-2* have three inputs labeled A, B, and C. However, some AND gates and OR gates may have many inputs. On the other hand, an inverter always has only *one input*, labeled A in the figure. Each gate and inverter has only *one output*, shown in the figure on the right of the symbol and labeled Q. The output bit is created by the gate and transmitted to the next gate.

Truth Table

In *Figure 8-2*, the logic function of each gate is shown in a table next to it. This table is called the *truth table* for that gate or function. A complete truth table shows *every possible combination* of input bits received by a logic function or circuit, and the state of the output transmitted for each input combination. For a three-input gate, there are eight (2^3) possible input combinations which correspond to the binary numbers from 000 (decimal 0) through 111 (decimal 7).

The output of the AND gate, shown in *Figure 8-2a*, is 1 only if *all the inputs are 1*. If any one or more inputs are 0, then the output is 0. This is called the AND gate because the output is 1 only when input A *AND* input B *AND* input C are all 1.

The output of the OR gate, shown in *Figure 8-2b*, is 1 if *at least one input is 1*. In other words, if input A *OR* input B *OR* input C is 1, *OR* if more than one input is 1, *OR* if all inputs are 1, then the output is 1. If all the inputs are 0, then the output is 0.

To invert a bit means to change it to the opposite bit, thus, the inverter output, shown in *Figure 8-2c*, is always the *opposite* of the input. If the input is 1, the output is 0; if the input is 0, the output is 1. (Stated another way, if the input is 1, the output is NOT 1; if the input is 0, the output is NOT 0.) The little circle at the output of the inverter symbol indicates inversion. An inverter could be called an inverting digital amplifier.

Relays Show How Logic Functions Can Be Performed Electrically

In a moment, we will see how these three simple decisions can be combined to make more complicated ones. But first, let's see how electrical circuits can perform the AND, OR, and NOT functions.

Gates are digital devices, but they don't have to be electronic. What makes a system digital is that it handles information by switching back and forth between different states that represent bits of information. There are even digital systems that use something other than electricity to represent the states—such as air pressure or the positions of mechanical parts. Probably the easiest way to understand electrical gates is by using *electromechanical relays* instead of transistors. Remember, an electromechanical relay is a mechanical switch that is operated by an electromagnet. Like a transistor in a switching circuit, the switch in a relay has two states—ON and OFF.

As shown in *Figure 8-3*, we can make a three-input AND gate out of three relays by connecting the normally-open contacts in *series*. A three-input OR gate is similar, but with the normally-open contacts connected in *parallel*. Using these concepts, we could obviously make AND gates and OR gates having any number of inputs. The inverter is a single relay with normally-closed contacts as the output.

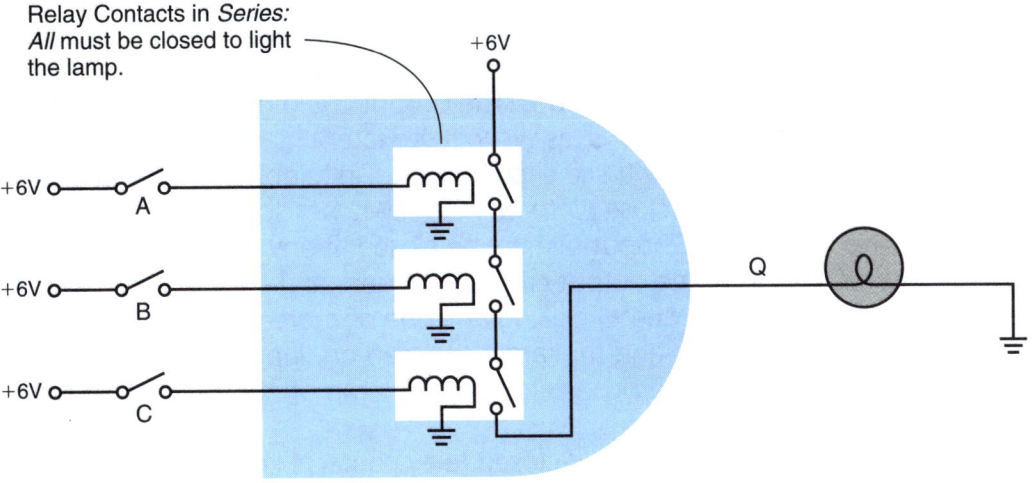
a. Positive "AND" Gate

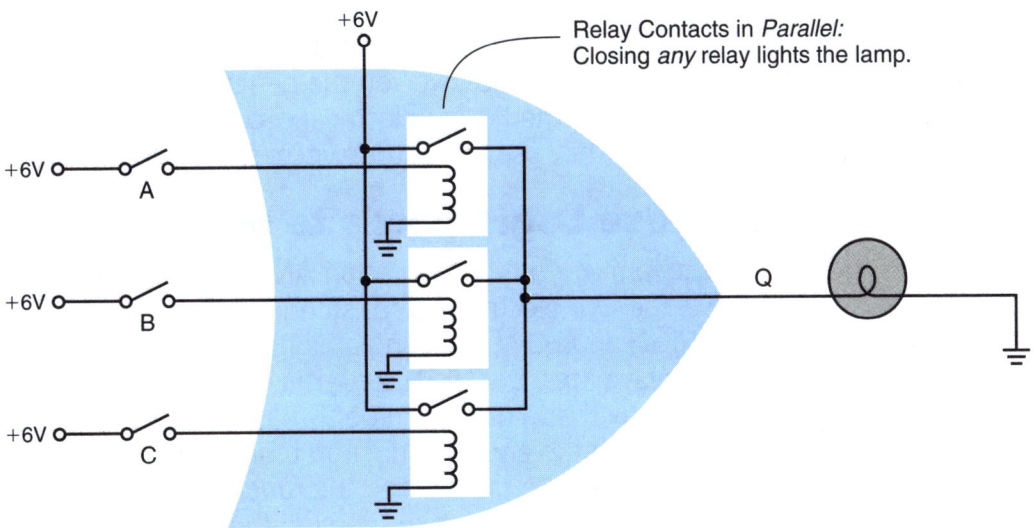
b. Positive "OR" Gate

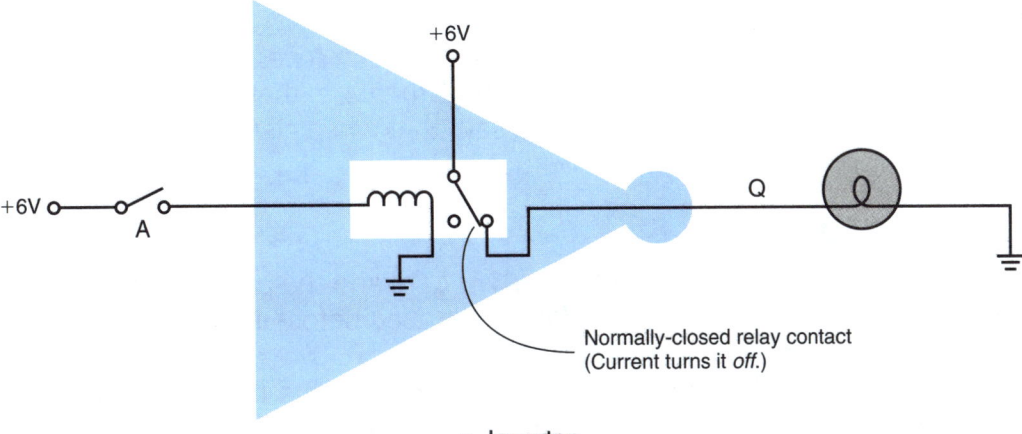
c. Inverter

Figure 8-3. It is easy to see how the three basic logic functions can be performed electrically when relays are used instead of transistors.

Let's assume that each relay is designed so that the contacts switch position when 6 volts is applied to the coil. In this example, input signals are provided by manual switches connected to a +6-volt power supply. The other side of each coil and the power supply is grounded (zero volts). The same +6 volts is supplied to the relay contacts in each gate. The output from each gate is connected to a 6-volt lamp which is grounded on the other side. (The output could go to the input of another gate made out of relays like this, as we will see later.)

You can easily see how these circuits work. First, look at the AND gate. To turn the lamp on, you have to close *all three* input switches so that all three relays close. Input A AND input B AND input C must be closed. If any input is open, the corresponding relay is open, the output circuit is broken, and the lamp is turned off.

In the case of the OR gate, closing *any one or more* of the input switches closes the corresponding relay(s) and the lamp is turned on. Input A OR input B OR input C or any combination may be closed. To turn the lamp off, all input switches must be open so that all the relays are open.

For the inverter, the normally-closed relay contacts keeps the lamp on when the input switch is open. Closing the input switch opens the relay contacts so the lamp is turned off.

In the 1930s, the first electrical digital systems used gates made out of relays in arrangements similar to these. There are other gate designs that use diodes, resistors, and relays. But the examples here should give you the general idea of how the three basic logic functions can be performed using electrical circuits. *Electronic* gate circuits work much the same way, but they use transistor switches instead of relays.

Positive or Negative Logic Needs to Be Specified

But wait—we're not finished making relays act as an AND gate or an OR gate. We have to decide which bits the two electrical states stand for. Remember, the electrical states are usually called HIGH and LOW, meaning a higher voltage and a lower voltage. For the example relay gates, the HIGH state is +6 volts, and the LOW state is zero volts. For now, we will say that a gate circuit only handles HIGH and LOW states.

Let's now connect the truth table with a function table. Any gate's electrical operation is completely described in a table called the *function table* for that circuit. A function table is very much like a truth table, showing the output state for every combination of input states. The difference is that a truth table shows *logic* states (1 and 0), but a function table shows *electrical* states; in this case, described as HIGH (H) and LOW (L)

The function table *(Figure 8-4)* for the relays in *series* says that when all three inputs are H (high), the output is H (high). Otherwise, the output is L (low). The function table for the relays in *parallel* says that when all three inputs are L (low), the output is L (low). Otherwise, the output is H (high).

Positive Logic

To make the series-relay circuit be an AND gate and the parallel-relay circuit be an OR gate, we have to choose *positive logic* (discussed before in Chapter 7). That is, the H (high) state means 1 and the L (low) state means 0. By writing 1 for H (high) and 0 for L (low) in each function table in *Figure 8-4,* we get a *truth* table for each circuit with positive logic. (In a moment, we will talk about the truth tables with negative logic.)

The truth table for the *series* circuit with *positive* logic says that the output is 1 only when all inputs are 1. That is the same truth table that we saw before for an AND gate. We would call this relay circuit not just an AND gate, but a *positive* AND gate. It is an AND gate only if we use positive logic.

FUNCTION TABLES

L = LOW VOLTAGE (GROUND)
H = HIGH VOLTAGE (+6 VOLTS)

RELAYS IN SERIES

A	B	C	Q
L	L	L	L
L	L	H	L
L	H	L	L
L	H	H	L
H	L	L	L
H	L	H	L
H	H	L	L
H	H	H	H

RELAYS IN PARALLEL

A	B	C	Q
L	L	L	L
L	L	H	H
L	H	L	H
L	H	H	H
H	L	L	H
H	L	H	H
H	H	L	H
H	H	H	H

TRUTH TABLES

with Positive Logic:
H = 1
L = 0

AND FUNCTION

A	B	C	Q
0	0	0	0
0	0	1	0
0	1	0	0
0	1	1	0
1	0	0	0
1	0	1	0
1	1	0	0
1	1	1	1

OR FUNCTION

A	B	C	Q
0	0	0	0
0	0	1	1
0	1	0	1
0	1	1	1
1	0	0	1
1	0	1	1
1	1	0	1
1	1	1	1

with Negative Logic:
H = 0
L = 1

OR FUNCTION

A	B	C	Q
1	1	1	1
1	1	0	1
1	0	1	1
1	0	0	1
0	1	1	1
0	1	0	1
0	0	1	1
0	0	0	0

AND FUNCTION

A	B	C	Q
1	1	1	1
1	1	0	0
1	0	1	0
1	0	0	0
0	1	1	0
0	1	0	0
0	0	1	0
0	0	0	0

Figure 8-4. We have to specify either positive or negative logic before we can say which logic functions are performed by the series and parallel gates in Figure 8-3.

For the *parallel* circuit with *positive* logic, its truth table shows an output of 0 only when all inputs are 0. That fits the truth table of an OR gate. So this particular circuit is a *positive* OR gate.

Negative Logic

The truth tables for these same two gates with *negative* logic show something very strange. Remember, negative logic means that a logic 1 state is represented by the least positive of the electrical information levels representing the bits 1 and 0. When the high voltage (+6 volts) stands for 0 and the low voltage (zero volts) stands for 1, the series-relay gate performs the OR function instead. A high voltage at all three inputs still causes a high voltage at the output, but all those "highs" are zeroes now. Similarly, the parallel-relay gate performs the AND function.

This shows an important general rule. *A positive AND gate acts as a negative OR gate, and a positive OR gate acts as a negative AND gate.* That fact often comes in very handy when digital systems are being designed. But an inverter is always an inverter, whether we use positive or negative logic.

Electronic Gates Use Transistors Instead of Relays

Dozens of different design concepts have been used for making electronic gate circuits with transistors instead of relays. For one example, look back at the diode-coupled amplifier circuit in *Figure 5-4*. If the signal inputs are at high and low levels that drive the amplifiers into saturation (low level) and cutoff (high level), you can now identify that NPN transistor circuit as an *inverter*. It has a single input, and, under the signal conditions specified, the output is always in the opposite electrical state.

Figure 8-5 shows some examples of other gate circuit concepts. To keep the drawings simple, only two inputs are shown on each gate. However, there could just as well be three or more inputs.

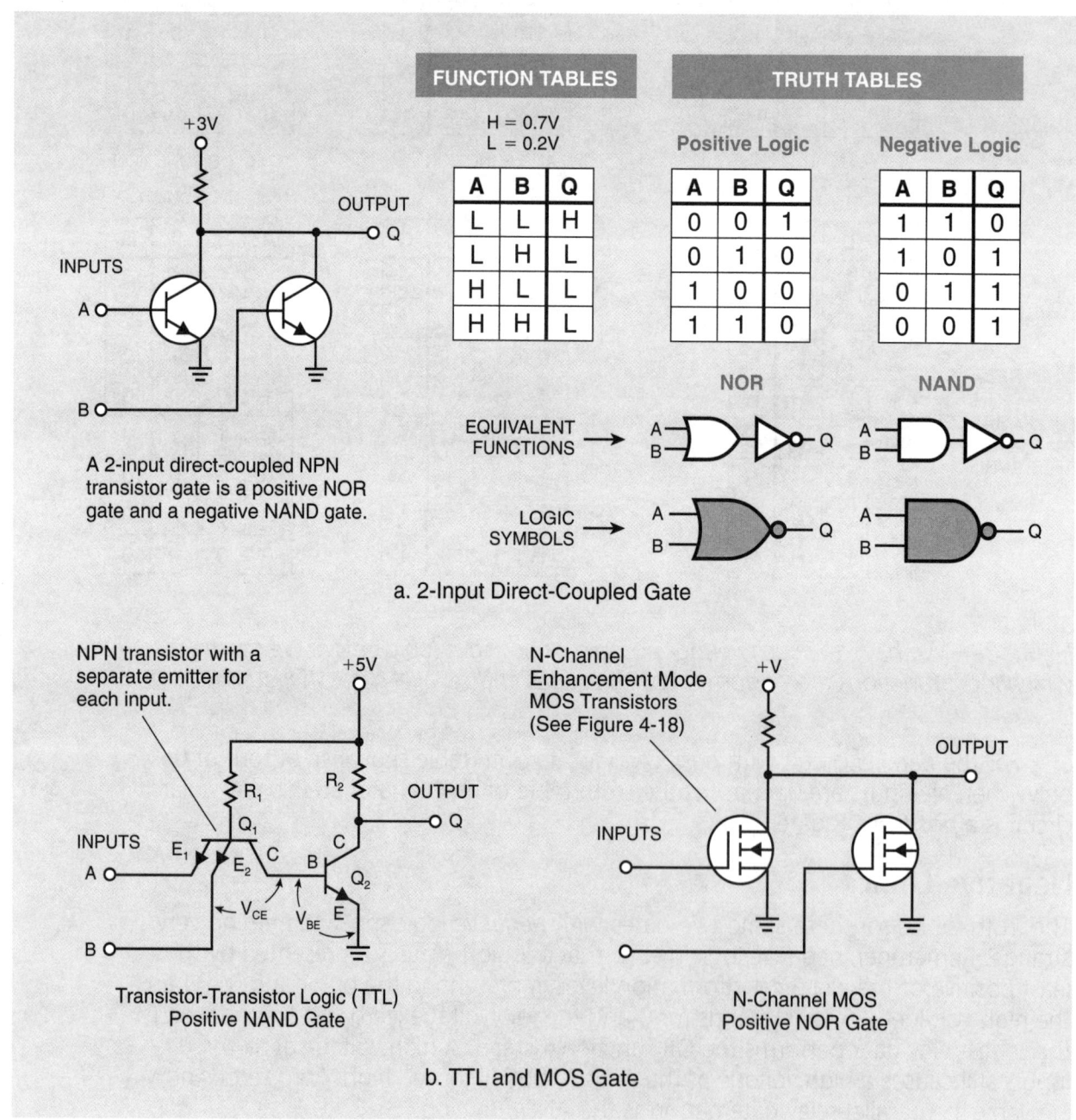

Figure 8-5. Typical electronic gates perform the NAND (NOT-AND) function or the NOR (NOT-OR) function instead of plain AND or OR.

A particularly simple and old-fashioned variety is a *direct-coupled gate* made of two NPN transistors in parallel as shown in *Figure 8-5a*. It's like two inverters that share a common resistor, called a pull-up resistor. When either input is high (about 0.7 volt in this case), its transistor saturates and pulls the output low to about 0.2 volt. This fact is summarized in the function table.

Two truth tables are shown for this gate—one with positive logic and the other with negative logic. You will see that neither of these truth tables represents the AND function or the OR function. Instead, this is what we call a positive NOR gate or a negative NAND gate.

A NOR gate acts as an OR gate followed by an inverter. Whenever one or more inputs are 1, the output is 0. The NOR symbol is a spearhead like an OR gate, with a little circle at the output point to indicate *inversion*. Identify the OR function using the input bits and follow the OR output with an inverter.

Similarly, *a NAND gate acts as an AND gate followed by an inverter*. Whenever all the inputs are 1, the output is 0. The NAND symbol is like an AND gate with a little circle at the output point to indicate *inversion*. Identify the AND function using the input bits and follow the AND output with an inverter.

Although the logical designs for digital networks are often worked out with AND and OR gates, the designs are mostly modified later to use NAND and NOR gates when they are actually built. This is because NAND and NOR gates are usually simpler and require fewer transistors. When a plain AND or OR gate is needed, designers typically use a NAND or NOR gate followed by an inverter.

The circuitry in digital systems nowadays consists almost entirely of *integrated circuit chips*. There may be hundreds of thousands of gates on a single chip. Nearly all the gates in a single chip use very similar circuit designs. There are several different groups of IC chips, called *logic families* or *technologies*, with names like MOS, CMOS, TTL, ECL, I²L, and so forth. All the chips of a given family use similar kinds of gate circuits. *Figure 8-5b* shows schematic diagrams of typical TTL and MOS gates.

Example 1. NAND Gate Function and Truth Tables

What is the function table and the positive-logic truth table for the TTL NAND gate of *Figure 8-5b*?

A multiple-emitter transistor Q_1 is coupled to an inverter transistor Q_2, thus, the name transistor-transistor logic (TTL). When an emitter of Q_1 is at the L (low) level, the base-emitter junction of Q_1 is forward-biased and conducts through R_1. This drives Q_1 into saturation and V_{CE} is very small. This causes the V_{BE} of Q_2 to be below the level (0.7V) for Q_2 to have forward conduction; therefore, Q_2 is cut off and output Q is at the H (high) level.

When all emitter inputs (A and B in this case) of Q_1 are at the H (high) level, its base-collector junction becomes a forward-biased diode, providing a forward-biased path through R_1 for the base-emitter junction of Q_2. This turns on Q_2 and drives it into saturation; therefore, output Q is at the L (low) level. From this information, we can construct the function table and the positive-logic truth table as shown below.

Function Table

A	B	Q
L	L	H
L	H	H
H	L	H
H	H	L

Positive Logic Truth Table

A	B	Q
0	0	1
0	1	1
1	0	1
1	1	0

The TTL gate is a positive-logic NAND. The output Q is inverted from a positive-logic AND gate.

Combining Little Decisions to Make Bigger Ones

Now that you know about gates, let's see how they can be put together to make more complex and useful decisions. As an example, consider the logic network for a *1-bit full adder*. *Figure 8-6* shows one particular design for this function. It uses seven AND gates, two OR gates, and three inverters. In developing and studying a logic diagram like this, it doesn't matter whether positive or negative logic will be used when the circuit is actually built. Either way, the logic diagram is the same.

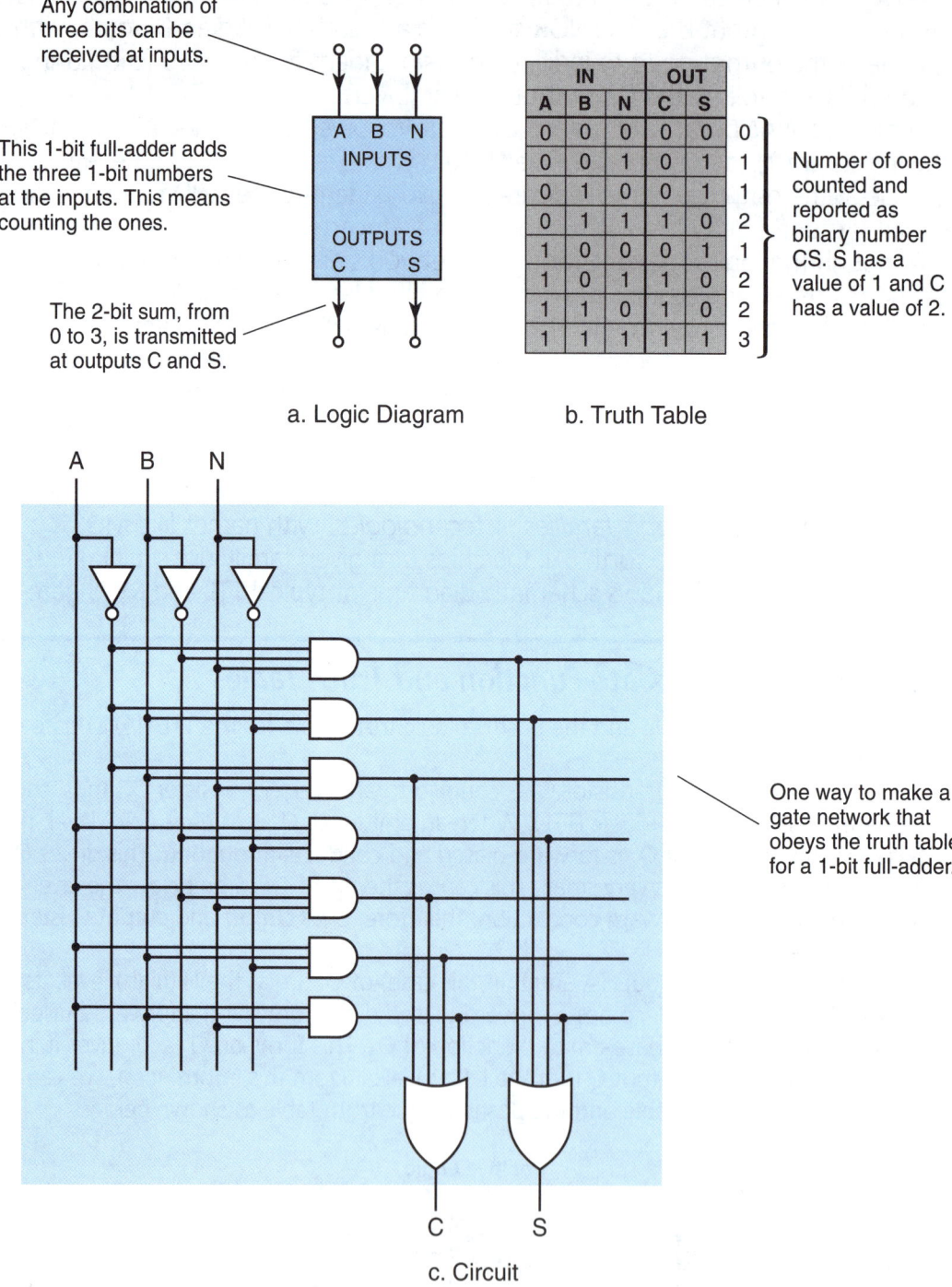

Figure 8-6. Logic gates can be connected together to make much more complex decisions, such as this network to add three bits together. This is an example of combinational logic—for every input combination, there is a certain output combination.

In a moment, we will see how several 1-bit full adders can be used to add binary numbers in a calculator or a computer. By itself, the adder in *Figure 8-6* simply *adds three bits to each other and produces a 2-bit binary sum*. That is, it counts the "ones" in three inputs. This function is shown in the truth table. Like most truth tables, this one shows all possible combinations of bits at the inputs (arranged as though they formed a binary number). It also shows the outputs that result from each input combination. The adder has two outputs instead of just one as in a single gate.

The three incoming bits are called A, B, and N. Think of them as three separate one-bit numbers. The two outgoing bits are called C and S. (Later, we will see what these letters stand for.) These output bits form a 2-bit binary number. Bit S is the least significant bit, having a value of one. Bit C is the most significant bit, having a value of two.

In the truth table, notice that if all three input bits are 0, the sum is 00. If only one of the input bits is 1, then the output is 01 (C is 0 and S is 1), which is binary for decimal 1. If any two of the input bits are 1, then the output is 10, which is binary for decimal 2. If all three inputs are 1, then the output is 11, which is binary for decimal 3. Thus, the three input bits are added to give an answer of 0, 1, 2, or 3. It's all done by logic gates connected to perform the AND, OR, and NOT functions.

Example 2. Determining Logic Diagram Outputs

For the following circuit, when are the output lines from the AND and OR gates at a 1 level? $\overline{A}$ is the opposite state from A, similarly with B and C. A dot between input signals means the signals are "ANDed" together. A plus sign between input signals means the signals are "ORed" together.

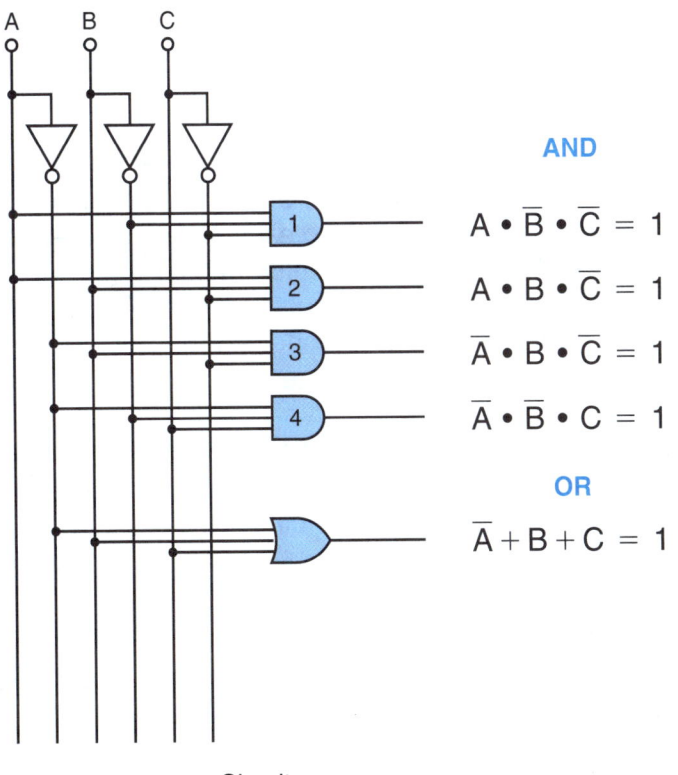

AND

$A \cdot \overline{B} \cdot \overline{C} = 1$

$A \cdot B \cdot \overline{C} = 1$

$\overline{A} \cdot B \cdot \overline{C} = 1$

$\overline{A} \cdot \overline{B} \cdot C = 1$

OR

$\overline{A} + B + C = 1$

Truth Table Positive Logic

	IN		OUT
A	B	C	
1	0	0	1
1	1	0	1
0	1	0	1
0	0	1	1

0	0	0	1
0	0	1	1
0	1	0	1
0	1	1	1
1	0	0	0
1	0	1	1
1	1	0	1
1	1	1	1

a. Circuit

b. Truth Table

Canonical Combinational Networks Come Straight from Truth Tables

The 1-bit full adder illustrates some important general principles about the design of digital systems. You can use these principles to design logic networks yourself, if you like—either on paper or using real integrated-circuit gates. A design of this general type starts with a truth table specifying what the network will do. For every *combination* of input bits shown on one line of the truth table, there is a certain prearranged combination of output bits. So this particular field of design is called *combinational logic*.

Any number of different combinational logic designs are possible for a given truth table. Designers usually try to find the simplest network that can be built with the most inexpensive kinds of gates that are available. But for each truth table, there is *one particular network* that is the easiest to design simply by looking at the table. It is called the *canonical* combinational network for that truth table. That name is pronounced ka-NON-ical, and it means "according to the rules."

The particular design that we have seen for the 1-bit full adder is the canonical one. Let's study it more closely and see how it works.

Connecting the AND Gates

A canonical network generally has a *separate AND gate for every input combination*; that is, there is one AND gate for each line in the truth table. Each of these AND gates recognizes one particular pattern of input bits, detailed by the line in the truth table, and has a *1 output when the input combination appears*. Each AND gate in *Figure 8-7* is labeled according to the input combination that it detects. For instance, when the input combination ABN is 000 (zero), Gate 0 has a 1 output, but all the other AND gates output 0. And when 001 is being received, only Gate 1 responds with a 1. All other gates have 0 as their outputs; therefore, when any AND gate is active, it will have a 1 at its output.

To permit the AND gates to detect zeroes in the inputs A, B, and N, an inverter is driven by each of these inputs. The inverter outputs are called the *inverse inputs*. They are labeled as $\bar{A}$, $\bar{B}$, and $\bar{N}$, called A-bar, B-bar, and N-bar. For instance, when Input A is receiving a 0, Inverse Input $\bar{A}$ is 1. That is, $\bar{A}$ is 1 whenever A is NOT 1.

Notice that each AND gate has one input for each of the three main inputs A, B, and N, or from their inverse input. That is, one input to each AND gate is either A or $\bar{A}$, another input is either B or $\bar{B}$, and the third input is either N or $\bar{N}$. If an AND gate is supposed to respond to a 1 on Input A, it is connected directly to A. But if this AND gate is supposed to respond to a 0 on Input A, it is connected to $\bar{A}$ instead. In the same way, the other two inputs to a particular AND gate are connected to detect either 0 or 1 on Inputs B and N.

For example, Gate 1 (which recognizes a combination of A=0, B=0, and N=1) is driven by $\bar{A}$, $\bar{B}$, and N. So when A is NOT 1, AND B is NOT 1, AND N is 1, Gate 1's AND gate output is 1. Similarly, Gate 2 recognizes combination 010 by being connected to $\bar{A}$, B, and $\bar{N}$.

Connecting the OR Gates

Finally, the two outputs C and S are produced by separate OR gates, labeled Gate C and Gate S. The inputs to Gate C and Gate S come from various AND gates. How do we determine which particular AND gates? Look at the output columns in the truth table of *Figure 8-7*. Output C is supposed to be 1 whenever we have input combination 3 OR 5 OR 6 OR 7. So we connect outputs from AND gates 3, 5, 6, and 7 to the OR gate inputs for output C. Similarly, output S is supposed to be 1 whenever we

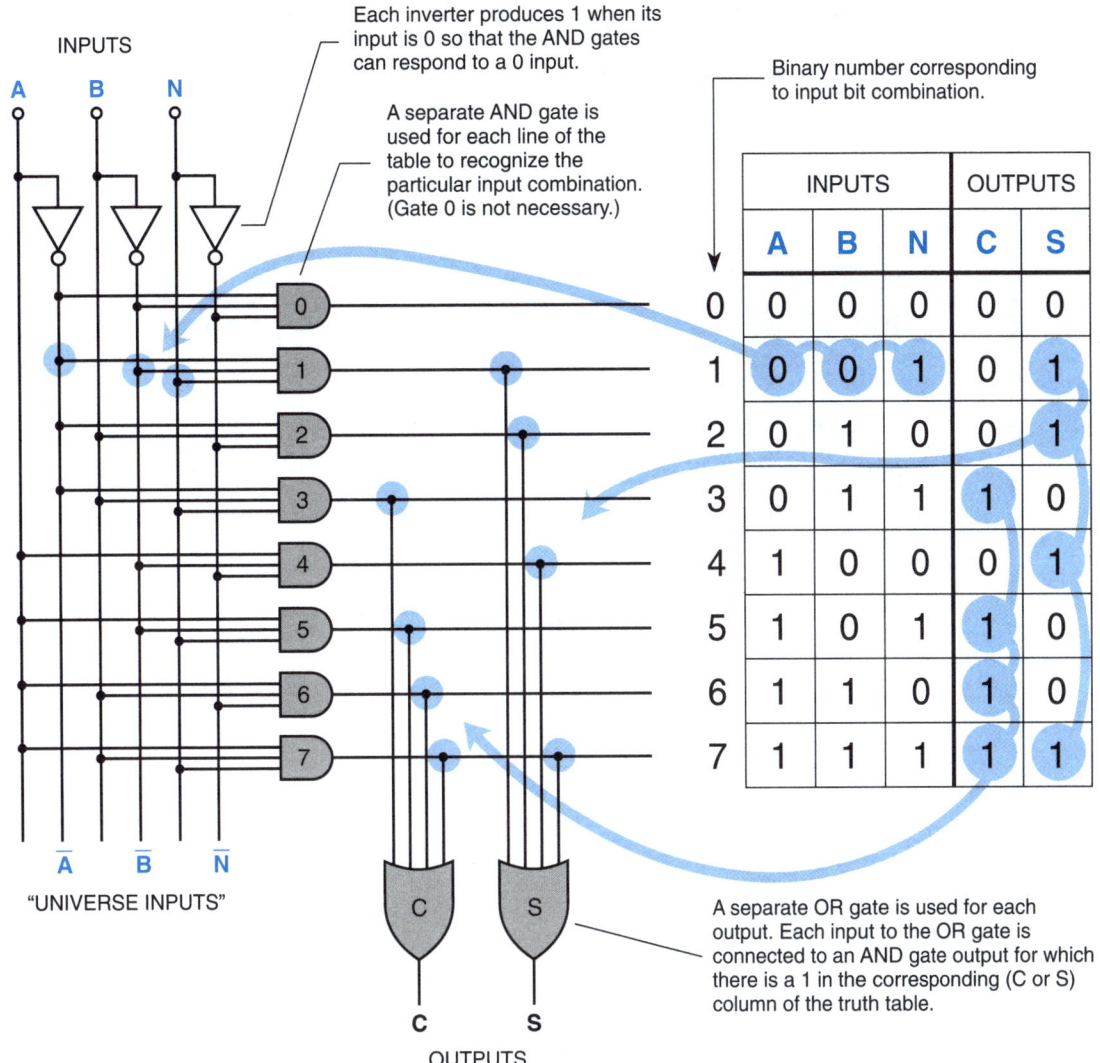

Figure 8-7. This 1-bit full-adder design is the canonical type that comes straight from the truth table.

have input combination 1 OR 2 OR 4 OR 7. So the inputs for the OR gate for output S are driven by the outputs from AND gates 1, 2, 4, and 7.

At this point, we see that we don't need AND gate 0. That's because combination zero in our truth table doesn't cause any ones in the outputs. So, for this case, gate 0 is omitted from the canonical design, leaving us with only seven AND gates.

The canonical network for a given truth table is hardly ever the simplest design; that is, one with the fewest possible gates and connections. The example in *Figure 8-7* would produce the same results if we left out certain connections, so that gates with fewer inputs can be used. But the canonical design is a good starting point for finding simpler versions.

Adding Binary Numbers with Several 1-Bit Adders

Finally, let's see how several 1-bit full adders can be used together for a more impressive job—adding two binary numbers. This further illustrates the fundamental principle of digital systems, which is breaking complex functions down into simple ones. Or, to put it another way, simpler circuits are used as *building blocks* in making a greater variety of more complicated circuits. Of course, we could design a canonical network or other special circuits to add two binary numbers, but doing the job with

Example 3. A Logic Diagram From a Truth Table

For the following truth table, what are the circuit connections of an AND-OR gate array?

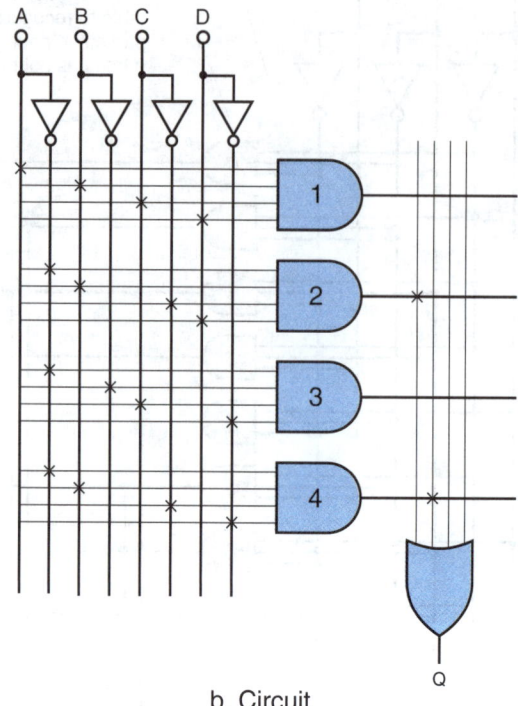

	IN				OUT
	A	B	C	D	Q
1	1	1	1	1	0
2	0	1	0	1	1
3	0	0	1	0	0
4	0	1	0	0	1

a. Truth Table

b. Circuit

1-bit adders is a simpler and more convenient way, and provides a good learning example.

Figure 8-8 shows the logic design of a *4-bit binary adder* which is made up of four 1-bit full adders of the type shown in *Figure 8-7*. The eight main inputs to this combinational network are two 4-bit binary numbers called A and B which are to be added together. These numbers are called the addends. Each of them can be anything from zero (0000) to fifteen (1111). The five outputs, in turn, form a 5-bit binary number S which is the sum of A and B. It ranges from zero (00000) to thirty-one (11111). Notice that the least significant "ones" bit of addend A is labeled A0 (A-zero), the "twos" bit is A1 (A-one), the "fours" bit is A2 (A-two), and the most significant "eights" bit is A3 (A-three). The bits of Addend B and Sum S are numbered in the same way. The Sum S also has a "sixteens" bit as its most significant bit, labeled S4 (S-four).

Notice that the four 1-bit adders are numbered from 0 to 3 like the bits of the addends. Addend bits A0 and B0 get added together as inputs A and B of adder 0. Output S of adder 0 provides bit S0 of the sum. Similarly, adder 1 adds bits A1 and B1 to produce sum bit S1, adder 2 adds bits A2 and B2 for sum bit S2, and adder 3 adds bit A3 and B3 for sum bit S3, and the carry bit S4.

Now here is an important point that makes the additions come out right. Output C of each 1-bit adder feeds input N of the next adder to the left. The C stands for "carry-out" from the 1-bit adder, and the N stands for "carry-in" to the adder. This is just like you do when adding, by hand, two ordinary decimal numbers that produce a sum greater than nine. If the sum of a column is two digits, you write down the least significant digit and carry the other to be added with the next column. As we have seen, output S from each 1-bit adder is the least significant bit of that adder's sum, and output C is the most significant. So C gets carried and added in with the next two addend bits to the left, as shown in the addition example in *Figure 8-8*.

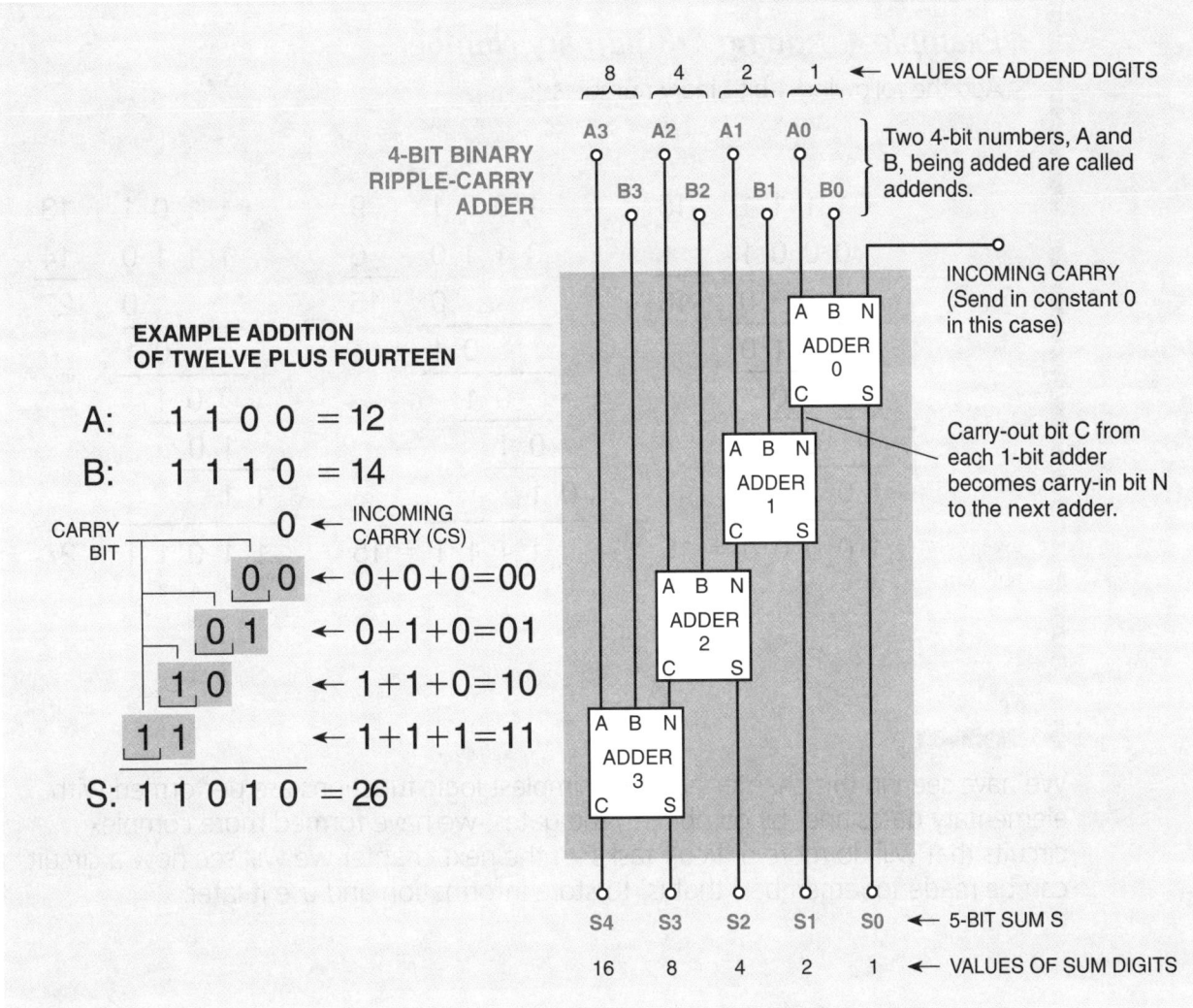

Figure 8-8. One-bit full adders can be chained together to make a ripple-carry adder that adds two binary numbers of any length. This is a combinational network, but not a canonical design.

Adder 3 doesn't have another adder on the left to receive its carry-out bit C, so that signal is brought out as the "sixteens" bit of the sum, S4. This is like writing down the final carry on the left end of the sum when doing an addition on paper.

Adder 0 doesn't have another adder on the right to provide a carry-in bit N, so we provide a constant 0 to that input. In this case, Adder 0 only has to add two bits, not three. It could be a simpler kind of 1-bit adder without a carry-in input N. Such a circuit is called a *half-adder*.

You can see that it would be possible to extend this string of 1-bit full adders on out to the left as far as we like. To add two 8-bit numbers, for instance, we could use eight 1-bit full adders. Each would receive a carry-in bit from its neighbor to the right. Some computers add pairs of binary numbers as long as 128 bits.

The adder design in *Figure 8-8* is called "ripple-carry," because the carry bits ripple through from right to left. Although it works well, additions take a lot more time than with some other adder designs. In a typical integrated circuit, each gate delays the signals passing through it by several nanoseconds (billionths of a second). Delays through many gates can slow the system down considerably. So when long binary numbers must be added as fast as possible, a ripple-carry design would not be used.

Example 4. Adding 4-Bit Binary Numbers

Add the following 4-bit binary numbers:

A.

```
              1 1 1 1     15
              0 0 0 1      1
INPUT
CARRY BIT →   →     0     16
              ─────────
              →   1 0
CARRY
BIT           →   1 0
              →   1 0
              → 1 0
              ─────────
              1 0 0 0 0 = 16
BIT VALUE →   16 8 4 2 1
```

B.

```
    1 0 0 1     9
    0 1 1 0     6
    ─────────
          0    15
    ─────────
        0 1
    ─────────
        0 1
    ─────────
      0 1
    ─────────
    0 1
    ─────────
    1 1 1 1 = 15
    8 4 2 1
```

C.

```
      1 1 0 1     13
      1 1 1 0     14
      ─────────
            0    27
      ─────────
          0 1
      ─────────
          0 1
      ─────────
        1 0
      ─────────
      1 1
      ─────────
      1 1 0 1 1 = 27
      16 8 4 2 1
```

Summary

We have seen in this chapter how the simplest logic functions are performed with elementary gates and, by combining the gates, we have formed more complex circuits that will do more difficult tasks. In the next chapter we will see how a circuit can be made to remember; that is, to store information and use it later.

Quiz for Chapter 8

1. The simplest circuits in a computer or other digital system are called:
   a. amplifiers
   b. logic gates
   c. rectifiers
   d. adders

2. A circuit that performs the NOT function is called:
   a. a converter
   b. a flip-flop
   c. an OR gate
   d. an inverter

3. The state of the output for each and every possible input combination is shown in:
   a. a truth table
   b. a logic diagram
   c. a flip-flop table
   d. a gate diagram

4. The output of the _____ gate is 1 *only* if all the inputs are 1.
   a. NOT
   b. NOR
   c. OR
   d. AND

5. If all the inputs are 0, then the output is 0 for a/an _____ gate.
   a. OR
   b. NOR
   c. AND
   d. two of the above

6. We can make an AND gate out of relays by connecting the relay switches in:
   a. parallel
   b. series
   c. ripple
   d. any combination

7. A single relay of the type that is normally-closed (on) when there is no current in the coil performs the logic function of:
   a. an OR gate
   b. an AND gate
   c. an inverter
   d. an EXCLUSIVE OR gate

8. By writing 1 for high and 0 for low in a *function* table, we get:
   a. a truth table with positive logic
   b. a truth table with negative logic
   c. a gate table
   d. an inverted function table

9. A positive AND gate acts as
   a. a negative NAND gate
   b. a negative NOR gate
   c. a negative OR gate
   d. a negative inverter

10. Two NPN transistors in parallel that share a common load (pull-up) resistor act as:
    a. a double inverter
    b. a positive AND gate
    c. a negative OR gate
    d. a negative NAND gate

11. An AND gate followed by an inverter acts as:
    a. a NAND gate
    b. a NOR gate
    c. a negative inverter
    d. an exclusive AND gate

12. The circuitry in digital systems today consists almost entirely of:
    a. vacuum tubes
    b. relays
    c. integrated circuit chips
    d. power switches

Answers: 1 b, 2 d, 3 a, 4 d, 5 d, 6 b, 7 c, 8 a, 9 c, 10 d, 11 a, 12 c

Questions for Chapter 8

1. What are the three basic digital decisions or logic functions called?

2. What is a circuit that performs the NOT function usually called?

3. What does a complete truth table show?

4. If an AND gate has the following bits for inputs, what will its output be in each case? (1101; 1000; 1111; 0000)

5. If an OR gate has the following bits for inputs, what will its output be in each case? (1101; 1000; 1111; 0000)

6. Do digital gates have to be electronic? Explain.

7. Explain the difference between a function table and a truth table.

8. How does it affect an inverter if a system is changed from positive logic to negative logic?

9. An OR gate followed by an inverter acts like a _____ gate.

10. An AND gate followed by an inverter acts like a _____ gate.

11. How are inverse inputs used to permit gates to detect zeros for inputs.

12. What is meant by the *canonical* combinational network for the truth table?

13. What is a disadvantage of a ripple-carry adder?

14. Of the three inputs to a full-adder (A, B, and N), which is not used for a half-adder and why?

15. What do the S and C outputs of a full-adder stand for?

CHAPTER 9
How Electronic Memories Operate

As we noted in the last chapter, the main functions of a digital system (other than inputting and outputting information) are concerned with *decisions* and *memory*. We have already talked about how decisions are made by logic gates connected together in various ways. In this chapter, we will talk about memory; that is, about how information is *stored* in digital systems.

This subject is divided into two parts, because there are two general ways to use memory. One of these uses is called *main memory*, and the other is called *sequential logic circuits*, or just sequential circuits. The meanings of these names will become clear to you as we go along.

Starting to Learn About Computers

First, let's look at an example that includes both sequential circuitry and main memory so you can see the difference. To do this, we will start talking about how a computer works, including an adder such as we discussed in Chapter 8. *Figure 9-1* shows the most important subsystems of a computer, leaving out a lot of details that don't concern us right now.

This would be called an 8-bit computer because it works on information in the form of 8-bit bytes. (Remember, a byte is a group of bits (usually eight) that are transmitted and worked on together; sometimes these groups are called "words," which we will discuss later.) Each of the broad arrows in the figure stands for eight conductors running along together, carrying all eight bits of a byte in parallel. A group of conductors carrying several bits in parallel is usually called a bus.

Routing Circuitry

The block labeled ROUTING CIRCUITRY stands for various connections that can be made among different parts of the computer. A byte coming into the routing circuitry along a particular bus can be routed to nearly any other bus coming out of this circuitry. The computer is forever shuttling bytes (all eight bits at once) from one place to another through the routing circuitry.

Controller

The subsystem called the CONTROLLER tells the routing circuitry where to send the various bytes and provides the timing of the transfers. The controller also tells all other parts of the computer what to do with the bytes they receive and when to do it. The control signals are carried on control buses connected between the controller and all the other parts. The control bus connections are not shown in *Figure 9-1* to simplify the diagram.

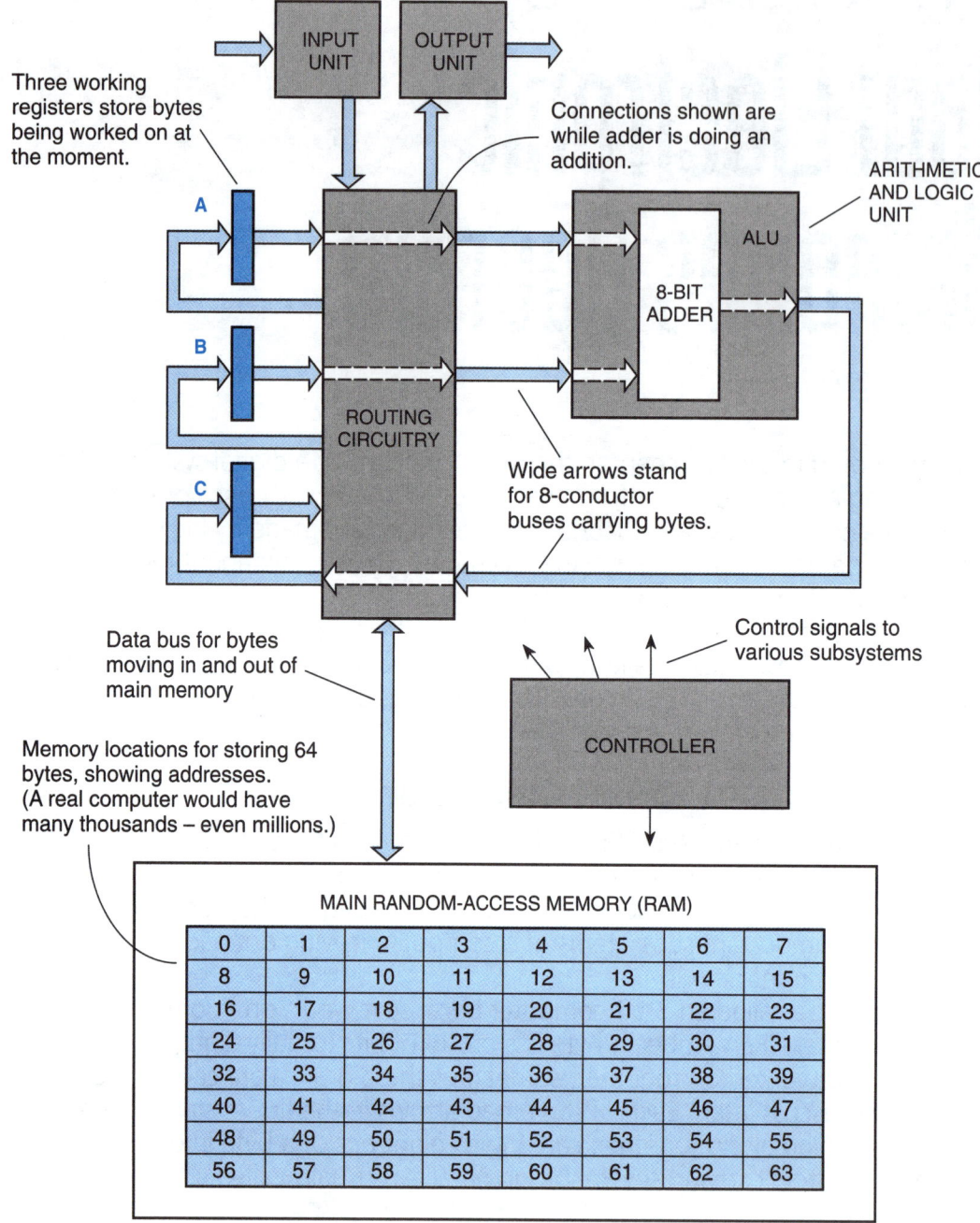

Figure 9-1. The main subsystems of a computer, showing the difference between sequential circuits (registers, in this case) and main memory (the main RAM).

Memory: A General Storehouse for Lots of Bytes

The main memory is where the computer stores bytes that are not being worked on at the moment. Each little rectangle in the memory diagram stands for a memory location, meaning a place where one byte can be stored. Each memory location is identified by a number called its address, like the address of a house on a street, or an apartment in a building. In a typical desktop computer, the main memory might hold 1,048,576 bytes (100000000000000000000 in binary form). But in this simple example, we are showing only 64 memory locations (1000000 binary). Memory addresses start with zero, so in this case the address numbers run from zero to 63 (111111).

A byte can be read or written at any of the main memory locations at any time, in any order. That makes this a *random-access memory*, or RAM (rhymes with ham). Random means not according to a definite sequence or pattern.

Bytes being transmitted in and out of the memory use the same bus. The controller tells the memory when to write (store) a byte coming in, when to read (retrieve) a stored byte and send out a copy, and which address to use.

Main and Mass Memory

In *Figure 9-1*, the memory shown is a random-access memory called a main memory. It is in this memory that the program instructions and data for the immediate task that the computer is running are stored. There is another memory called mass memory. It is used to store programs, results, text, pictures, and data for use at a later time by the computer. There may be indefinite periods of time before the use of this information. A mass memory unit is a subsystem (a major, separate part of a system) where fairly *large amounts* of information are stored for *indefinite time periods*. The most common mass memory used today is a magnetic disk. Whenever you hear or see the word "memory" used by itself, it can mean the fast, main memory that is self-contained in the computer or the large, stable, long-term mass memory used to load the main memory.

Example 1. Designing a Memory Decoder

Design a decode network that would place a 1 on each of eight column lines to read out eight bits of a word from a RAM. Start with a truth table, then design the circuit.

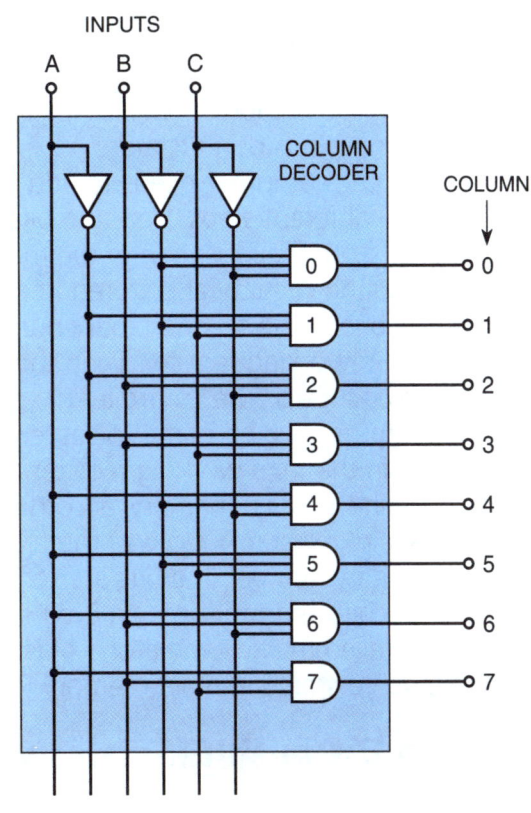

COL. NO.	INPUT A	B	C	OUT
0	0	0	0	1
1	0	0	1	1
2	0	1	0	1
3	0	1	1	1
4	1	0	0	1
5	1	0	1	1
6	1	1	0	1
7	1	1	1	1

a. Truth Table

b. Circuit

Sequential Circuits: Storing Bits Where They Are Needed

A system's main memory isn't the only place where information can be stored. *Figure 9-1* shows three extra storage places for single bytes, called *registers*. A register is a unit that stores a group of bits. Generally, the bits come into the register by one bus, are held for a while, and go out by another bus. Registers are used in many different places in digital systems. They receive bits, hold them for awhile, and then release them; that is, the bits they receive are not stored permanently. The particular registers we are discussing are called the computer's *working registers*. They provide temporary storage for a few bytes while the computer is working on them.

There are registers of various kinds. One we will discuss is called a *sequential circuit*. Sequential circuits are logic circuits that form circuits, many for storage of bits, that operate by moving bits in a particular timed sequence. The information is stored right in among the circuitry that makes decisions, instead of being sent to a separate main memory subsystem. Later, we will look closer at why such circuits are called sequential.

Adding Two Binary Numbers

To help show the difference between the uses of sequential circuitry and main memory, we will see what happens when our computer adds two 8-bit binary numbers. The two numbers to be added have already been stored as bytes at certain locations in the main memory.

The adding is done by an 8-bit adder which is very similar to the 4-bit adder that we studied in the last chapter. It is part of a very important subsystem called the *arithmetic and logic unit*, or ALU. (In this case, "arithmetic" is pronounced arith-MET-ic instead of a-RITH-me-tic; ALU is pronounced by saying the letters A-L-U.) Keep in mind that the adder itself cannot store any of the information that flows into it. This is because it is just a combinational logic network. Its output bits stay constant only as long as the input bits don't change.

Whenever the computer needs to add two numbers, the controller makes it go through several different steps. First, the bits of one of the two binary numbers to be added are read from their location in the main memory RAM and transmitted through the routing circuitry and stored in one of the working registers. Next, in the same way, the bits of the second binary number (addend) are read from memory and stored in another working register. Then these two registers are connected to the adder's two 8-bit inputs. The controller gives the signal to perform the addition. After the adder's outputs have had time to settle down in their final pattern, the resulting bits are stored in the third working register. From there, an 8-bit byte can be copied into a location in the main memory, according to signals from the controller. Stating the process again, register A receives the first 8-bit number, register B receives the second 8-bit binary number, the bits are sent at the same time to the 8-bit adder, which adds the bits and produces an 8-bit sum, which is stored in register C. Register C, sometimes called the accumulator, holds the 8-bit sum until the controller directs the sum bits to be stored in main memory RAM at a particular address.

Storing a Bit with Electricity: The Basic Flip-Flop

In this chapter and the next, we will continue to look at other details of how a computer works. But right now, let's talk more about the storage of bits. Several different methods are used for storing a bit. Probably the most important to learn about is the flip-flop, also called a bistable (bi-stable) element or static memory element. This concept for storing one bit is used in many sequential circuits and some main memories.

Figure 9-2 shows the basic idea of a flip-flop. It's just two inverters with the output of each one forming the input to the other. The inverters are said to be *cross-coupled* to each other. Also shown is an electrical schematic of such a memory element, using inverters of the direct-coupled NPN type. As usual, we assume positive logic in using this electrical circuit, so that the higher voltage stands for 1 and the lower voltage means 0.

This circuit can be in *either of two states or conditions*. One of these states is shown in *Figure 9-2a*. The left inverter is transmitting a 1 (Q=1), and the right inverter is transmitting a 0 ($\overline{Q}$=0). The left transistor is cut off (Q=1), and current from R1 of the left inverter turns on the right transistor so $\overline{Q}$ is 0. In *Figure 9-2b*, we see the opposite state. The *right* transistor is now cut off ($\overline{Q}$=1), and current from R2 of the right inverter turns on the left transistor so Q=0. Now the right inverter is transmitting a 1, and the left inverter is transmitting a 0.

When we say something is stable, that means it will stay that way until something comes along and changes it. This circuit is called bistable because it is stable in *either of two states*. You could think of many familiar examples like that. One would be the stiff cover of a book lying on a table. You can either flip the upper cover to the right or flop it to the left, and it will stay there. It won't stay balanced up in the air. An electronic bistable element behaves very much like that, which is why it is called a flip-flop.

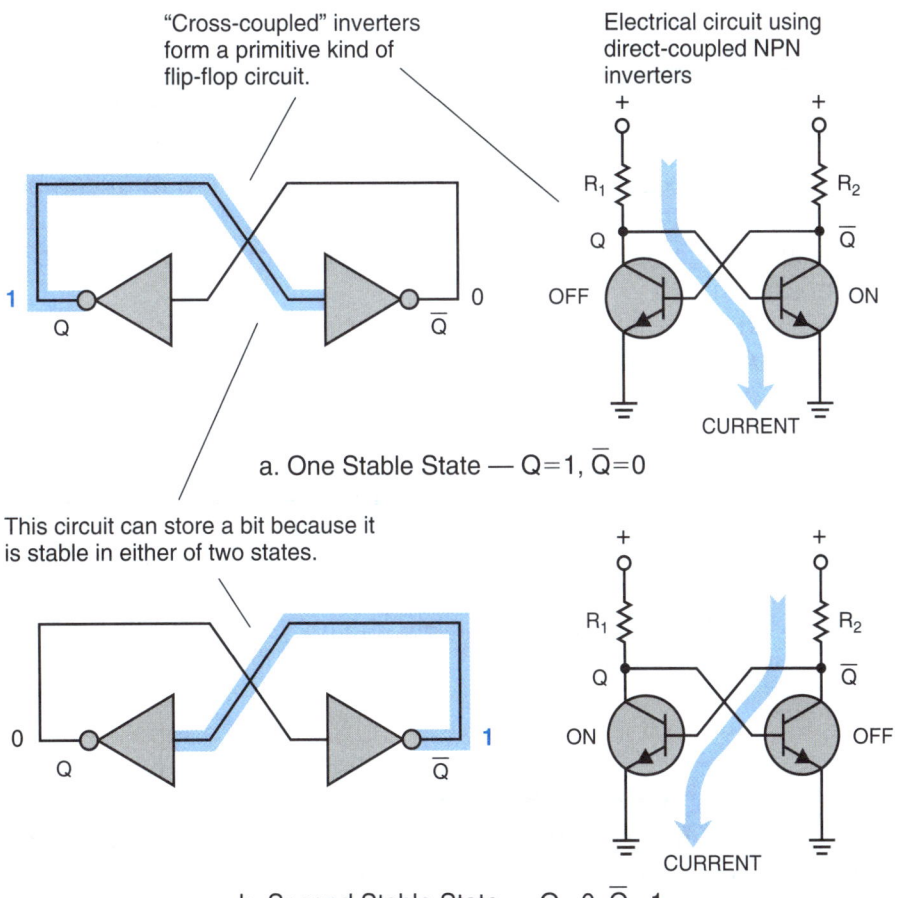

Figure 9-2. Basic concept of the flip-flop, also called a bistable element or a static memory element.

To use a flip-flop as a storage place for one bit, we simply let one state stand for 0 and the other state for 1. To store a 1, we flip the circuit to that state, and it stays there until changed as long as power is supplied. To change the bit and store a 0, we flop the circuit to its other state.

The R-S Latch Has Inputs for Flipping and Flopping

The flip-flop circuit of *Figure 9-2* isn't complete. It doesn't have any inputs or outputs. There is no way to make the circuit flip and flop. How do we solve that problem?

One way, if we are using direct-coupled NPN circuitry as in *Figure 9-2*, is to add a second transistor in parallel with each inverter as shown in *Figure 9-3a*. By momentarily turning on one or the other of these extra transistors, we can flip or flop the circuit to the desired state.

As shown in *Figure 9-3b*, forcing one inverter into a particular state forces the other inverter to be in the opposite state. The extra transistors turn the cross-coupled inverters into positive NOR gates. Remember, a NOR gate is like an OR gate followed by an inverter. Supplying a 1 at either input makes the output of the inverter a 0. The result (two cross-coupled NOR gates) is called an R-S latch. It is a still a very primitive flip-flop—but a complete and useful one. It is called a latch because, when it is set to a particular state, it stays there—latching in the bit state until it is changed.

The R input to the left NOR gate in *Figure 9-3* is called the *reset* input. This input is used to flip the bistable element to store a 0. (To reset a stored bit means to make it a 0.) The left gate's output, labeled Q, is used as the main output for describing the state of the bit stored in the latch. It is called the *true* output. Resetting forces Q=0.

The S input to the right gate is called the *set* input. It is used to flip the bistable element so Q is a 1. (To set a bit means to make it a 1.) With Q=1, the inverter with the set input will have an output of 0. The additional opposite output can be handy in some applications, so its output is also provided from the flip-flop. It is called the *complement* output and is labeled $\overline{Q}$ (Q-bar). Complement is another word for inverse or opposite.

By the way, you may sometimes come across circuit diagrams showing latches made of positive NAND gates instead of NOR gates. That sort of latch works much the same way, but it is usually named an $\overline{S}$-$\overline{R}$ latch, called "S-bar R-bar" latch. The simplest circuit is shown in *Figure 9-3c*.

Circuits with Memory are Sequential, not Combinational

Since a flip-flop has ordinary digital inputs and outputs, its operation can be described in a truth table. *Figure 9-4* shows a truth table for the R-S latch.

As in most truth tables, there is a separate line for each possible input combination. For any circuit with two inputs (R and S in this case) the combinations are 00, 01, 10, and 11.

In logic diagrams, the two input signals R and S would be supplied by some other circuitry in the system, not shown in *Figure 9-4a*. Normally, both R and S are kept at 0. When the inputs are 00, the outputs are designated in the truth table as Q_0 and $\overline{Q}_0$ (Q-sub-zero and Q-bar-sub-zero). The Q_0 state is the state Q ended in after being set or reset the last time. $\overline{Q}_0$ is, of course, the opposite state. It is the initial state when R=0 and S=0. When the latch is set, R=0 and S=1, and from the truth table, we see that Q=1 and $\overline{Q}$=0. When the latch is reset, R=1 and S=0, and Q=0 and $\overline{Q}$=1. Now note the input combination R=1 and S=1. If both R and S are 1 at the same time, Q and $\overline{Q}$ are both driven to 0, but when R and S are returned to 0, the state that Q and $\overline{Q}$ end up in cannot be predicted reliably. Therefore, in a logic circuit, having R=1 and S=1 at the same time *is not allowed*.

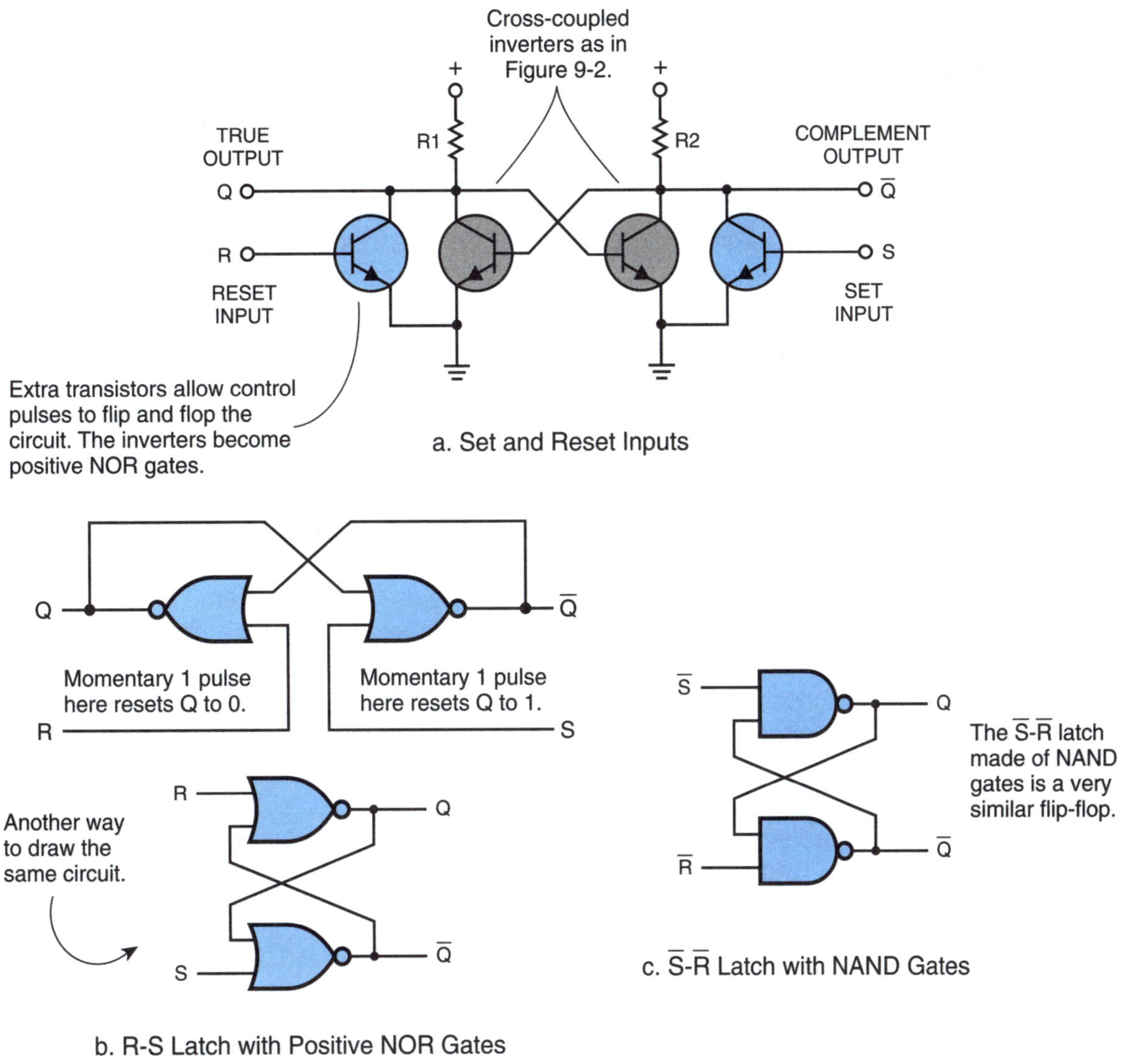

Cross-coupled
inverters as in
Figure 9-2.
+
+
R1
R2
TRUE
OUTPUT
Q
COMPLEMENT
OUTPUT
Q̄
R
S
RESET
INPUT
SET
INPUT
Extra transistors allow control pulses to flip and flop the circuit. The inverters become positive NOR gates.
a. Set and Reset Inputs
Q
Q̄
Momentary 1 pulse here resets Q to 0.
Momentary 1 pulse here resets Q to 1.
R
S
Another way to draw the same circuit.
R
Q
S
Q̄
b. R-S Latch with Positive NOR Gates
S̄
Q
R̄
Q̄
The S̄-R̄ latch made of NAND gates is a very similar flip-flop.
c. S̄-R̄ Latch with NAND Gates

Figure 9-3. The R-S latch is a primitive flip-flop made of two cross-coupled NOR gates.

R
Q
S
Q̄

a. Logic Diagram

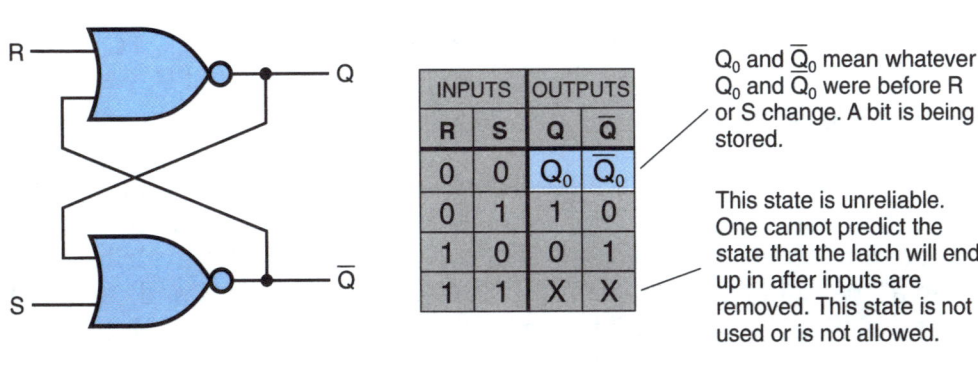

INPUTS		OUTPUTS	
R	S	Q	$\overline{Q}$
0	0	Q_0	$\overline{Q}_0$
0	1	1	0
1	0	0	1
1	1	X	X

Q_0 and $\overline{Q}_0$ mean whatever Q_0 and $\overline{Q}_0$ were before R or S change. A bit is being stored.

This state is unreliable. One cannot predict the state that the latch will end up in after inputs are removed. This state is not used or is not allowed.

b. Truth Table for R-S Latch

Figure 9-4. The truth table for an R-S latch shows that a flip-flop is a sequential circuit, not a combinational circuit.

The input combination of R=0 and S=0 and output states Q_0 and $\overline{Q}_0$ remind us that a flip-flop is *not a combinational circuit* like a gate or adder. Instead, it is a *sequential* logic circuit or network. The word "sequential" refers to things that happen in a sequence, meaning that the state of the circuit depends on what happened in the past; that is, what happened in the sequence of events that occurred just before the present event occurs.

A sequential logic circuit is similar to a combinational logic circuit, in that it has several bits going in and several coming out. What is different is that some output combinations depend on what happened in the sequence of different input combinations just before the input combination that presently is to be applied.

Registers Use Gated Latch Flip-Flops

Next, let's see how flip-flops could be used to make registers for our computer example in *Figure 9-1*. There are a number of different kinds of flip-flops that are more complex and more useful than the R-S latch. All of them store one bit, and all are made from one or two R-S latches (or $\overline{S}$-$\overline{R}$ latches consisting of NAND gates, mentioned earlier). A typical simple register might be made of a type of flip-flop called the *gated latch*. *Figure 9-5a* shows one way to make a gated latch out of a single R-S latch and *Figure 9-5c* how it could be used in a register.

The flip-flop in *Figure 9-5a* has one input, labeled D for data, for a bit to be stored. It has one output Q for the bit which is presently being stored. A second input called the clock input (labeled CK), controls the storage process. The clock is normally at 0. To store the new bit coming in at D, the clock input is quickly pulsed to 1 and back to 0. (By the way, the clock signal in a gated latch is often called the gating signal instead.)

The truth table in *Figure 9-5b* shows what happens more exactly. When the clock is 0, the output Q is D_0, meaning whatever D was as a result of the last clocking. Whether D=1 or D=0, because the clock is zero, there is no change in the state of Q. When the clock is 1, the output Q goes to the state of D gated in by the clock.

Note that if D=1, then $\overline{D}$=0, and when CK=1, the set gate has both inputs at 1 which sets Q to 1. If D=0, then $\overline{D}$=1, and when CK=1, the reset gate has both inputs at 1 which resets Q to 0. Thus, Q follows D when gated by CK. If CK=0, Q is not affected by the state of D. The gating action ends when CK goes back to 0 from 1.

Logic gates work like a gate in a fence which can be opened or shut. To gate a digital signal means to control it by blocking it or passing it through. When the clock feeding each AND gate is at 1, the gates are open. When the clock is changed to 0, the gates are shut. Thus, to store the bit being presented at input D, we unlatch the circuit for a very brief time with a 1 in the clock line, and then latch it again by sending the clock back to 0.

To make an 8-bit register, we simply use eight gated latches controlled by the same clock signal as shown in *Figure 9-5c*. This is called a *parallel* register, because it receives and transmits all bits in parallel at the same time. To store a byte being received at the eight inputs, we briefly pulse the clock input to 1 and back to 0. The result at the output is the state of each bit that was at the input when the CK signal went to 1 and then returned to 0.

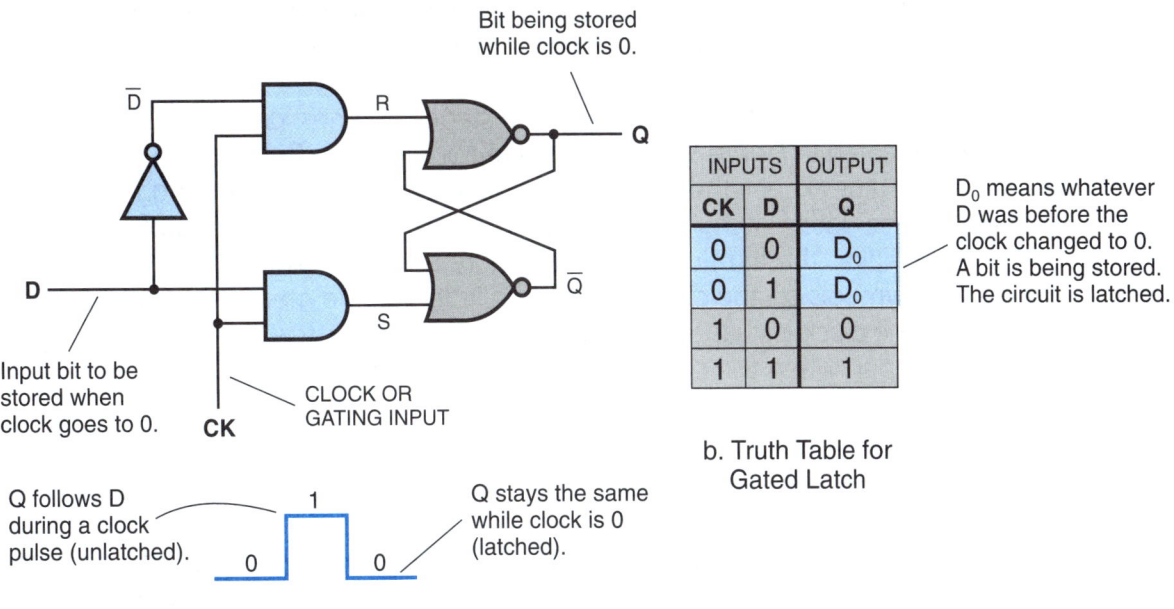

a. Gated Latch Using R-S Latch and Logic Circuits

b. Truth Table for Gated Latch

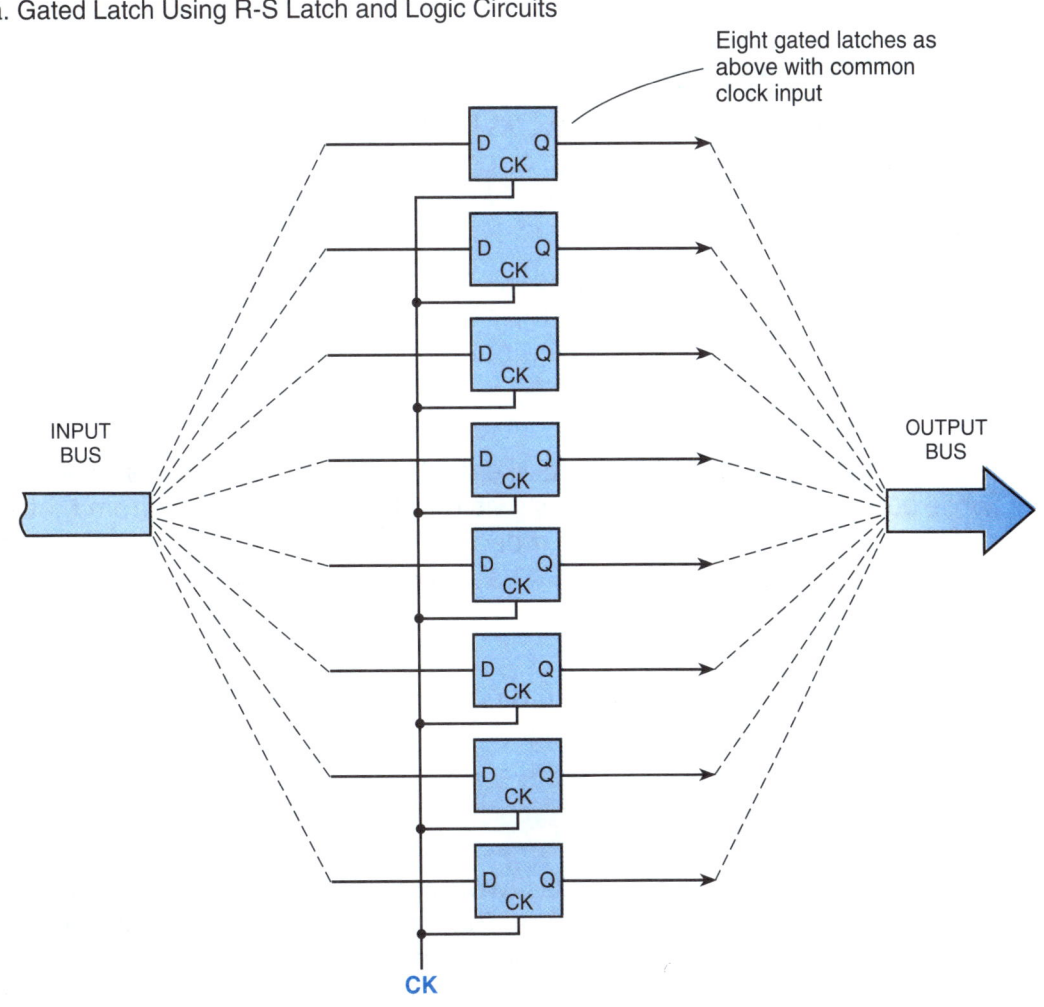

c. 8-Bit Gated Latch Parallel Register Logic Diagram

Figure 9-5. A parallel register can be made of gated latches.

Other Flip-Flops Use Variations of the Same Principles

If you continue learning about digital systems, you will run into other kinds of flip-flops called R-S (not quite the same as the R-S latch), D, T, and J-K. *Figure 9-6* shows how these units are typically represented in logic diagrams, along with a truth table for each one.

Each of these flip-flops consists of two stages in series, called the *master* and the *slave*, as indicated in *Figure 9-6a*. Each stage contains one R-S latch (or $\overline{S}$-$\overline{R}$ latch). Having two stages keeps the outputs from following the inputs during the clock pulse, as in the gated latch. The outputs change only at the very end of the clock pulse. That makes the flip-flops useful for more purposes. This behavior is indicated by the two symbols on the clock inputs. The little triangle means the outputs change only when the clock is changing, and the little inversion circle means those output changes happen when the clock changes from 1 to 0.

No Clock Column

To go along with this use of the clock, the truth tables for these flip-flops are a little bit different than we have shown before. They don't include a column for the clock input. Instead, the left side of each table shows any other inputs *before a complete clock pulse*, and the right side shows the true output Q *after* a clock pulse. Let's look at each truth table more closely.

Clocked D-Type Flip-Flop

The D-type flip-flop *(Figure 9-6b)* works like the gated latch, the true output Q follows the data input D during the clock pulse.

Clocked R-S Flip-Flop

The R-S flip-flop *(Figure 9-6c)* works like the previous simple R-S latch, except that the outputs change only at the end of the clock pulse. (As before, when both R and S are 1 at the time of clocking, the true output is uncertain and would not be used.)

J-K Flip-Flop

The J-K flip-flop *(Figure 9-6d)* works like the R-S flip-flop, except that when clocked while both inputs are 1, the outputs switch to the opposite states from that existing before the clocking. So if Q was 1, it becomes 0, or vice versa. This is called *toggling*. To toggle something means switch it back and forth between states.

T Flip-Flop

Finally, the T flip-flop *(Figure 9-6e)* has no inputs for a bit to be stored. Instead, a clock pulse simply makes the outputs toggle to the opposite states. It's like a J-K flip-flop with a permanent 1 at both inputs.

Preset and Clear

Figure 9-6f shows a flip-flop that has two extra inputs called preset and clear. Any kind of flip-flop can have these inputs. They are separate ways of putting a bit into the flip-flop, *between clock pulses*. The preset input is like the set input on an R-S latch, and the clear input is like the reset input. Both are normally at 0. A momentary 1 at the preset input "presets" the flip-flop to 1 at the Q output. A momentary 1 at the clear input "clears" the Q output to 0.

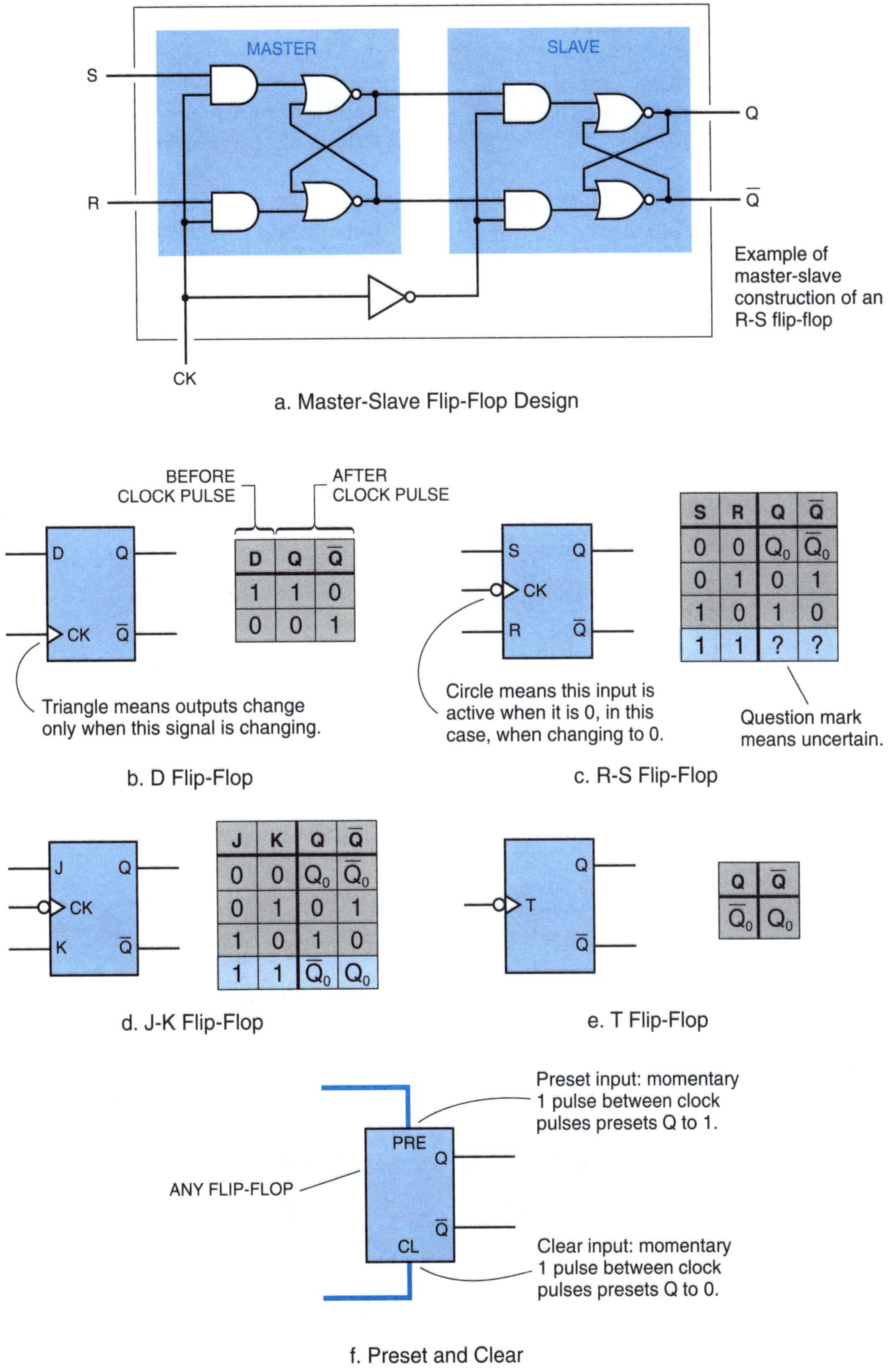

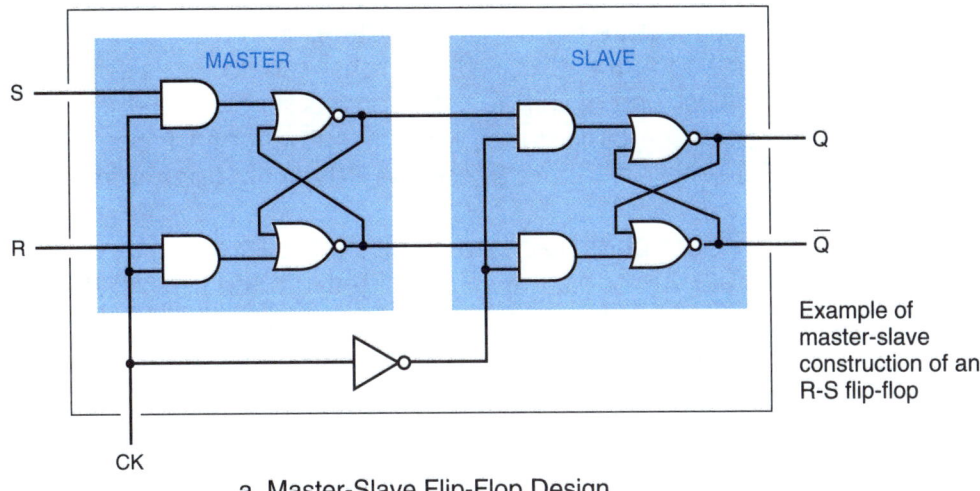

Figure 9-6. Several kinds of flip-flop that are widely used. They typically consist of a master stage followed by a slave stage. Any design can be provided with preset and clear inputs.

A Shift Register Passes Bits from One Flip-Flop to the Next

One very handy feature to have in a register is the ability to *shift all the stored bits from one flip-flop to the next*. A register that can do this is called a *shift register*.

The simplest kind of shift register consists simply of a string of master-slave flip-flops clocked by the same signal, with the output of each one feeding the input of the next. For simplicity, *Figure 9-7* shows D-type flip-flops, with one data input D and one true output Q.

At every clock pulse, each stored bit shifts over to the next flip-flop (toward the right in *Figure 9-7a*), and a new bit is shifted into the first flip-flop from some other circuitry. An 8-bit byte would take eight clock pulses to be shifted in completely. This shift register also has external connections from the outputs of the eight flip-flops. That allows all the stored bits to be read out in parallel after they have been shifted in.

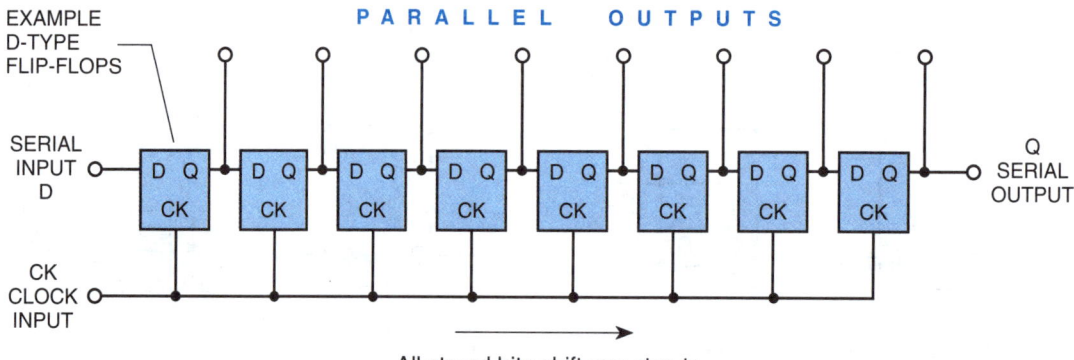

a. Serial Shift Register with Parallel Outputs

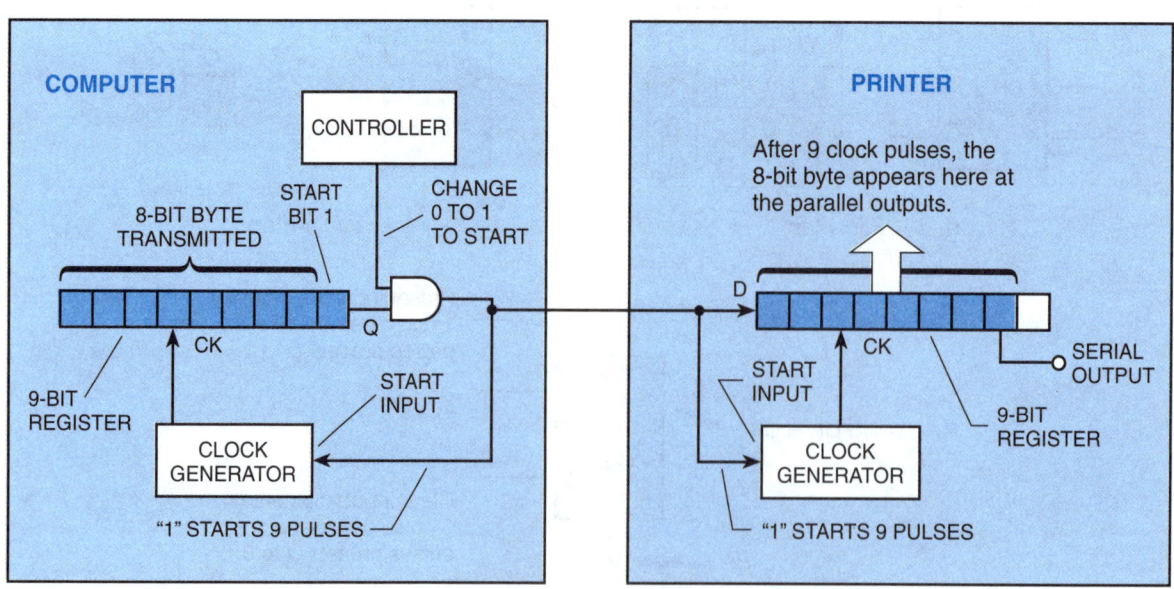

b. Serial Shift Register Application

Figure 9-7. In a shift register, the output of each flip-flop feeds the input of the next one. Data can be sent over a single wire using serial data transfer.

One application of shift registers, shown in *Figure 9-7b*, is to send and receive bytes serially, meaning one bit at a time in a single conductor. A computer is sending a character to a serial-input printer as we discussed in Chapter 7. The 7-bit character is part of an 8-bit byte.

To shift bits from one register to the other, the clock signals to the registers must be synchronized. *Figure 9-7b* shows one way to do this. There is a separate clock generator circuit at each end of the transmission line. The line is connected to a "start" input on both clock generators. Changing that input from 0 to 1 starts the clock. Both clocks are started at the same instant, by a "start" bit of 1 that is loaded into the computer output register ahead of the 8-bit byte. (We talked about the start bit in Chapter 7.) Both clocks stop automatically after nine pulses.

Notice that the signal from the computer's output register is gated by an AND gate which is controlled by a signal from the controller. Before sending the byte, this control signal is 0.

To send a byte to the printer, the computer loads the 8-bit byte and the start bit into the output register. (We will see how in a moment.) To begin the shifting process, the controller switches the control signal to 1. The gate opens and transmits the start bit (a 1) which is sitting in the first bit position of the shift register's serial output. Both clocks "see" the start bit and begin clocking. All nine bits shift merrily through the line, and the clocks stop. The computer switches the control signal to 0 while it loads the next byte into its output register. The 8-bit byte that was received by the printer register is taken from the register's parallel output and sent out on a parallel data bus to other parts of the printer.

Example 2. Outputs After Clocking a Shift Register

In the shift register of *Figure 9-7a*, what will the parallel outputs read if all bits of the shift register are cleared to 0 and then the byte 00001111 is right shifted in with six clock pulses?

What is the output bit from the LSB stage?

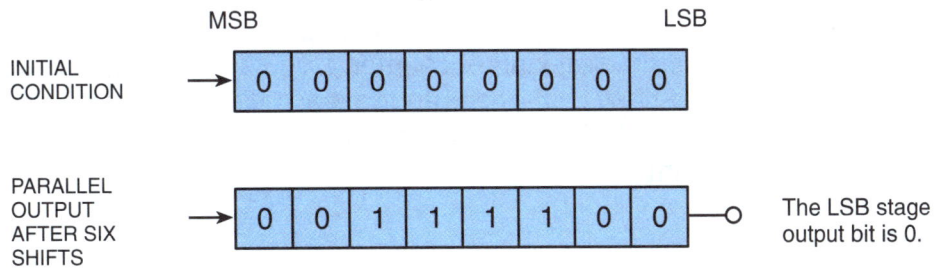

What will be the parallel output if the register is shifted 11 pulses with a second byte of 01010101 being shifted in?

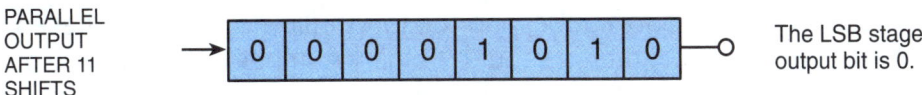

What is the output bit from the LSB stage? The LSB stage output bit is 0.

Some Shift Registers Can Be Loaded In Parallel

How does the computer load an outgoing byte into its output register? By parallel loading through the parallel inputs that some shift registers have. *Figure 9-8a* shows one way to do it. Each flip-flop in the register must have a clear and preset input.

The parallel loading is done by means of gates feeding the preset and clear inputs on each flip-flop in the shift register. In this case, parallel loading takes place when the shift register receives a 1 on a "load control" input. A 1 on the parallel input is gated to the preset input of the flip-flop and sets the Q output to a 1. A 0 on the parallel input is gated to the clear input of the flip-flop and resets (clears) the Q output to a 0.

In the application of *Figure 9-8b,* to load a byte into the register in parallel, the controller tells the RAM to fetch the byte from a certain address and put it on the memory data bus. At the same time, the routing circuitry connects the memory data bus to the parallel inputs of the output register. Then the controller pulses the load control input on the output register and stores the byte.

The ninth flip-flop at the front end of the register is not connected to the parallel loading circuitry of the other eight flip-flops. Instead, its preset input is connected directly to the transmit control line from the controller. When the transmit line goes to a 1, not only does the controller start the clock, but it also presets a 1 start bit into the front bit of the register. The clock generators are designed to wait an instant before sending the first clock pulse.

Counters Are Made Out of Registers

For an introduction to another kind of sequential circuit, consider the clock generator in the computer and the printer in the serial transfer application of *Figure 9-7*. The clock generator has to send nine clock pulses for each byte, and then stop. But how would this circuit know when nine clock pulses had been sent? That is, how would the clock generator *count* the pulses?

That is a typical job for a type of sequential circuit, or building block, called a *counter*. There are many different kinds of counters. They can be thought of as registers with special features. The number stored in the register changes in a regular, prearranged pattern every time an input pulse is received.

Figure 9-9a shows the simplest kind of counter, called a *binary ripple counter*. It is just a string of T-type flip-flops, with the output of each flip-flop feeding the *clock input* of the next one to the left. Pulses being counted are fed to the clock input of the first flip-flop on the right.

Counting

As mentioned earlier, an input pulse makes a T-type flip-flop toggle to the opposite state. These particular flip-flops toggle when their inputs go from 1 to 0. The first flip-flop on the right toggles every time the input pulses goes from 1 to 0. Its output goes from 1 to 0 every *second* pulse, toggling the second flip-flop. Similarly, the third flip-flop toggles at every *fourth* pulse, and the fourth flip-flop toggles at every *eighth* pulse. The output of the counter is a four-bit binary number formed by the outputs of the flip-flops in parallel. Because of the toggling, this output number increases by one at every pulse. The next number after fifteen (1111) is zero. If we were to add a fifth flip-flop, the counter would count up to 11111 (31), and so on.

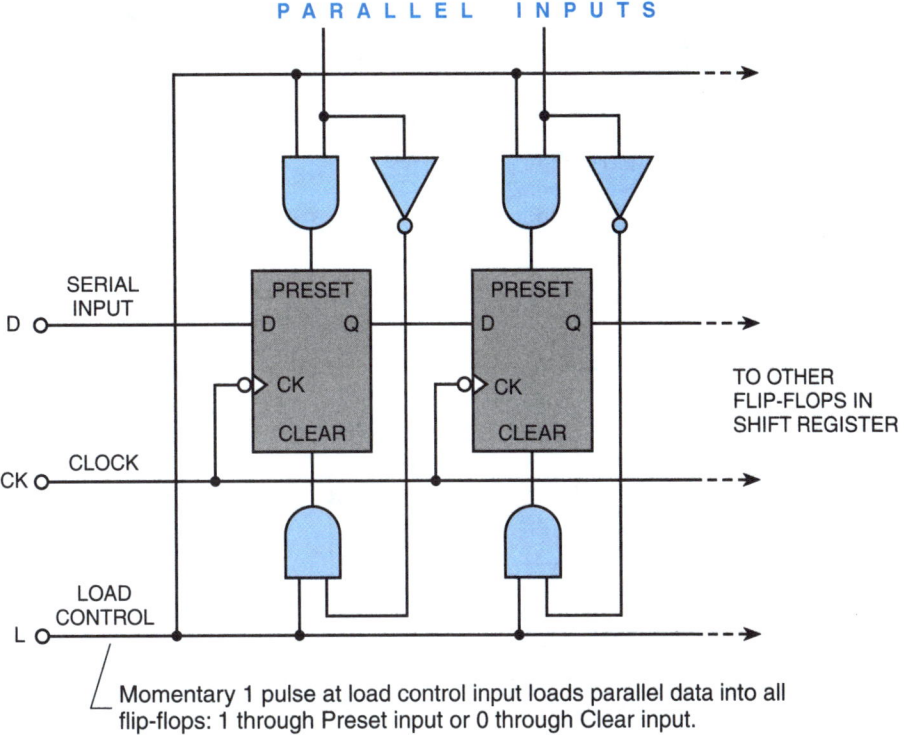

a. Serial Shift Register with Parallel Inputs

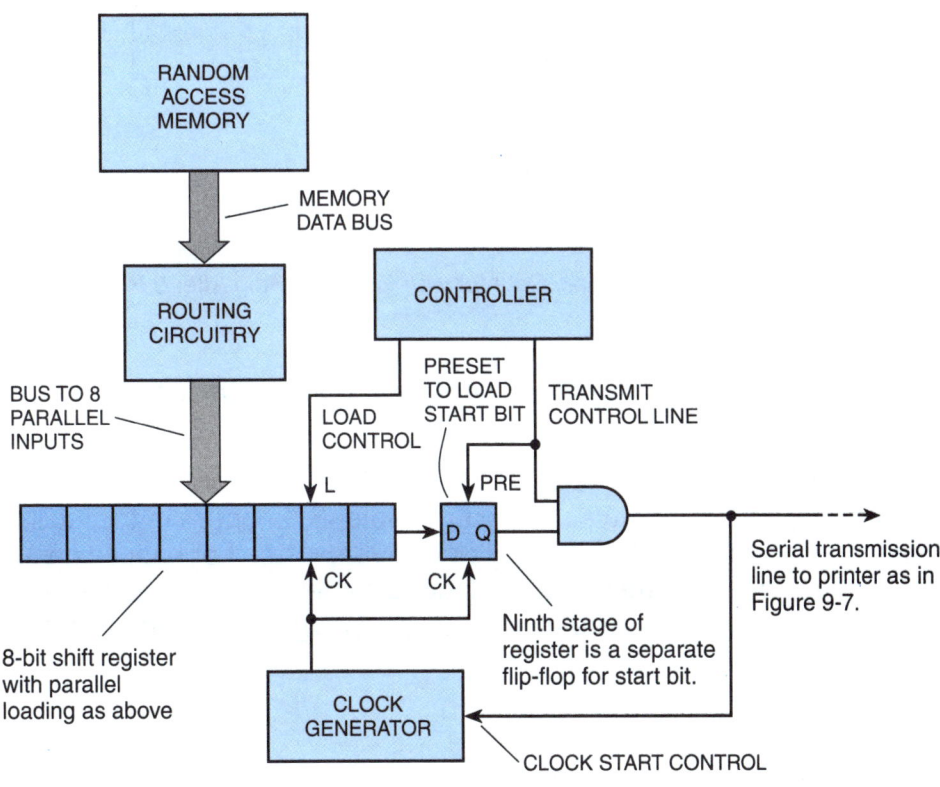

b. Application Using Serial Shift Register with Parallel Inputs

Figure 9-8. The serial output register in Figure 9-7 would be loaded through parallel inputs.

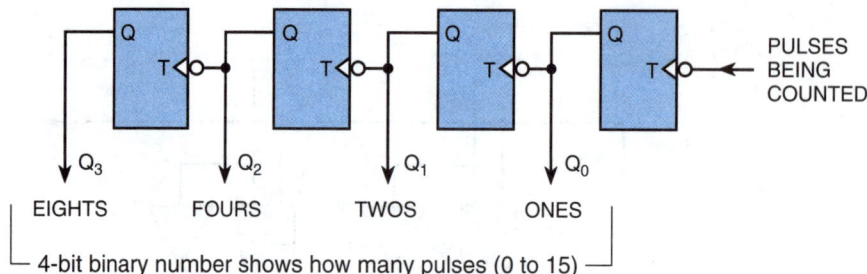

a. 4-Bit Ripple Counter

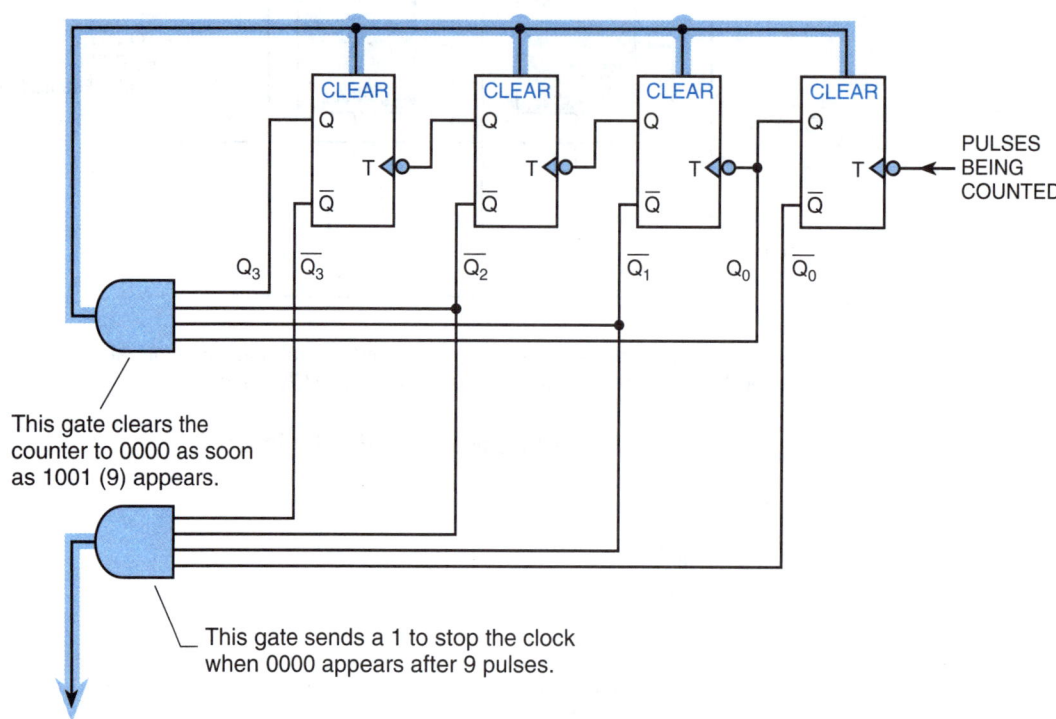

b. 4-Bit Ripple Counter with Gating to Stop After 9 Pulses

Figure 9-9. Examples of counters – another important kind of sequential circuit.

This particular counter is called a "ripple" counter for the same reason as the "ripple-carry" adder that we studied in Chapter 8. Changes in the registered number ripple through the circuit after every clock pulse, instead of happening at exactly the same time. There are other ways to design counters so that the flip-flops change at the same instant. Circuits like that are called *synchronous* counters instead of ripple counters.

Gating Nine Pulses

Figure 9-9b shows a way to modify a four-bit binary ripple counter to make it go back to zero and send an output signal after every nine pulses. That's what we need in our nine-pulse clock generator. The modification requires two AND gates. This would be called a *modulo-9 counter*. A counter's modulus is the number of different states through which it counts. In the counter required, the modulus is nine, representing numbers from zero to eight.

The upper AND gate in *Figure 9-9b* receives the Q (true) outputs of the first and fourth flip-flop and the $\overline{Q}$ (complement) outputs of the second and third flip-flops. This gate puts out a 1 only when the register outputs a count of 1001 (nine) on its Q outputs. We say that the AND gate recognizes or decodes that combination. Its output is connected to the clear input on all four flip-flops. Remember, a 1 at the clear input resets a flip-flop to 0. Thus, as soon as the ninth pulse is received, the AND gate forces the count back to 0000. The 1001 remains for just a brief instant before going to 0000.

The lower AND gate, in turn, is fed by the complement outputs of all four flip-flops. So it decodes the count of 0000, putting out a 1 in that case. This provides an output to the rest of the clock generator circuitry, telling it that nine pulses have been received and stopping the clock generator until the controller starts it again.

Example 3. Modulus of a Counter
What is the modulus of the following counter?

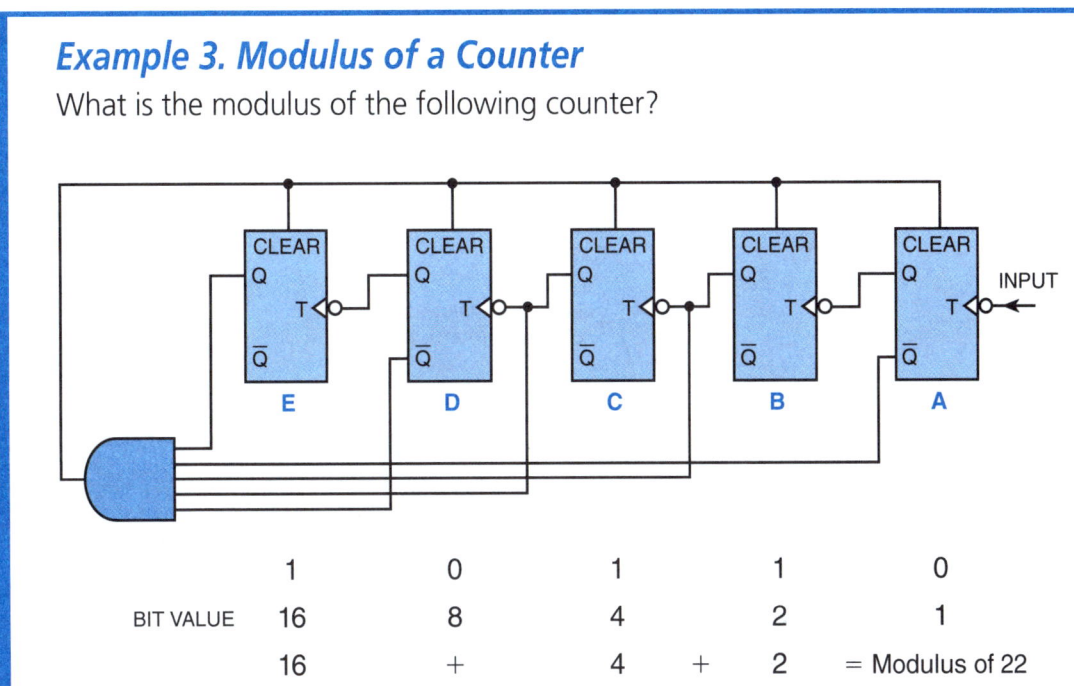

MOS Dynamic Memory Elements Can Take the Place of Flip-Flops

Instead of flip-flops, some integrated circuits use another method for storing one bit. This circuit is called the *dynamic memory element*. It involves storing electric charge on capacitors. The method only works using MOS integrated circuits; that is, ICs with circuits which use MOS transistors (MOSFETs) instead of bipolar transistors. You may want to review the information about MOSFETs in Chapter 4, and notice the MOS-type NOR gate shown in *Figure 8-5b*.

This time, to demonstrate the method, we will use a dynamic shift register as shown in *Figure 9-10a*. This dynamic shift register uses N-channel enhancement-mode MOS transistors. In the place of master-slave flip-flops, we use master-slave dynamic memory elements. Each of these consists of two MOS inverters with an extra MOS transistor in the input to each inverter. The extra transistors are called *gating transistors*. They act simply as on-off switches. Remember, when the control signal to a MOSFET of this type is high (1), it turns on. And when the control signal is low (0), the transistor turns off.

In this circuit, we have not one, but two, separate clock signals that go to all the storage elements in the shift register. These signals are called Phase 1 and Phase 2, abbreviated as φ1 and φ2. (The symbol φ is the Greek letter for F, called "phi,"

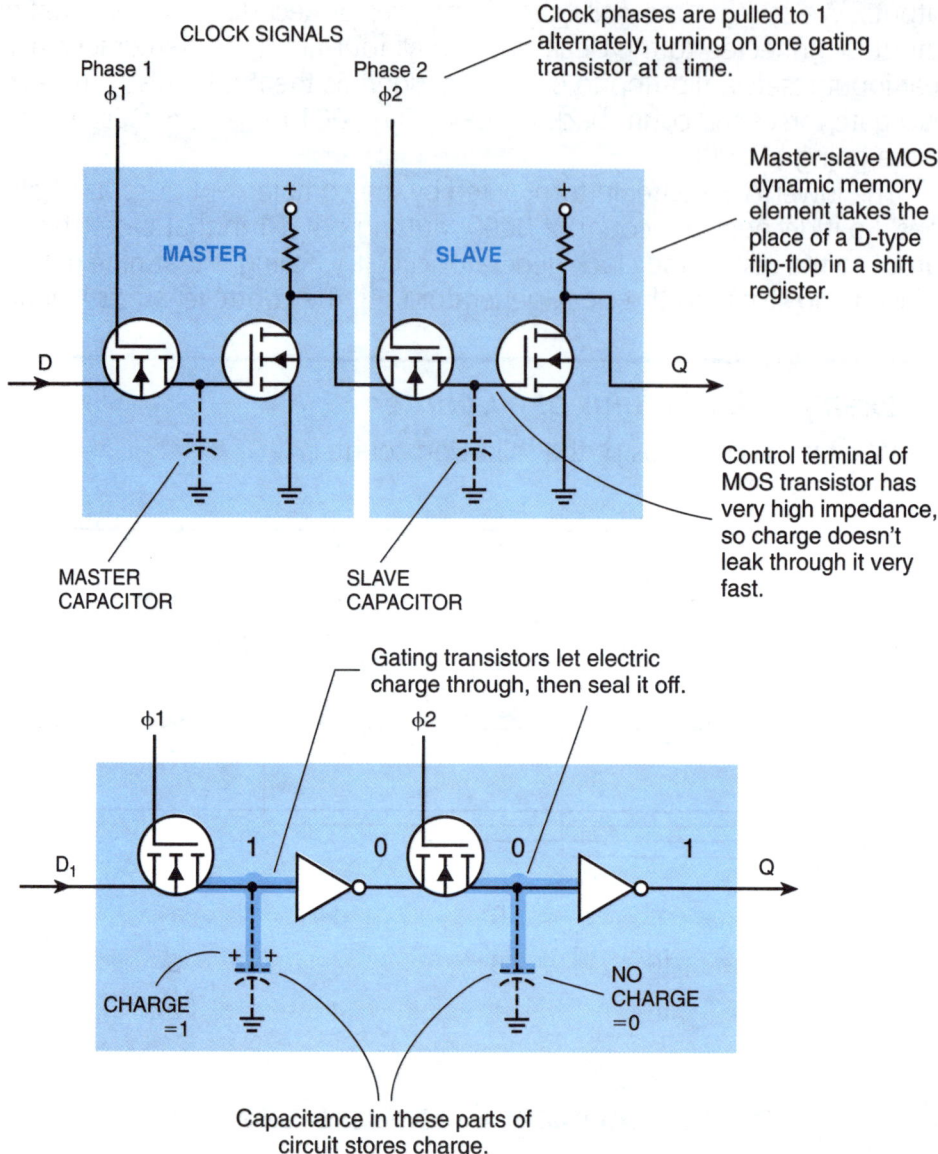

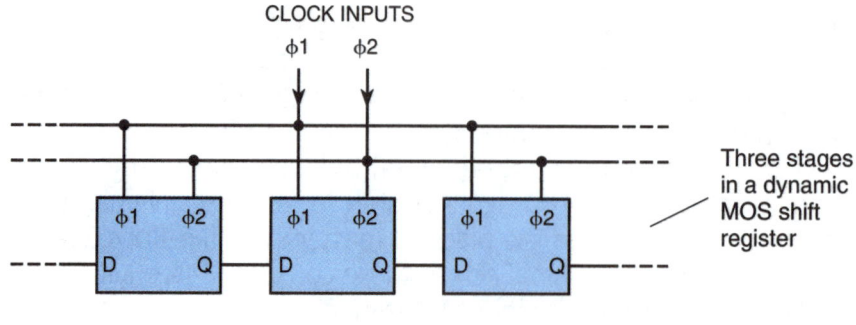

Figure 9-10. Dynamic memory elements store bits as electric charges instead of states in a flip-flop. The charges won't stay very long so this type of shift register must be clocked at a minimum frequency.

pronounced either like fee or fie.) Pulses in these clock lines alternate, first a Phase-1 pulse, then a Phase-2 pulse, then a Phase-1 pulse, and so on. These pulses turn on the two gating transistors—Phase 1 turns on the first one, Phase 2 turns on the second one.

This storage method depends on the basic principle that the input conductor and transistor input of each inverter have a certain amount of capacitance. As shown in *Figure 9-10a*, the circuit operates as if there were a capacitor connected between the inverter transistor's input and ground.

When Phase 1 is pulsed while a 1 is being received at the input (D) to the storage element, the voltage level of the 1 signal pushes a tiny positive charge into the master capacitor. When the master gating transistor turns off, the charge cannot leak off very fast. This is because the gate (control terminal) of a MOS transistor has very high input impedance, as discussed in Chapter 4. As a result, the input voltage to the master inverter stays very close to the original 1 level. In response, the master inverter output is at a 0 voltage that doesn't change at all.

Similarly, when Phase 1 is pulsed while a 0 is being received, the master inverter's input voltage stays pretty close to the 0 level after the gate transistor turns off. Thus the master inverter output is at a constant 1 voltage.

The same thing happens at the slave inverter input when Phase 2 is pulsed. The slave capacitor holds the input voltage after the pulse is over, and the slave inverter transmits the opposite bit as a constant voltage. This is the same bit that was stored in the master stage during the preceding Phase-1 pulse. Thus, every time both phases are pulsed, all the bits stored in the shift register shift over to the next master-slave memory element *(Figure 9-10b)*.

The reason why dynamic storage elements are widely used is that they take up considerably less space on an IC chip than a flip-flop. The main drawback is that the stored charges eventually leak out—in a few milliseconds. So to keep from losing the stored data, a dynamic shift register has to be clocked faster than some minimum frequency.

This is where we get the name "dynamic" for this kind of storage element. Dynamic means moving or working. To save the stored information, we have to keep the bits moving every so often. On the other hand, static means stationary. We can leave a static memory element alone, and it will hold data as long as the power supply is not interrupted.

Random-Access Memories Use Arrays of Storage Elements

Finally, let's come back to random-access main memories and see how they use static or dynamic storage elements. The general idea is to lay out a large number of storage elements on an integrated-circuit chip, packed closely together in rows and columns. This pattern is called a rectangular array. The storage elements, whether static or dynamic, are usually called cells. There are conductors running along the rows and columns. Each conductor, called a row line or a column line, makes contact with all the cells in a particular row or column. These conductors are used for selecting cells at certain addresses, and reading or writing the bits.

For example, the general RAM example shown in *Figure 9-11* has 256 cells, in a square array with 16 bits on each side. These bits are grouped into 32 eight-bit bytes, with two bytes in each horizontal row. However, a group of bits at the same address in a RAM is usually called a *word* instead. For example, some systems transfer data back and forth in 8-bit bytes, but the data is stored and worked on in words consisting of two or more bytes apiece. This example system uses a byte as a word.

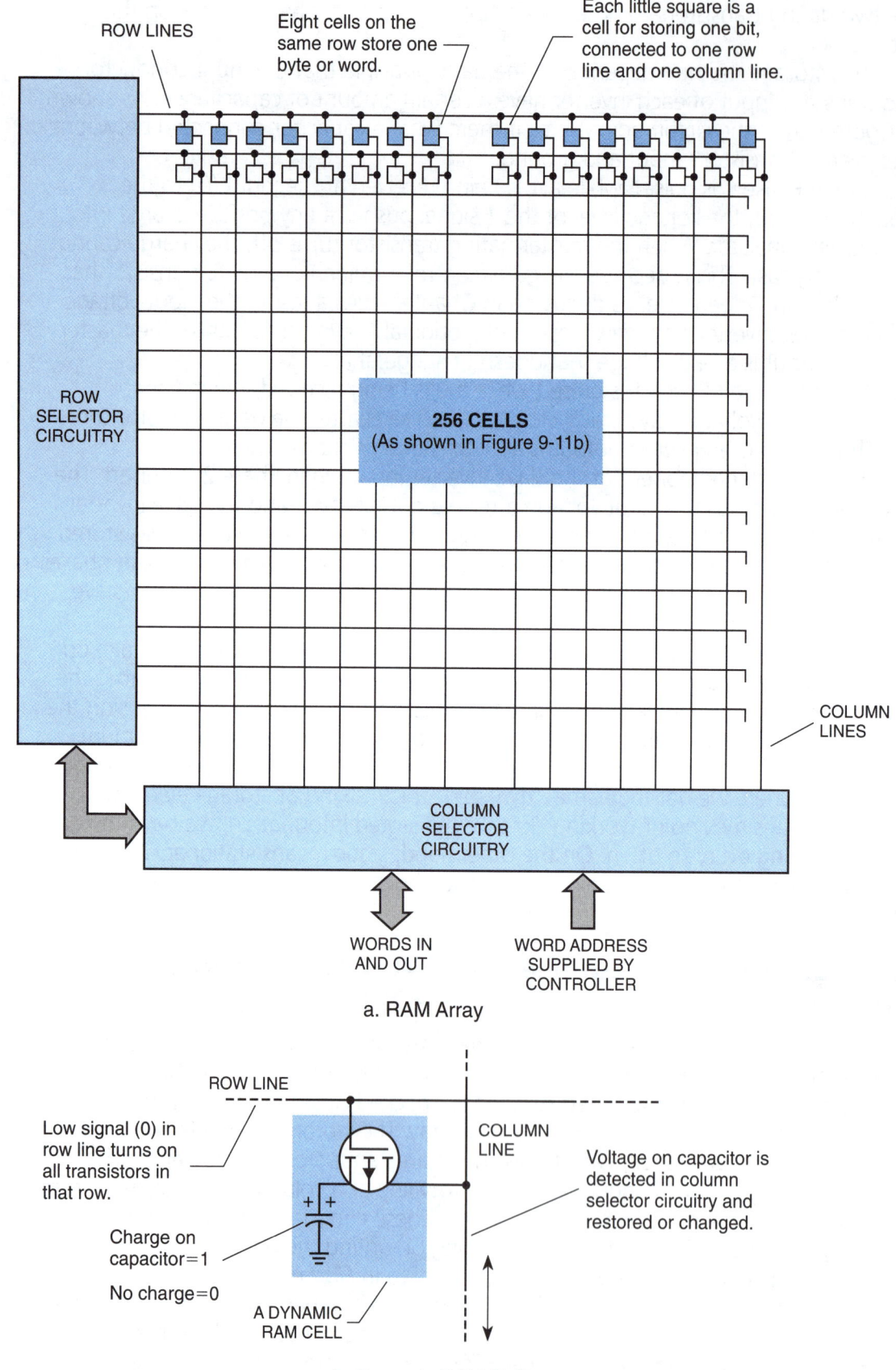

Figure 9-11. Concept of a random-access memory and a dynamic RAM cell

In the general RAM example, all the cells along one row would be selected by a signal in that row line from the ROW SELECTOR CIRCUITRY. The particular word desired in that row would be selected by making contact only with the specific eight column lines, through the COLUMN SELECTOR CIRCUITRY.

Figure 9-11b shows one way to make a dynamic RAM. Each cell consists merely of a p-channel enhancement-mode MOS transistor with a tiny capacitor on one main terminal (the source). The other main terminal (the drain) is connected to the column line. The transistor's control terminal (the gate) is connected to the row line. As in the dynamic shift register example, a charge on the capacitor means 1, and no charge means 0.

To read a stored word, a 0 is applied to its row line. All the transistors in that row turn on, connecting the capacitors to the column lines. All the words on that row are copied into a long parallel register in the column selector. The selected word is transmitted to the output bus. Meantime, all the words in the register are copied right back into their memory cells. This strengthens or refreshes the stored charges in that row. Every so often, the column selector refreshes all the rows in this way, to keep the stored data from being lost. Words are written into the memory by a similar process.

Figure 9-12 shows a type of static RAM cell. Each cell is a primitive flip-flop made of two direct-coupled NPN inverters as in *Figure 9-2*. Each column of cells has two separate column lines connected to the cells through diodes. Each row line serves as a ground connection for all the cells in that row. Normally, the voltage of a row line is above ground voltage, barely low enough to keep the inverters working. The column lines are normally at ground, keeping the diodes reverse-biased so there is no current through them.

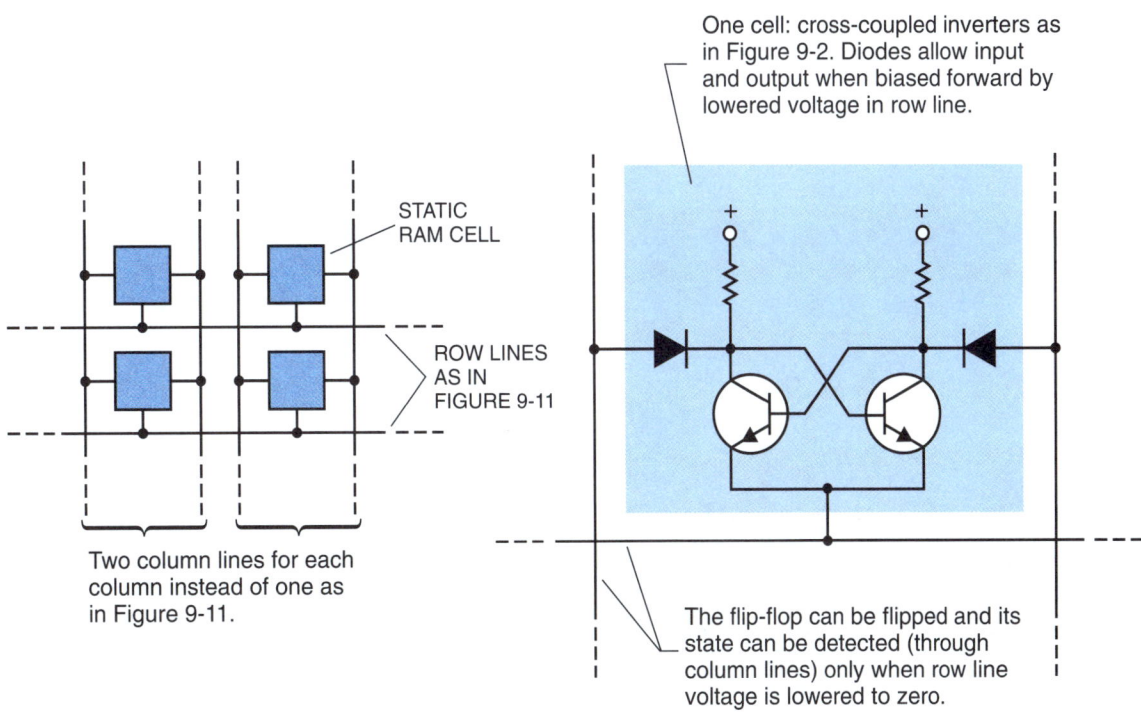

Figure 9-12. Each cell in a static RAM is a primitive flip-flop such as this example.

To store a bit, a row is selected by pulling the row-line voltage down to ground. This enables a pulse of about 1.4 volts in either column line to forward-bias its diode on the selected cell, setting the flip-flop to either the 1 or 0 state. Other cells on the same column line are not affected because their diodes remain reverse-biased. A cell is read by pulling its row line to ground and connecting both column lines to a 1.4-volt source through large resistances. The flip-flop connection in the 0 state will draw a tiny current—not enough to flip the flip-flop, but enough to detect the state present in the cell.

Static RAMs made with bipolar transistors (NPN or PNP) can be made to read and write much faster than dynamic RAMs. But their cells are considerably larger, and the IC manufacturing process is more complicated. That means fewer cells can be put on a single IC chip, and the cost per bit of storage is much higher. Bipolar static RAMs also use more power than dynamic RAMs. For these reasons, dynamic RAMs are presently used more widely than static RAMs.

Summary

Throughout these last three chapters we have learned about digital circuits, both combinational and sequential. We have learned that simple logic circuits are combined to make complex circuitry, and we have seen how the complex circuitry is combined into input, decision making, memory and output circuits to make digital systems. In the next chapter, we will learn more about digital systems.

Quiz for Chapter 9

1. The computer stores bytes that are not being worked on at the moment in the:
   a. Arithmetic Logic Unit
   b. bus
   c. Read-Only Memory
   d. RAM

2. Each memory location is identified by a number called its:
   a. file
   b. address
   c. directory
   d. access number

3. Circuits that provide temporary storage for a few bytes while the computer is working on them are called:
   a. working registers
   b. working ROM
   c. storage counters
   d. main controllers

4. Logic circuits that have storage places for bits built into them are called:
   a. sequential circuits
   b. gates
   c. combination circuits
   d. inverters

5. The part of a computer that performs operations such as addition is the:
   a. RAM
   b. ROM
   c. ALU
   d. BUS

6. Probably the most important static memory element used in many sequential circuits and some mass memories is the:
   a. gate
   b. inverter
   c. flip-flop
   d. amplifier

7. A flip-flop's main output for the bit stored in it is labeled Q and called the:
   a. main output
   b. true output
   c. toggle output
   d. memory output

8. To set a bit means to make it a:
   a. 1
   b. 0
   c. value opposite to what it was
   d. value equal to what it was

9. To store a new bit in a gated-latch, we unlatch the circuit for a very brief time with the clock while the bit is on the terminal labeled:
   a. R
   b. D
   c. Q
   d. T

10. The little triangle on the clock pin of a flip-flop means the outputs change only when the clock:
    a. is changing
    b. is high
    c. is low
    d. stopped

11. A way of putting a bit into the flip-flop between clock pulses is the:
    a. Q pin
    b. latch pin
    c. toggle pin
    d. preset pin

12. Instead of flip-flops, some integrated circuits called dynamic memory elements store:
    a. pictures instead of bits
    b. bits in gates
    c. electric charges on capacitors
    d. electric current in resistors

Answers: 1 d, 2 b, 3 a, 4 a, 5 c, 6 c, 7 b, 8 a, 9 b, 10 a, 11 d, 12 c

Questions for Chapter 9

1. A group of conductors carrying several bits in parallel is usually called a _____.

2. The subsystem called the _____ tells the routing circuitry where to send the various bytes.

3. The _____ is where the computer stores bytes that are not being worked on at the moment.

4. What does *random-access memory* mean?

5. What three things does the controller tell memory?

6. Storage places in the computer for single bytes are called _____.

7. How can two inverters be used to make a flip-flop?

8. What is meant by the term *complement* output?

9. Logic gates are combinational, flip-flops are _____.

10. To _____ a digital signal means to control it by blocking it or passing it through.

11. What is the name of the circuit that is formed of flip-flops and passes bits along from one flip-flop to another?

12. What does the term *toggle* mean?

13. How would a computer load an outgoing byte into an output register?

14. What is the simplest kind of binary counter?

15. What is meant by a counter's modulus?

CHAPTER 10
Computers and Other Digital Systems

In earlier chapters about digital electronics, we have used various parts of computer systems as examples. Now let's fill in some of the gaps we have left in your understanding of electronic circuitry used in computers and similar equipment. This book is about electronics, not computers, but to understand *digital* circuitry, you need to see how it is used.

In talking about computers here, bear in mind that we are using that word in a very general way. There are personal computers of many kinds, sometimes called microcomputers. Each one has a keyboard, a video display, at least one magnetic disk unit, and usually a printer. There are larger computers used in offices and factories, also known as minicomputers. They typically have several stations, called terminals, that resemble separate personal computers. And there are even larger computers (sometimes called mainframes), which may be connected to hundreds of terminals. They handle huge amounts of information, perhaps for a fairly large business.

We are also speaking of computers that are *built into other systems*. Nowadays, it seems that almost everything electronic contains at least one computer. The computer size may not be such that it fits in a room, or on a desk, or in a breadbox, but may be on one or two integrated circuit (IC) chips that are included in video games, video cassette recorders, compact disk players, or ignition control systems for cars. Such a built-in computer that serves a highly specialized function is said to be dedicated to that purpose. A dedicated computer usually doesn't have its own keyboard, display, disk units, or printer. As mentioned previously, in some cases it consists only of one or two IC chips.

Central Processing Unit, Memory, and Input/Output

First, let's complete our general picture of the main internal parts of an 8-bit computer, which we first looked at in *Figure 9-1*. *Figure 10-1* is a repeat of *Figure 9-1*. It shows more details than before, yet it still glosses over a lot of points that make one computer different from another. Many of these differences are in what we call the routing circuitry. The routing circuitry is the circuitry controlled by the controller that switches the information pipelines (the buses) and the system interconnections so the system operates in the proper sequence. Other differences occur due to the number of inputs and outputs, the memory sizes, and the informal working circuits, such as clocks, registers, size of buses, etc. To understand the function of some these in more detail, and how they work together, we will use *Figure 10-1* as a reference.

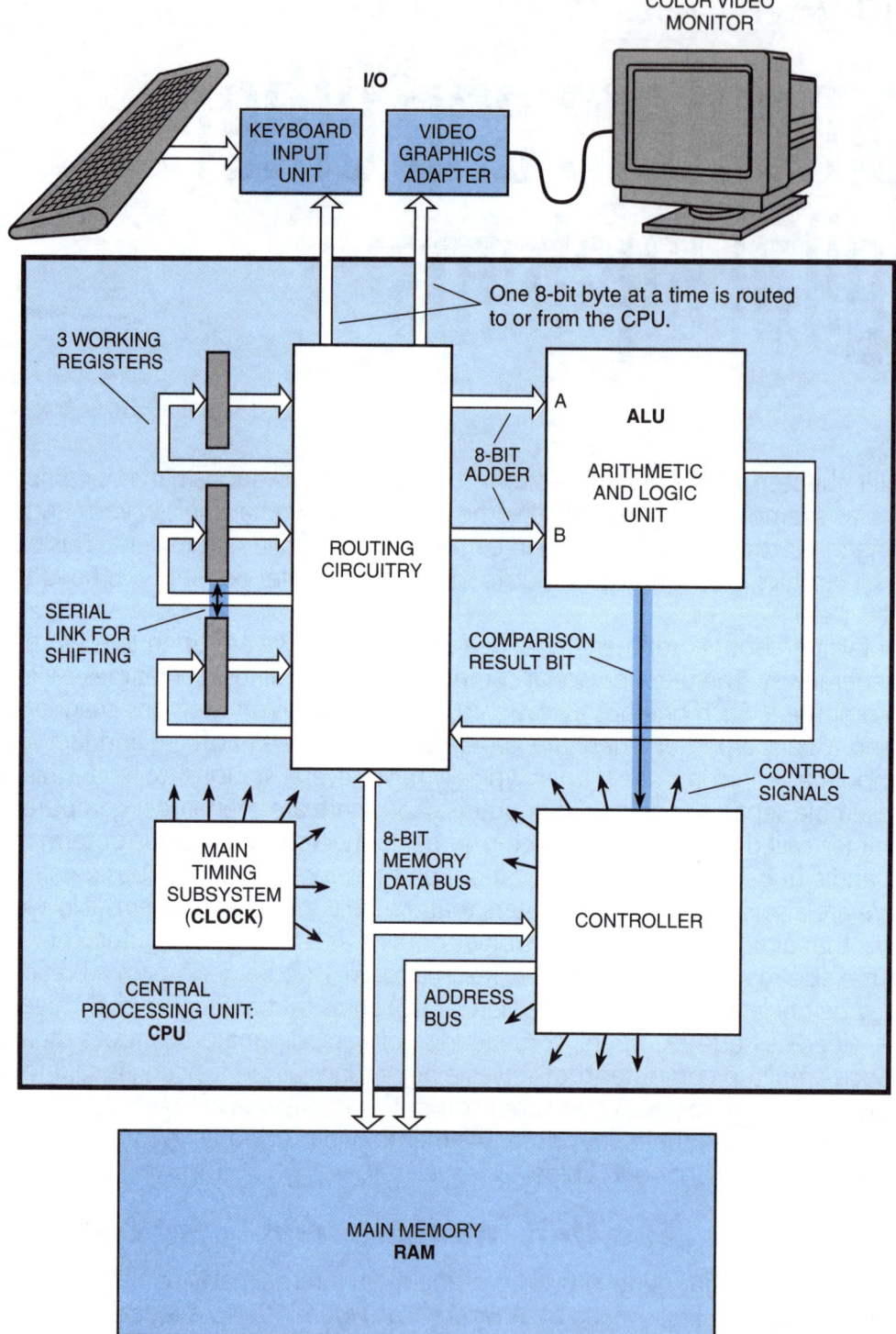

Figure 10-1. The main subsystems of a computer are classified as input/output (I/O), central processing unit (CPU), and random access memory (RAM).

The ALU, working registers, controller, and routing circuitry—all together—are called the *central processing unit* (CPU) of the computer. The diagram also includes a central clock or timing subsystem. This refers to circuitry that sends signals throughout the computer system to keep in time (synchronize) everything that happens. It can be considered to be part of the CPU. For an 8-bit computer like this, the entire CPU would probably be contained in one IC chip.

One detail added in *Figure 10-1* is the *address bus*. The controller uses it to send an address to the RAM when a byte is to be written or read. Those bytes are carried to the RAM on the *memory data bus*, usually just called the data bus. Notice that the data bus is connected to the controller so that bytes stored at particular addresses in RAM can be sent to the controller.

Also, notice that a keyboard is provided to input information. The keyboard converts each keystroke into one byte. Bytes from this unit are routed to the RAM and are stored at certain addresses furnished to the RAM by the controller on the address bus.

The output of the computer in *Figure 10-1* goes to a color video monitor for display on its screen. A video monitor is basically like a TV receiver, but it doesn't have a tuner to receive picture and sound information from TV broadcast stations or cable TV systems. It is designed specifically to display data from a computer. However, it needs a special interface circuit board called a *video graphics adapter* between it and the CPU. This adapter is a separate, specialized computer with its own memory that contains thousands, even millions, of bytes. The video graphics adapter creates a display on the monitor screen from bytes stored in its memory. The CPU changes the display by sending selected bytes from the main memory to the video graphics adapter memory. These bytes are stored at selected addresses in the video graphics adapter memory. All the byte transfers and byte storage are controlled in sequence by the controller.

The input and output functions of a computer are so similar that they are usually discussed as a single subject called input/output (I/O). As a result, we can think of a computer system as consisting of three main sections or subsystems: CPU, RAM, and I/O. Let's look at each of these in more detail. We'll begin with the CPU.

Functions of the ALU and Working Registers

In the CPU of *Figure 10-1*, the routing circuitry is shown between the three working registers and the ALU. Two of the working registers are *shift* registers with a serial link between them. Besides receiving and transmitting bytes in parallel, these particular registers can shift the stored bits in either direction between them. The shifting, like the storage of a byte in a RAM or a register, is governed by signals from the controller. If both registers are shifted at the same time in the same direction, bits shift from one register to the other. This can be very useful in performing various steps in arithmetic calculations.

Under control of the signals from the controller, an ALU can do any of several things to the two bytes that it receives *(Figure 10-2a)*. The ALU's operations are listed in *Figure 10-2b*. The ALU can add two input binary numbers represented by Byte A and Byte B, subtract Byte A from Byte B, or subtract Byte B from Byte A. In the subtraction process, the ALU sends a bit directly to the controller. The bit tells the controller that Byte A is larger or smaller than Byte B. In other words, it compares Byte A and Byte B. This comparison is important in determining the output of the subtraction.

The ALU can also perform AND and OR functions, in parallel, on corresponding bits of the two input bytes. The results form the corresponding bits of the output byte. Let's use the AND function as an example. As shown in *Figure 10-2c*, bits A_0 through A_7 of Byte A are fed as one set of inputs to an AND gate. At the same time, bits B_0 through B_7 of Byte B are fed as the second set of inputs to the same AND gate. The corresponding Q_0 through Q_7 bits of the output will be a 1 if both input bits of Byte A and Byte B are a 1; for example, if bit A_7 is a 1 and bit B_7 is a 1, then bit Q_7 will be a 1.

Another useful function performed by the ALU is called an exclusive-OR (XOR) function. For the XOR function, the output bit is 1 *if, and only if, one* of the two

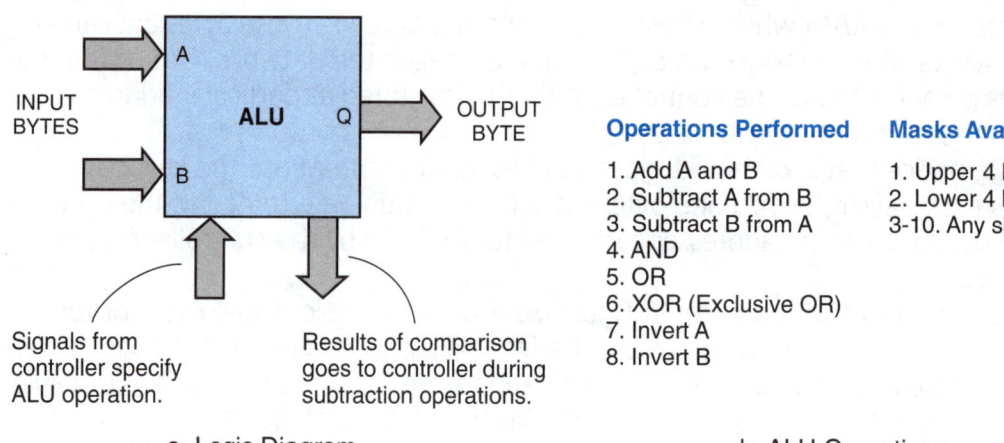
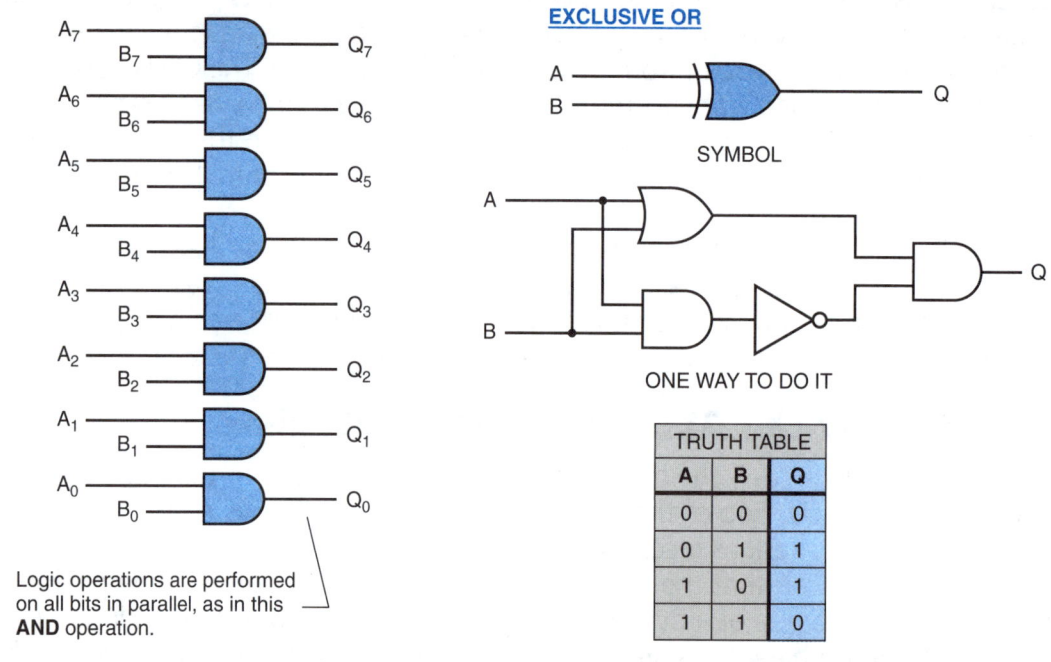

Figure 10-2. A computer's ALU can add, subtract, and do logic operations on all or some bits of a byte.

corresponding input bits is 1. *Figure 10-2d* shows the symbol, truth table, and a circuit to perform the XOR function. The symbol for the XOR function is an OR gate symbol with a double line for the back of the spearhead. The circuit shown is made up of a combination of AND, OR, and NOT gates.

The last operation listed in *Figure 10-2b* that the ALU performs is to *invert* all the bits of either input byte.

The ALU can *mask* certain bits of the input bytes fed to it. Masking means to cover up or ignore some of the bits, so that the operation is performed only on the unmasked bits. For instance, as shown in *Figure 10-2b*, the upper four bits, or the lower four bits, or any single bit of a byte can be ignored in an operation by the ALU.

A computer is not restricted to just shifting, adding, subtracting, and performing logic functions. It can do a multitude of different functions by using the simple ALU operations in sequence in a certain order. For instance, multiplication is done by a series of additions and shifts, and division is done by a series of subtractions and shifts. As we will see in a moment, the sequence is called a program.

Example 1. ALU Operations

Do the following ALU operations on Byte A and Byte B and show the ALU output results:

ALU Operation	OR	AND	XOR	NOT	NOT
Byte A = 11011010	11011010	11011010	11011010	11011010	
Byte B = 10111100	10111100	10111100	10111100		10111100
ALU Output	11111110	10011000	01100110	00100101	01000011

Now perform the ADD operation. Check yourself with decimal equivalents.

ADD	Binary	Decimal
A	11011010	218
B	10111100	188
Sum	110010110	406

Now perform the subtraction of B from A. To perform subtraction, the B input to the ALU is inverted to provide a "1s" complement. By providing a 1 at the carry input, it converts B to a "2s" complement. Adding a "2s" complement of B to A subtracts B from A.

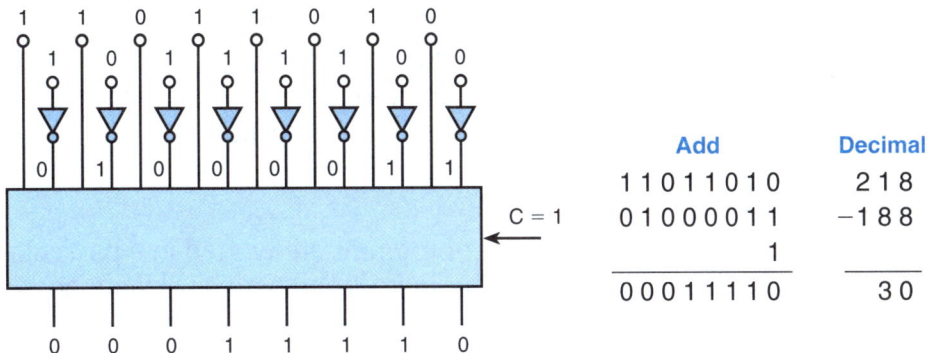

```
        Add         Decimal
     11011010         218
     01000011        -188
            1
     --------        -----
     00011110          30
```

Example 2. Exclusive-NOR Truth Table

What is the truth table for a 2-input positive exclusive-NOR logic circuit?

EXCLUSIVE-OR

A	B	Q
0	0	0
0	1	1
1	0	1
1	1	0

EXCLUSIVE-NOR

A	B	Q
0	0	1
0	1	0
1	0	0
1	1	1

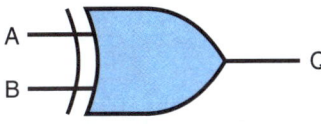
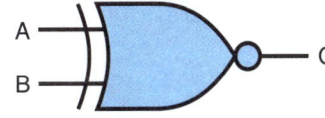

Example 3. Masking Bits

Do an ALU OR operation on Bytes A and B. First, mask off the upper four bits of each byte. Second, mask off the lower four bits of each byte.

ALU Operation	OR (Mask Upper)	OR (Mask Lower)
Byte A 11011010	11011010	11011010
Byte B 10111100	10111100	10111100
ALU Output	11111110	11111110

The Controller: Following Instructions in a Program

We have said that signals from the controller tell all the other parts of the computer system what to do at various times. For example, as shown in *Figure 10-3,* it tells the routing circuitry where to send various bytes that it receives; it clocks the working registers, causing them to store new bytes or shift the stored bytes; it tells the ALU which operations to perform and which masks to use; and it sends an address for the RAM to use while it tells the RAM to read or write a byte.

The Computer's Program

To do all these things, the controller follows a *program* that someone has written and stored in the RAM. A program is a number of bytes called *instructions*, organized in sequence so that one instruction follows another in a definite order. In the RAM, each instruction is stored at the next higher address than the preceding instruction in the sequence. The RAM may contain several different programs for various purposes. Each program begins at a certain address for the first instruction.

High-Level Languages

You may have heard or know that computer programs are written in a particular language such as Basic, Fortran, C or Pascal. These are called high-level languages and have instructions that you can read, like "Print Register A." However, the language we are talking about that is used for the controller instructions is called the *machine language* for this particular computer design. Each instruction is just a different binary number. Higher-level language programming uses other programs to convert the high-level instructions to machine-level language for the computer on which they are run. In running a Basic program, for instance, a computer uses another program called a Basic interpreter. Whenever a Basic instruction is executed (obeyed), the interpreter program quickly creates a series of machine-language instructions to execute the higher-level language instruction.

Machine Language

The bits of a machine-language instruction tell the controller what to do during a period of time called an *instruction cycle*. One instruction cycle usually consists of a certain number of cycles of the main clock. In a typical desktop computer, the main clock frequency might be 20 megahertz (20,000,000 clock cycles per second). Several hundred thousand instructions would be executed every second. Let's look at *Figure 10-3* and follow through an instruction cycle.

Instruction Cycle

At the beginning of each instruction cycle, the controller fetches (reads) an instruction from the RAM. The instruction's address comes from a counter in the controller called the *program counter*. In the controller, the instruction goes into a register

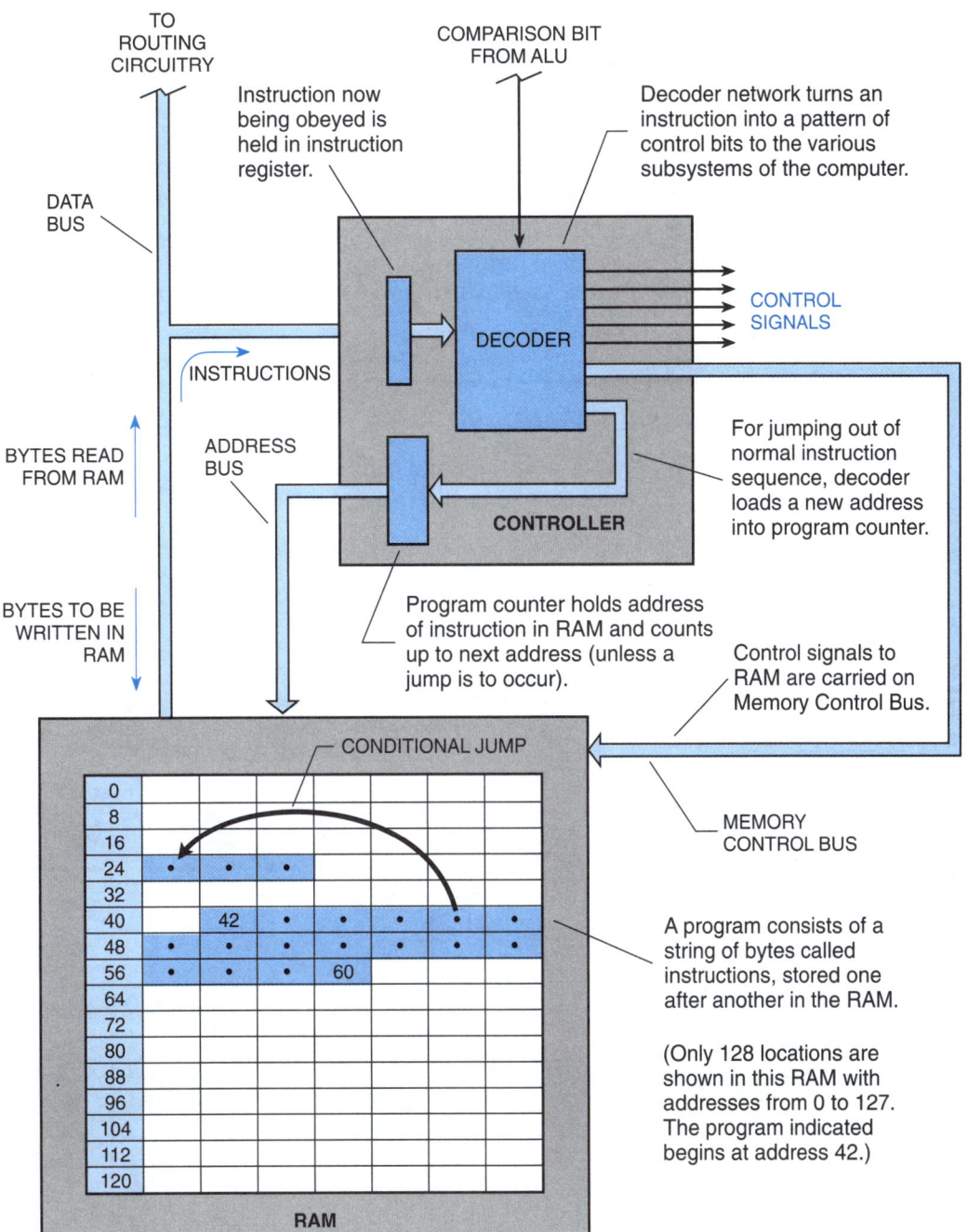

Figure 10-3. The controller operates by decoding one instruction at a time and providing an address for the next instruction to be read from the RAM.

called the *instruction register*. It holds the present instruction being obeyed. A combinational network called a *decoder* receives the instruction bits (along with the comparison bit from the ALU and possibly some other information.) It creates a pattern of control bits that go to the various subsystems. The control signals to the memory are carried in a bus called the *memory control bus*.

Based on information it receives during an instruction cycle, the controller decides which instruction to fetch for the next cycle. In most cases, the next instruction will simply be the one at the next higher RAM address. In that case, the decoder tells the program counter to step to the next higher number. Otherwise, the present instruction must tell the decoder what address to use for the next instruction.

If it is different from the next higher RAM address, the new address gets loaded into the program counter. Going to an instruction other than the next one in the sequence is called *jumping*.

Conditional Jump

There is a particularly useful type of instruction called a *conditional jump*. An example is shown in *Figure 10-3* for the instruction located at address 46; it tells the controller, "Compare Byte A to Byte B; if Byte A is smaller, then jump to Address 25; else continue with the next instruction." Conditional jump instructions let the programmer vary what the computer does according to what happens, instead of doing the same thing every time the program is executed.

Example 4. Addressing Memory

What is a simple decode network for a 16-word by 8-bit RAM memory?
What address would the controller send to this RAM to obtain word 11? **1011**

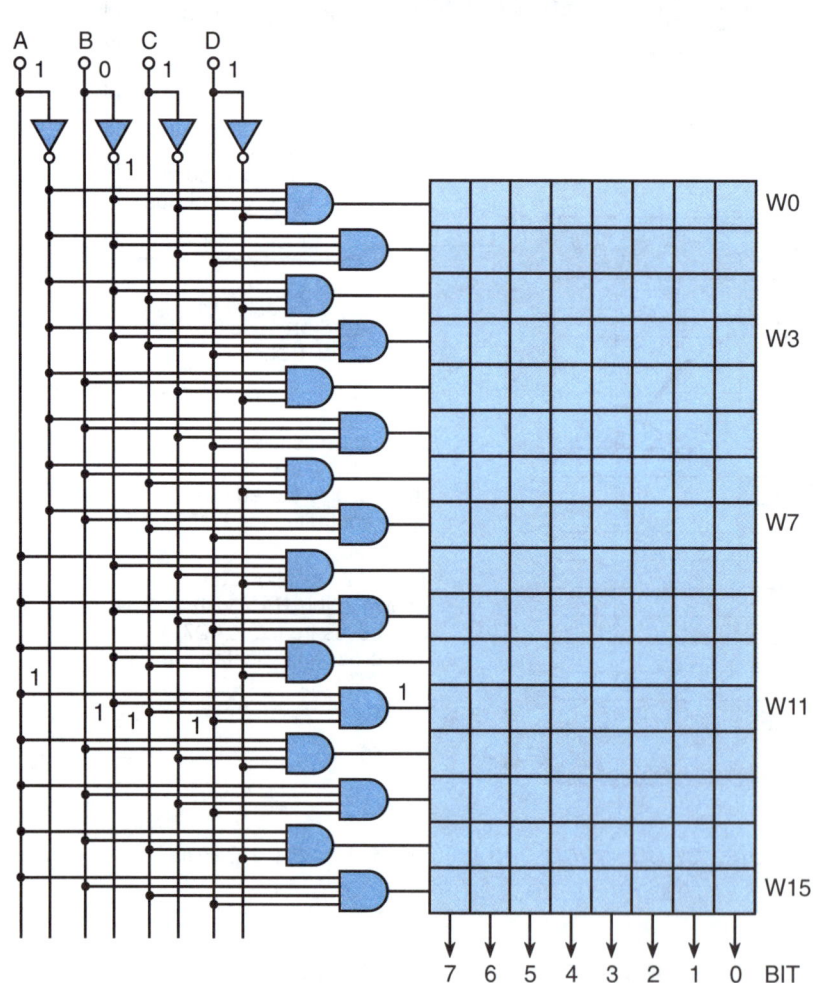

Selecting each word line will read out all 8 bits of word.

Truth Table

INPUT				W	WORD
A	B	C	D		LINE
0	0	0	0	1	0
0	0	0	1	1	1
0	0	1	0	1	2
0	0	1	1	1	3
0	1	0	0	1	4
0	1	0	1	1	5
0	1	1	0	1	6
0	1	1	1	1	7
1	0	0	0	1	8
1	0	0	1	1	9
1	0	1	0	1	10
1	0	1	1	1	11
1	1	0	0	1	12
1	1	0	1	1	13
1	1	1	0	1	14
1	1	1	1	1	15

The machine-language programs for even some of the simplest tasks can get pretty complicated. For instance, the routine for entering a single character from the keyboard may include dozens of machine-language instructions, with jumps at many different points to handle various possibilities. But remember that each instruction takes not much more than a millionth of a second, even in a fairly slow computer.

A Read-Only Memory for Permanent Programs

Programs for many of the most basic things a computer does—like handling the keyboard and monitor—are permanently built into a special section of the RAM. This section is called a *read-only memory* or ROM (rhymes with Tom). It is built differently from the rest of the memory. The computer can read bytes from any address in the ROM section, but it cannot write or change any of those bytes. The rest of the main memory would be called the ordinary RAM section, or the read-write section. A combination RAM and ROM main memory is shown in *Figure 10-4*. One advantage of ROM over ordinary RAM is that the stored data doesn't disappear when the power is turned off.

The ROM section of the memory consists of one or more separate ICs. All of them are connected to the same data bus, address bus, and memory control bus as the RAM chips. Each ROM chip and RAM chip is assigned a different set of addresses for its memory locations.

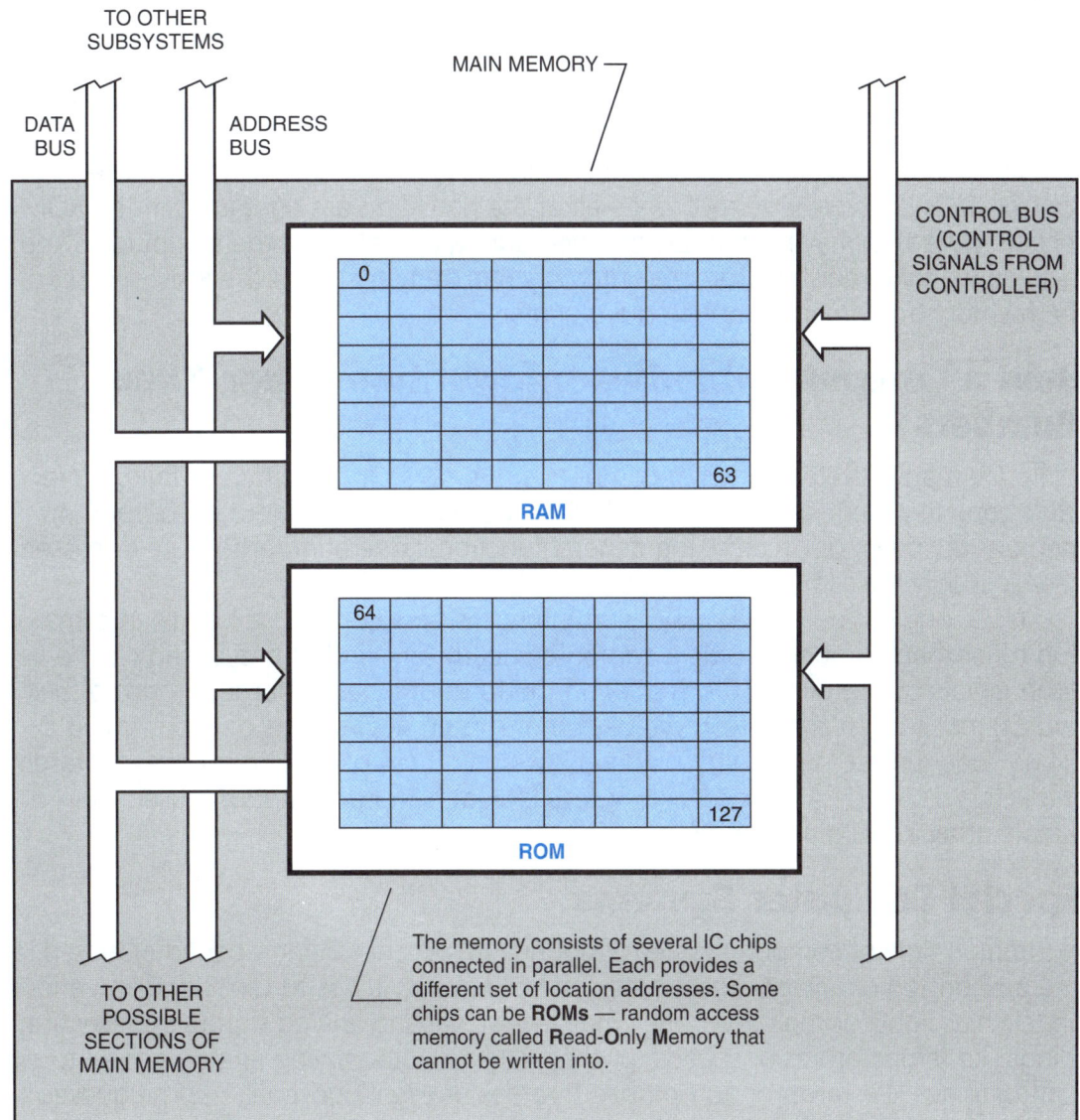

Figure 10-4. Part of computer's main memory is usually read-only memory (ROM), containing permanently stored programs that cannot be erased or written into.

PROMs

There are ROM chips of several different kinds, some of them allowing data to be changed by special methods. Most of these varieties are classified as programmable ROMs or PROMs.

With a permanent ROM, the stored data (1s and 0s) are built in by the manufacturer according to a customer's order. With a PROM, a 1 is initially programmed into every memory bit location during manufacture. Then a user can *permanently* program data by burning out fused links to program a 0 at particular bit locations while leaving a 1 at all other locations. The programming is accomplished by plugging the IC into a special computer system, sometimes called a PROM burner. To change the PROM memory to a new program, you have to remove the chip and plug in a different PROM having the new program.

Other PROMs allow stored data to be changed by electrical charge storage. These have various names such as EPROM, EEPROM, and EAROM. In most cases, to reprogram one of these chips with different data, you have to take it out of the system, possibly expose it to ultraviolet light to erase it, and plug it into a special programmer. Some types can be reprogrammed in the computer where the chip is used, but writing a new byte into one of these PROMs is a different process than writing into an ordinary RAM location. It takes more time and electrical energy and requires some special circuitry.

In a computer system which is dedicated (built in) to a special application—say, controlling a video cassette recorder—all of the programs are contained in the ROM section. Data that may have to be changed during operation of the computer is kept in some working registers (for small memory requirements) or in the RAM section of the memory (for larger memory requirements).

How a Computer Handles Information Other Than Numbers

There are many different kinds of computers that do so many different things. We can't hope to cover every type, but whether it be variations in input, or output, or memory, or speed, or circuitry, the general functional operating procedure discussed here is usually the same.

Now, one last point that may be of interest is how computers handle information other than numbers. Inside a computer, nearly any kind of processing can be done rapidly on the digital information in the form of bytes (or larger groups of bits (words) made of several bytes). We've demonstrated how strings of bits forming a binary code represent numbers. And we have seen how other binary codes represent characters using the ASCII code. Now, let's look at how computers are used to handle other kinds of information.

Special Computer Systems

A common type of computer system responds to switches being opened or closed, as in a pushbutton or a joystick; or some systems have switches to turn things on and off, such as lights or motors; or they may receive varying analog signals representing things like temperatures or speeds; or they may produce varying analog signals to control things like valves or automobile throttles or even loudspeakers. The question is: How does a digital computer handle these functions?

Figure 10-5 shows the general concept of how a computer system handles switched and analog signals by illustrating various portions of different types of systems:
- A system with a joystick input or a temperature sensor input.
- A system that needs something turned ON or OFF at the output.
- A system that has an analog output, like music or voice to a speaker.

It's all a matter of I/O and programming. Basically, the inputs come into the input registers that the computer is programmed to read every so often. The computer program analyzes the inputs and initiates action based on what the input indicates. Outputs to accomplish the action are then generated by the computer and loaded at the appropriate time into the output registers.

In *Figure 10-5,* the input and output registers accept 8-bit bytes just like the computer. Inside the computer, these registers are connected to the routing circuitry in the same way as other I/O units, such as the keyboard input unit or video graphics adapter in *Figure 10-1*. The program instructions makes the computer fetch the bytes from the input registers and bring them into the computer to act on them, and then store bytes in the output registers to produce the required action. It resembles the same program steps taken to write data to RAM memory and read data from RAM memory.

Joystick Input

The upper input register of *Figure 10-5* can receive digital input bits from as many as eight switches. In this example, only five are used to receive inputs from the joystick for a computer game. The five bits come from the five switches on the joystick. The joystick itself operates four of the switches when it is moved up, down, right, or left. The fifth switch is the pushbutton on the joystick unit. When the switches are open (as shown), a resistor pulls each signal line up to 5 volts for a bit value of 1. Closing any switch produces a bit value of 0 on that line only. For instance, moving the joystick to the right closes the RT switch, sending a 0 bit in that line. Several times each second, the computer latches the input register and reads its byte. Inside the computer, the program makes the computer look at each bit separately. It finds the 0 on the RT input bit line and executes an action to move an image on a video game screen to the right.

ON-OFF Output

Similarly, the digital output register of *Figure 10-5* puts out eight bits. Each output line can control a relay or electronic switching circuit to turn electric power ON and OFF in a lamp, a motor, a heating element, etc. To turn on the output lamp shown in *Figure 10-5,* the computer latches a new byte in the output register such that output line 0 has a 1 on its output. The 1 turns on the transistor and lights the lamp. The other bits in the byte may remain the same at that time, or they also may have been changed, depending on what action the computer program required as it processed the information from the inputs

Analog Input

An input register for an analog input (an input that varies continuously in a smooth fashion, not in discrete digital levels) must receive its input as digital information. Therefore, the input analog signal must be a converted from an analog signal to a digital signal with eight bits. The unit to do the conversion is called an analog-to-digital converter (ADC).

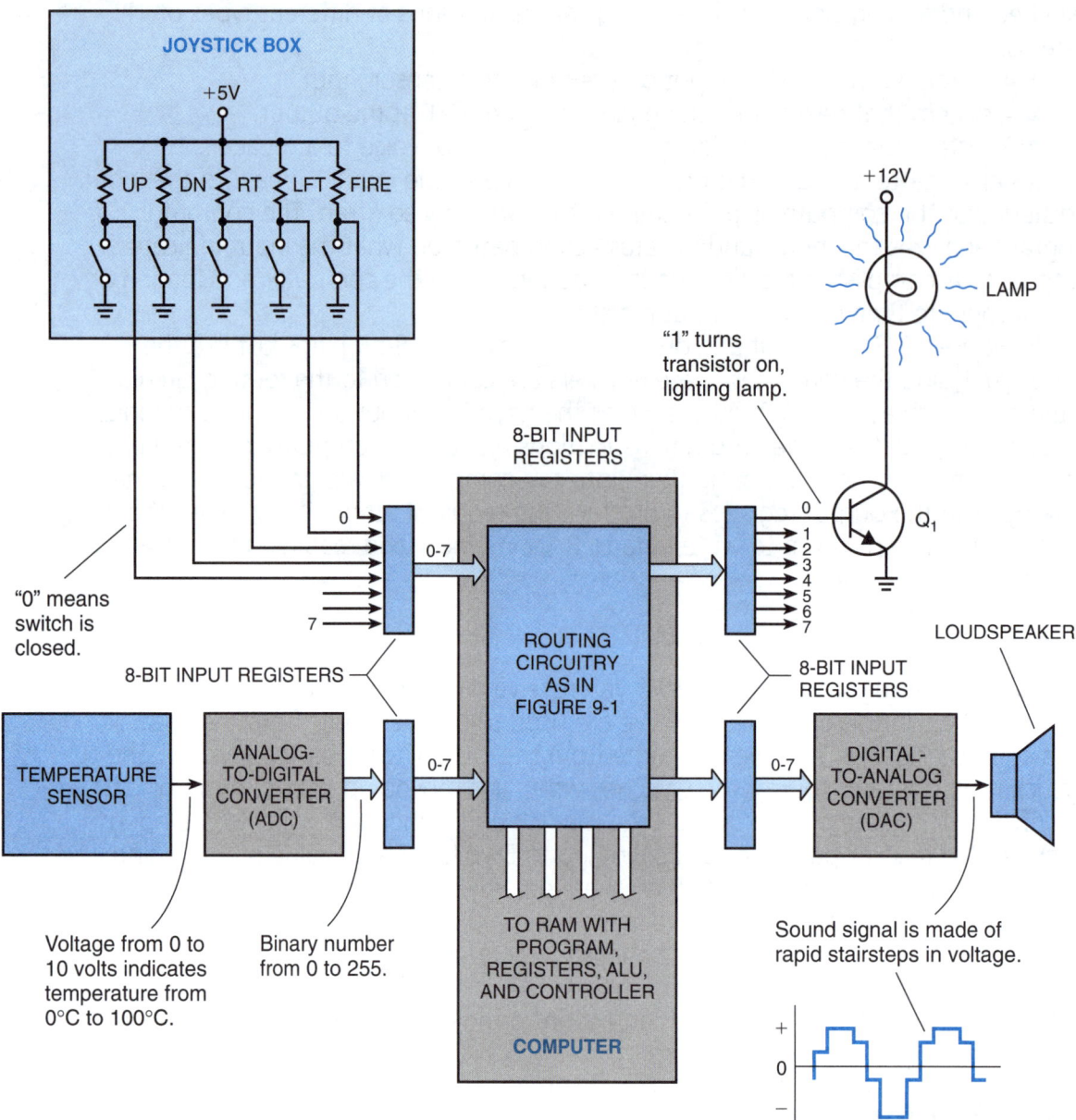

Figure 10-5. Nearly any kind of information can be put in and out of a computer through registers connected to the routing circuitry. The computer is programmed to read and write bytes in these registers as rapidly as required.

The analog signal in this example comes from an electronic temperature sensor. The sensor produces a voltage ranging from 0 to 10 volts to indicate the measured temperature over a specific range. Let's say 0 volts means 0°C and 10 volts means 100°C. The ADC puts out an 8-bit binary code that ranges from 0 to 255 (in decimal). An output number of 0 means 0 volts received by the ADC (indicating 0°C), and an output number of 255 means 10 volts received by the ADC (indicating 100°C). By calibrating the ADC 8-bit code, any temperature from 0 to 100°C can be indicated by the computer. At any time the program instructs it, the computer can read the ADC output into the input register, route the information to a decoder, and feed the decoder output to the output registers. The output could control a relay to turn on or off a heater and show the temperature on a display device.

Example 5. Analog to Digital Conversion

If the ADC of *Figure 10-5* has a binary code of 00000000 as an output for 0°C and a code of 11111111 as an output for 100°C, what is the binary code for each of the following temperatures in °C: 0, 20, 40, 60, 80, 100?

Temp °C	0°	20°	40°	60°	80°	100°
Decimal Equiv. of Binary Code	0	51	102	154	205	255
		32 1	64 1	128 1	128 1	128 1
		19	38	26	77	127
		16 1	32 1	16 1	64 1	64 1
		3	6	10	13	63
		2 1	4 1	8 1	8 1	32 1
		1	2	2	5	31
		1 1	2 1	2 1	4 1	16 1
		0	0	0	1	15
					1 1	8 1
					0	7
						4 1
						3
						2 1
						1
						1 1
						0

Bit Value	128	64	32	16	8	4	2	1
0°	0	0	0	0	0	0	0	0
20°	0	0	1	1	0	0	1	1
40°	0	1	1	0	0	1	1	0
60°	1	0	0	1	1	0	1	0
80°	1	1	0	0	1	1	0	1
100°	1	1	1	1	1	1	1	1

This example demonstrates how to convert a decimal number to a binary code.

Example 6. Digital to Analog Conversion

A DAC similar to the one in *Figure 10-5* receives the following sequence of bytes at 0.001 second intervals. Plot the code values against time to look at the analog signal. The decimal values of the codes are determined first, and then the waveform is plotted. What is the frequency of the waveform?

$$f = \frac{1}{0.012 \text{ sec}} = 83.3 \text{ Hz}$$

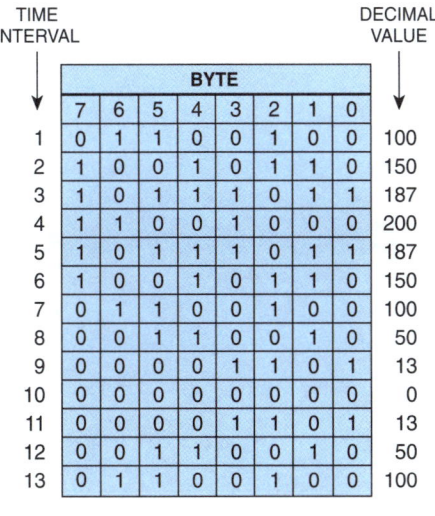
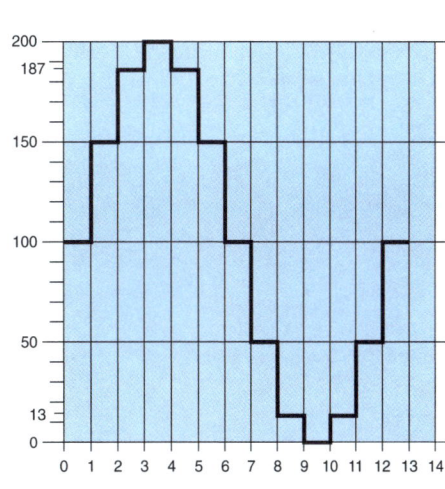

Analog Output

An analog output works the same way, but in reverse. Its output register feeds a unit called a digital-to-analog converter (DAC). The DAC takes an 8-bit digital code from the register and changes it to a corresponding voltage level. Each time a new code is read in, a new voltage level is generated. The voltage output from the DAC follows the number in the register. In *Figure 10-5,* the analog output voltage changes rapidly enough to create sound in a loudspeaker.

Summary

A computer or digital system can handle many kinds of information if it has the proper inputs, programming, and operating speed. It can produce outputs to display information and control many kinds of devices. It's basically a matter of putting together hundreds of thousands of gates and flip-flops in a logical manner, timing them to operate in sequence, and applying a set of program instructions to make them accomplish a specific task—all following procedures we have talked about in this book.

Quiz for Chapter 10

1. The ALU, working registers, controller, and routing circuitry, all together, are called the:
   a. read only memory
   b. write only memory
   c. central processing unit
   d. input and output device

2. To receive stored bytes from the RAM, the controller is connected to the:
   a. data bus
   b. I/O register
   c. printer
   d. keyboard

3. The circuitry that sends signals throughout the computer system to synchronize everything that happens is the:
   a. address bus
   b. central clock
   c. ALU
   d. RAM

4. The output unit for the monitor is a circuit board called:
   a. picture circuit board
   b. video memory board
   c. monitor memory board
   d. graphics adapter board

5. We can think of the main part of a computer system as consisting of three sections or groups of subsystems:
   a. CPU, RAM, and I/O
   b. RAM, ROM, and registers
   c. CPU, ALU, and keyboard
   d. monitor, keyboard, and memory

6. When doing subtracting, the ALU may determine which of the numbers is the smaller by a determination called:
   a. flagging
   b. comparison
   c. bit stuffing
   d. gating

7. Covering up or ignoring some of the bits is:
   a. shading
   b. inverting
   c. masking
   d. shifting

8. A series of additions and shifts are used to perform:
   a. multiplication
   b. ANDing
   c. inverting
   d. comparison

9. A program is a sequence of a number of bytes called:
   a. characters
   b. instructions
   c. assemblers
   d. linkers

10. Whenever a Basic instruction is executed (obeyed), a series of machine-language instructions is quickly created by the:
    a. program manager
    b. linker program
    c. interpreter program
    d. monitor

11. The section of the memory that contains permanently stored data that is not changed during operation of the computer is:
    a. BIOS
    b. RAM
    c. ROM
    d. EPROM

12. Going to an instruction other than the next one in the sequence is called
    a. sequencing
    b. paralleling
    c. jumping
    d. assembling

Answers:
1 c, 2 a, 3 b, 4 d, 5 a, 6 b, 7 c, 8 a, 9 b, 10 c, 11 c, 12 c

Questions for Chapter 10

1. What is a dedicated computer?

2. Name the three main sections or groups of subsystems of any computer.

3. Covering up or ignoring some of the bits so that an operation is performed only on certain bits of a byte is called _____.

4. The _____ tells the ALU which operation to perform.

5. The bits of a machine-language instruction tell the controller what to do during a period of time called an _____ .

6. What is the name of the counter that keeps up with an instruction's address?

7. What decides which instruction to fetch for the next cycle?

8. What instruction lets the programmer vary what the computer does according to what happens?

9. Explain the difference between ROM and RAM.

10. What kind of memory is used to contain all the programs in a dedicated computer, such as that in a VCR?

11. How does the ALU perform the operation of multiplication?   ... of division?

12. What does the interpreter program do to higher-level instructions written, for example, in Basic?

13. Explain what the five bits from a joystick would represent.

14. What is used to allow information in the form of a continuously varying voltage to be entered to the computer?

15. How is an analog output, such as to drive a speaker, produced by a computer?

CHAPTER 11
How Photoelectric Devices Work

To complete your basic understanding of electronics, there is one last topic that we need to cover. This is photoelectric semiconductor devices. "Photo" used as part of a word means having to do with light. Photoelectric semiconductor devices interact with light. There are three general categories:
1. Photodiodes and phototransistors *sense* (detect) light, using light to control current passing through them. These devices act as variable resistors controlled by light. They are said to be photoconductive, meaning that their resistance decreases when light shines on them.
2. Photovoltaic diodes (pronounced PHOto-vol-TAY-ic), also called solar cells, *convert light energy* into electric energy. They use light to generate an EMF or a voltage difference which, when connected in a circuit, produces an electric current, like a battery.
3. Light-emitting diodes, called LEDs, *create* light when current passes through them. (LED is pronounced by spelling the letters, like L-E-D.)

In this chapter, we will see how these photoelectric devices work and some ways in which they are used.

Photodiodes: Light Increases Reverse Leakage In Any Junction Diode

It will be easier to see how photoelectric semiconductor devices work if we start with photodiodes. Actually, any p-n junction diode can act as a photodiode. That is, *all p-n junction diodes are sensitive to light*.

Reverse-Biased Diode Under Test

Figure 11-1 shows a test setup that demonstrates the effect. In the circuit, there is a small silicon p-n diode which is reverse-biased by a flashlight battery (1.5V). That is, the battery is trying to force current backward through the diode, the direction in which the diode blocks current. The circuit includes a sensitive meter to measure the reverse current. There is also a resistor to limit the current and protect the diode from overheating when light strikes the diode and reduces its resistance. We are using a type of diode whose covering is a metal can instead of solid plastic.

At first, the meter shows a very tiny reverse current. No diode is perfect in performing its job as a one-way valve. Like a faucet dripping water when it is shut off, the diode will leak a little bit, however small. This is called *reverse leakage current*.

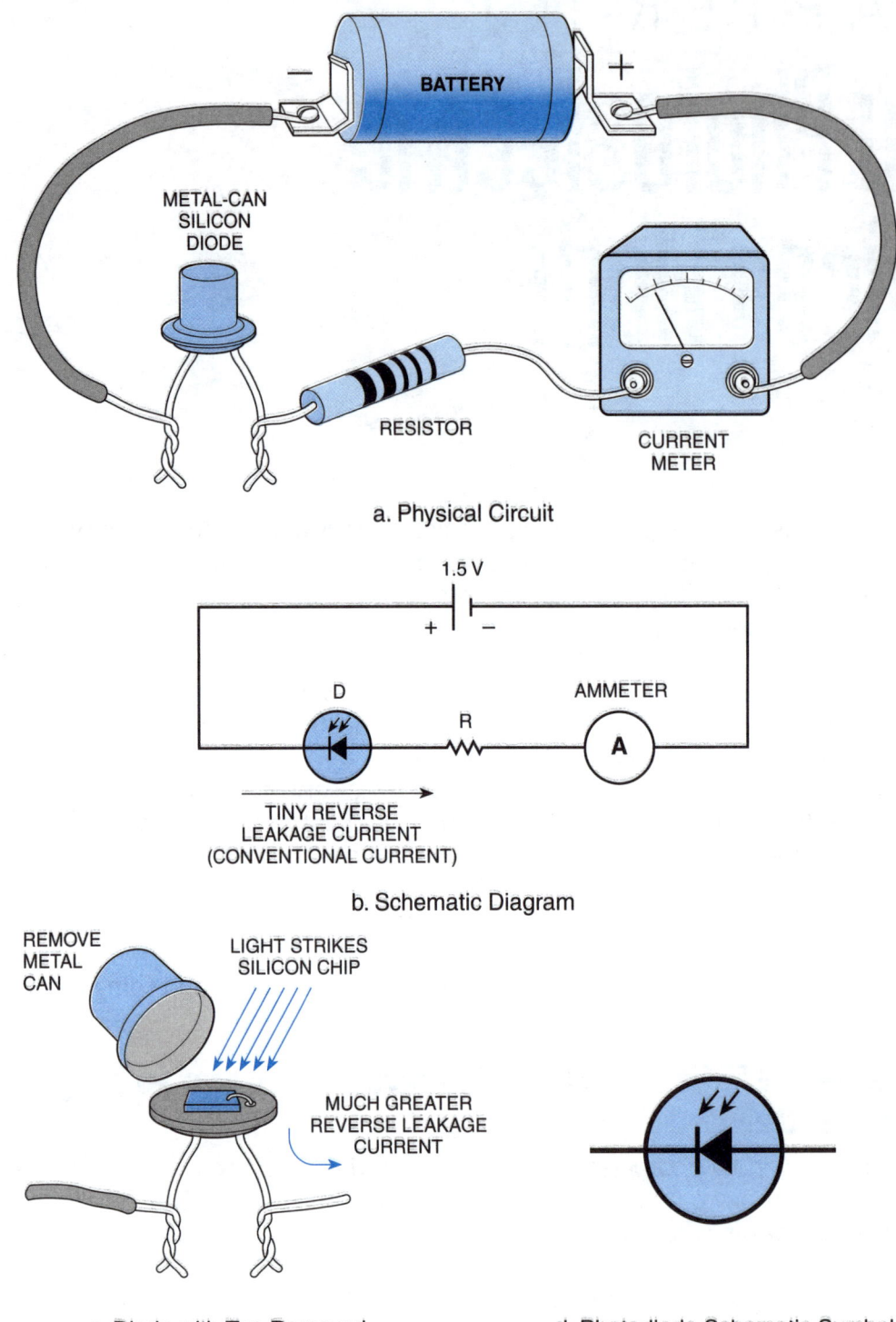

Figure 11-1. A photodiode is like an ordinary p-n junction diode with a window to admit light. The light greatly increases reverse leakage current.

Adding Light

Now if the top of the can is removed, light from the room is allowed to strike the silicon chip. As a result, the meter will show a considerable increase in reverse leakage current. The stronger the light, the greater the current. If you shine a bright light on the chip, and move your hand back and forth between the light and the diode, the meter needle will swing very low whenever the shadow hits the chip.

This is what happens with any photodiode. A photodiode is basically like any other p-n junction diode, but the chip is specially designed so that light will create as much reverse leakage as possible. And it has a special transparent window in its top so that light can strike the chip. The chip design mainly involves special shapes and layouts for the p and n regions and the metal conductors on top of the chip connecting the various regions.

The schematic symbol for a photodiode shown in *Figure 11-1d* describes this device fairly accurately. It is an ordinary diode symbol with one or two little slanting arrows pointing at it to represent light.

Effect of Heat Energy On Normal Reverse-Biased Diode

The way light increases reverse leakage current will help you understand later how solar cells and LEDs work. But before continuing, you may want to turn back to Chapter 4 and review how diodes are made and how they work.

To begin, we will talk about how reverse leakage current happens *without* light. *Figure 11-2a,* like *Figure 4-12,* shows a diode chip reverse-biased in a circuit. The reverse bias is trying to force electrons to the left through the chip. In the n region on the left, negative free electrons are trying to drift further left. And in the p region on the right, positive holes are trying to drift further right. In between, there is a depletion zone on both sides of the junction, where there are no free electrons or holes to carry current.

Atoms in the Depletion Zone

Let's consider the silicon atoms in the depletion zone shown in *Figure 11-2b*. The shaded rings in the picture represent the two outermost shells of permitted electron orbits. The orbits themselves are not shown—just the regions or bands where orbits are permitted. As we discussed in previous chapters, the electron shells are called *bands* when discussing the energy of the electrons. The lower band is called the valence band and the upper band is called the conduction band.

Remember that in this kind of diagram, we pretend that all electron orbits are circular and lie in the same plane, when actually the bands are really spherical surfaces. The radius of an orbit in the diagram would represent not only the average height of an electron above the nucleus but also the *energy* of the electron. This is because the positively charged core of the atom (the nucleus and inner electrons) attracts the electron as gravity pulls a satellite toward the earth. So it takes energy to lift an electron up to a higher orbit.

That energy is stored in the lifted electron as an increase in the electron's potential energy. This increase, or difference of potential energy between two orbits for an electron, is measured in volts. Remember, voltage is the potential energy of electrons. The height, or radius, of an orbit from the nucleus in this kind of diagram could be measured in fractions of an inch or in thousandths of a meter, but instead is measured in the volts of potential energy it takes to lift the electron to the respective orbit.

Now let's look at the normal silicon atom in the depletion zone as shown in *Figure 11-2b*. The lower valence band of orbits or energy levels is filled with its permitted limit of eight electrons. These electrons are busy forming covalent bonds with four neighboring atoms. The covalent bonds are actually pairs of electrons—four of the valence electrons belong to this atom, and the other four belong to four other atoms. Each covalent bond is shared with a neighboring atom. If this atom were a "hole", there would be a missing electron among the eight valence electrons—a place where an electron would fit—if the atom could get one. And as for the upper, conduction band in this atom, the permitted orbits there are empty. If this atom had a free electron, the conduction band is where the free electron would be orbiting.

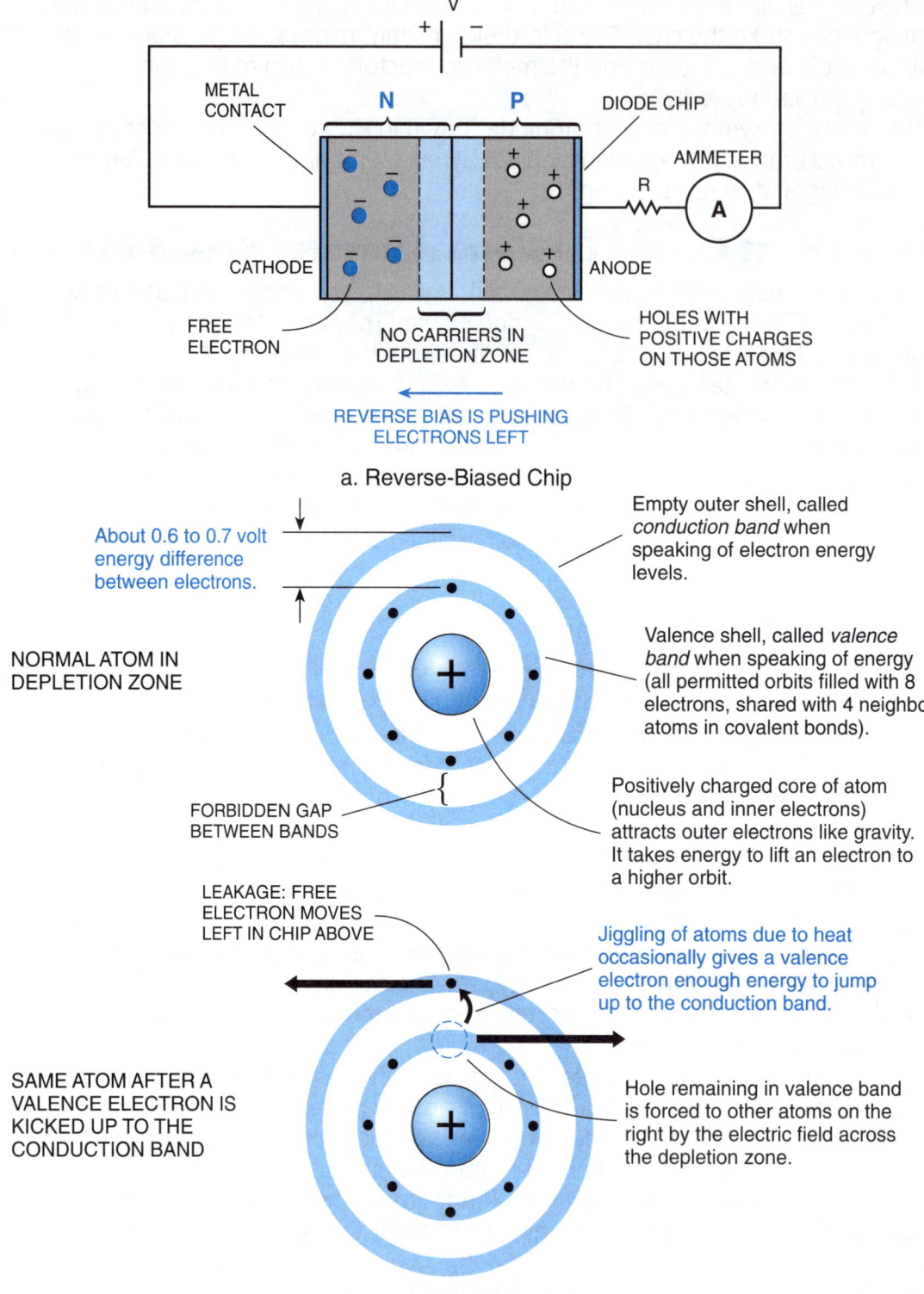

Figure 11-2. Heat energy causes reverse leakage current by knocking a few valence electrons up to the conduction band. This constantly supplies some free electrons and holes in the depletion zone.

The space between the two bands is called the forbidden gap. Because of the rules of quantum mechanics, no electrons are allowed to have energy levels in these gaps. At any point in a typical doped *silicon* crystal, *electrons in the conduction band are about 0.6 to 0.7 volt higher in potential* (on the average) than electrons in the valence band. This difference varies slightly because the bands have some width—not all the electrons in a band have the same energy.

Heat Effect

When there is no light, the main reason for reverse leakage current is that the silicon atoms are forever jiggling around in their places. This is due to heat energy in the crystal. The higher the crystal's temperature, the faster and more violently the atoms vibrate and bounce, pulling and pushing their flexible covalent bonds. Occasionally, two atoms jiggle against each other hard enough to knock a valence electron up into the conduction band as shown in the other atom of *Figure 11-2b*. Up there, the electron is a free one, and it leaves a hole behind in the valence band. And so, *heat energy constantly creates a few free electrons and holes throughout the chip*.

When this happens in the depletion zone, the result is two carriers (a free electron and a hole) where there are not supposed to be any. The electric field forces the free electron to the left and the hole to the right in the chip diagram of *Figure 11-2a*. (Silicon atoms with holes have positive charges, so we say holes carry positive charge.) The result is as though one electron had moved all the way across the depletion zone. The higher the chip's temperature, the more frequently valence electrons get jiggled up to the conduction band. This means more carriers in the depletion zone and more reverse leakage current. One consequence is that semiconductor devices don't operate properly when the temperature gets too high.

Photons of Light Also Knock Valence Electrons Free

Now let's bring light into the picture. Light is a form of energy. It behaves both as particles called photons and as very short, high-frequency radio waves. The energy of a photon depends on the frequency or wavelength of the waves, which also determines the color. The shorter the wavelength — the higher the frequency — the more energy the photon has.

As shown in *Figure 11-3a,* white light, which is a combination of all colors, can be separated into its frequency (and color) spectrum with a prism, so you see the individual colors as you do in a rainbow. In light which is visible to human eyes, photons which we see as purple (violet) have the shortest wavelength and highest energy. Photons of red light, on the opposite end of the rainbow spectrum, have the longest wavelength and lowest energy. A photon of barely visible red light, with a wavelength of about 70 micrometers (millionths of a meter) has enough energy to raise the potential of one electron by about 0.6 volt

Let's look at *Figure 11-3b,* where we have a reverse-biased photodiode. You can see what we are leading up to. *A photon of light can knock an electron from the valence band up to the conduction band (Figure 11-3c).* When light strikes the depletion zone of a reverse-biased diode, some photons hit valence electrons, transferring their energy to the electrons. When a valence electron absorbs a photon whose energy is high enough, the electron is instantly kicked up into an orbit in the conduction band. As in the case of heat energy, *this creates one free electron and one hole*. Brighter light produces more photons, which produces more carriers in the depletion zone, which results in more reverse leakage current. Thus, as light on the photodiode is increased, the resistance of the photodiode is decreased.

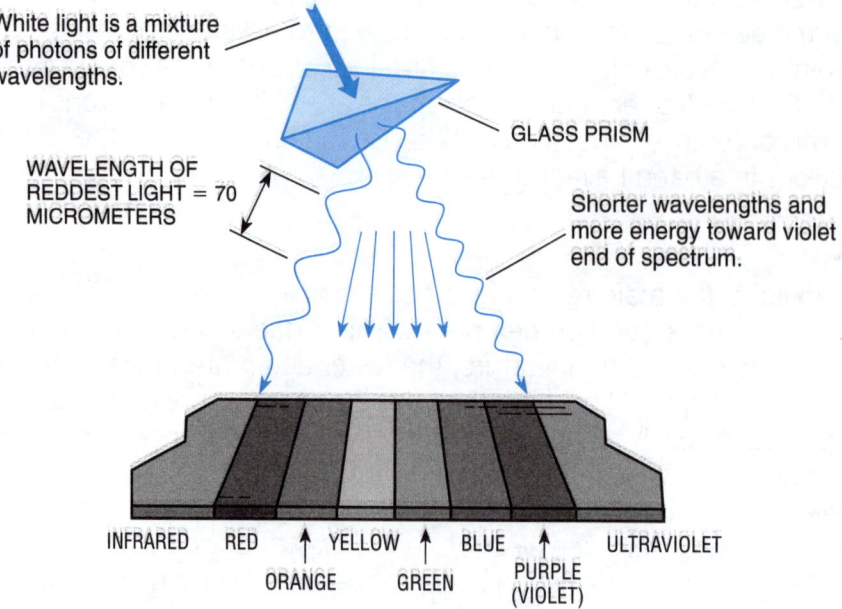

a. The Spectrum – A Sequence of Colors and Wavelengths

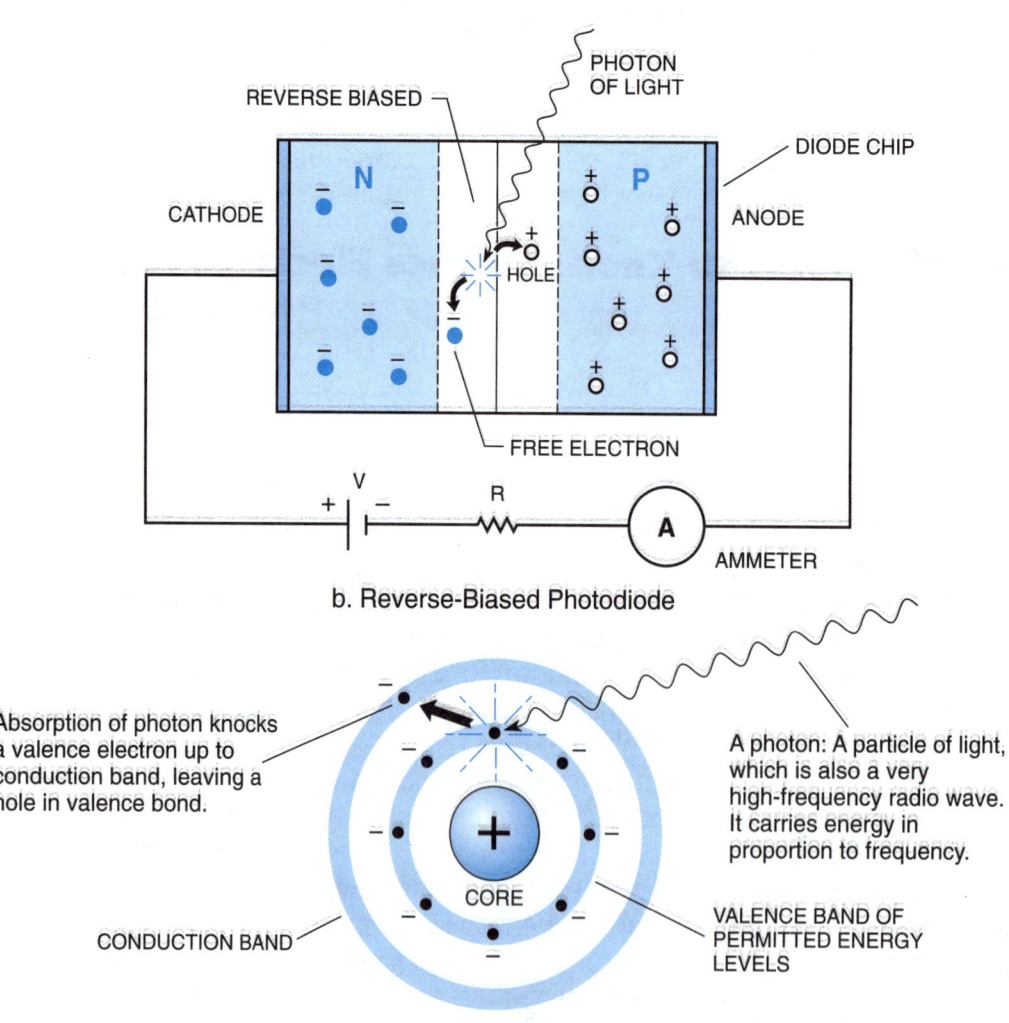

b. Reverse-Biased Photodiode

c. An Atom in the Depletion Zone

Figure 11-3. How a photodiode works: Photons absorbed in a reverse-biased depletion zone produce free electrons and holes which flow to form a large reverse leakage current.

Example 1. Wavelength in Angstroms for Light Frequencies

We know the relationship between wavelength and frequency for electromagnetic waves as:

$$\lambda \text{ (in meters)} = \frac{300{,}000{,}000}{f \text{ (in hertz)}}$$

Convert the following frequencies to wavelength in Angstroms when we know that one millionth of a meter (a micron) is 10,000 Angstroms.

f (in MHz)	Radio Waves 300	Micro Waves 3×10^4	Infrared 3×10^7	Visible Light 1×10^9	Ultra Violet 3×10^9
f (in MHz)	λ (in meters)	$\times 10^6 =$ λ (in microns)		$\times 10^4 =$ λ (in Angstroms)	
300	1	10^6		10^{10}	
3×10^4	0.01	10^4		10^8	
3×10^7	10^{-5}	10		10^5	
1×10^9	3×10^{-7}	0.3		3000	
3×10^9	10^{-7}	0.1		1000	

Phototransistors: Using the Collector-Base Junction as a Photodiode

In most applications of photodiodes, the current from the device is amplified by a transistor. *Figure 11-4a* shows a typical simplified amplifier circuit for this purpose, using a photodiode and an NPN transistor. Base current for the transistor comes through the reverse-biased photodiode. Changes in this current due to changes in light intensity at the photodiode are amplified by the transistor. Variations in the amplified collector current through the load transistor cause voltage variations at the output. A fairly small change in light intensity can cause a substantial step in the output voltage.

This leads us into understanding phototransistors. A semiconductor material version of the circuit is shown in *Figure 11-4b*. Notice that the p-n photodiode is connected in parallel with the p-n regions formed by the base and collector of the transistor. The p-type anode is connected to the p-type base, and the n-type cathode is connected to the n-type collector.

As shown in *Figure 11-4c,* the photodiode chip with its p-n junction could just as well be fastened to the base and collector regions of the transistor chip, without using any wires for the connections. In that case, the two devices would continue to work just as before, with the photodiode providing conventional current into the base. Going one step further with this reasoning, we would get the same results if the photodiode chip were not even a separate crystal. It could actually be made as part of the base and collector regions of the transistor chip. The transistor's base region would also act as the anode region of a photodiode, and the collector would serve as the cathode.

The Phototransistor

That is what a phototransistor is, in principle. As shown in *Figure 11-4d,* it is an ordinary NPN or PNP transistor without any terminal connected to the base region with a transparent window to let in light. The schematic symbol suggests this construction. It is a transistor symbol without a base connection, with one or two slanted arrows to indicate light coming in.

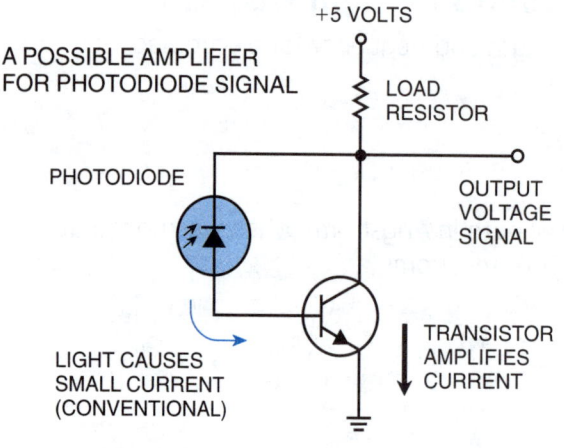
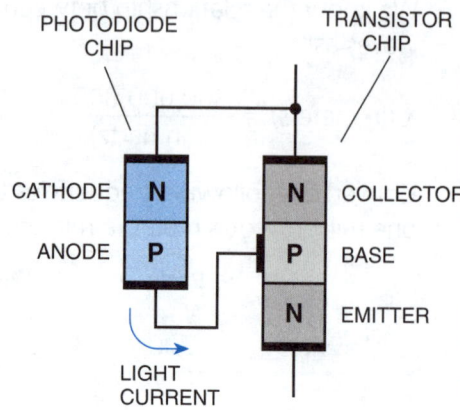

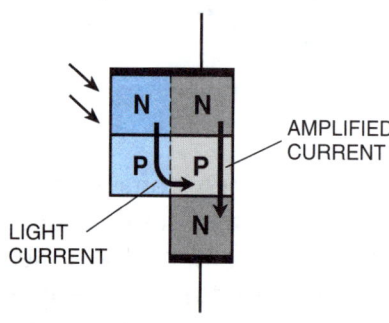
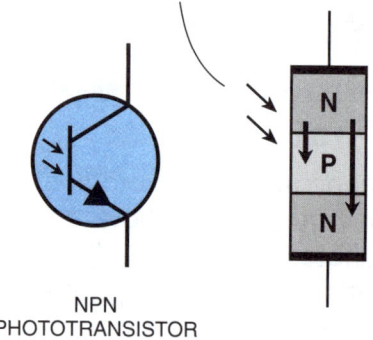

a. Photodiode Separate from Transistor

b. Semiconductor Material Version

c. Chips Combined

d. Phototransistor

Figure 11-4. A phototransistor is an ordinary NPN or PNP transistor designed so that the collector-base junction acts as a photodiode. Transistor action amplifies the current caused by light. (Currents shown are conventional current.)

The collector-base junction is designed to work as a photodiode, being spread out to catch as much light as possible. Light causes an increased reverse leakage current across this junction. This leakage current has the same effect as current being supplied to a base terminal. It causes a much greater current between the emitter and collector by transistor action.

As a result, you could say that a phototransistor is a photodiode with a built-in current amplifier. It has two terminals and works much like a photodiode alone. But the same amount of light and applied bias voltage produce much more current.

Solar Cells Use the Potential Difference Between Electron Bands

Next, let's talk about photovoltaic diodes, more commonly known as solar cells. These devices convert light energy into electrical energy by generating a voltage difference and causing current in a connecting circuit. Once again, this is done by a p-n junction that is basically the same as in any other diode; the difference is that the chip and the junction are specially designed for the particular purpose.

Example 2. Matching Light Source and Light Detector

Using light detector A, what is the relative response from the detector when light-emitting diodes made from the listed materials are used as light sources?

Light Sources		Relative Response of Light Detector A
Material	λ (Å)	
Indium Phosphide	9850	0.6
Gallium Arsenide	8980	0.98
Gallium Arsenide Phosphide	6500	0.33
Gallium Phosphide	5650	0.15
Gallium Nitride	4000	0

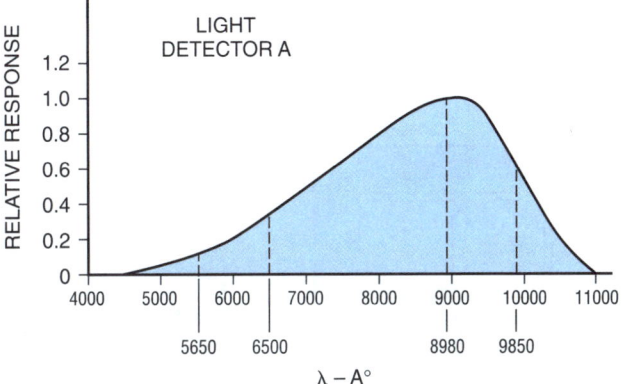

Figure 11-5a shows a possible concept for a solar cell. The wafer of n-type silicon is about three inches across and has a very thin p-type layer over its top surface. The bottom surface (left side in the diagram) is plated with a thin metal film which is the cathode terminal. The anode terminal, which contacts the thin p-type layer, is a network of very narrow metal strips so that they do not block much of the light. Light passing between the strips shines right through the silicon p region, which is nearly transparent. The light is mainly absorbed in the depletion zone on both sides of the junction.

The pictorial diagram of an experimental setup where a solar cell is operating a small electric fan is shown. The schematic diagram of the circuit is shown in *Figure 11-5b*. (Notice that the schematic symbol for a photovoltaic diode includes a little letter "v" to indicate voltage generation.) The circuit includes instruments to indicate the generated voltage and current. We will consider the anode voltage to be zero (ground). The generated voltage at the cathode will then be *negative*. Under normal indoor lighting, it might be about -0.3 volt. A little motor like this might draw about 20 milliamps. If we take the circuit outside in bright sunlight, the negative voltage at the cathode increases, and so does the current and the fan speed. But the voltage difference never gets much greater than about 0.6 volt. In dim light, the voltage and current fall nearly to zero.

The solar cell is making use of the effective voltage difference between electrons in the conduction band and the valence band of silicon atoms. As photons strike the valence electrons, they raise their potential by the voltage difference. The more light, the more photons, the more electrons, and the more current in the external circuit at the potential difference of the band. Since the voltage is approximately the same, and the current increases, the power delivered to the circuit increases.

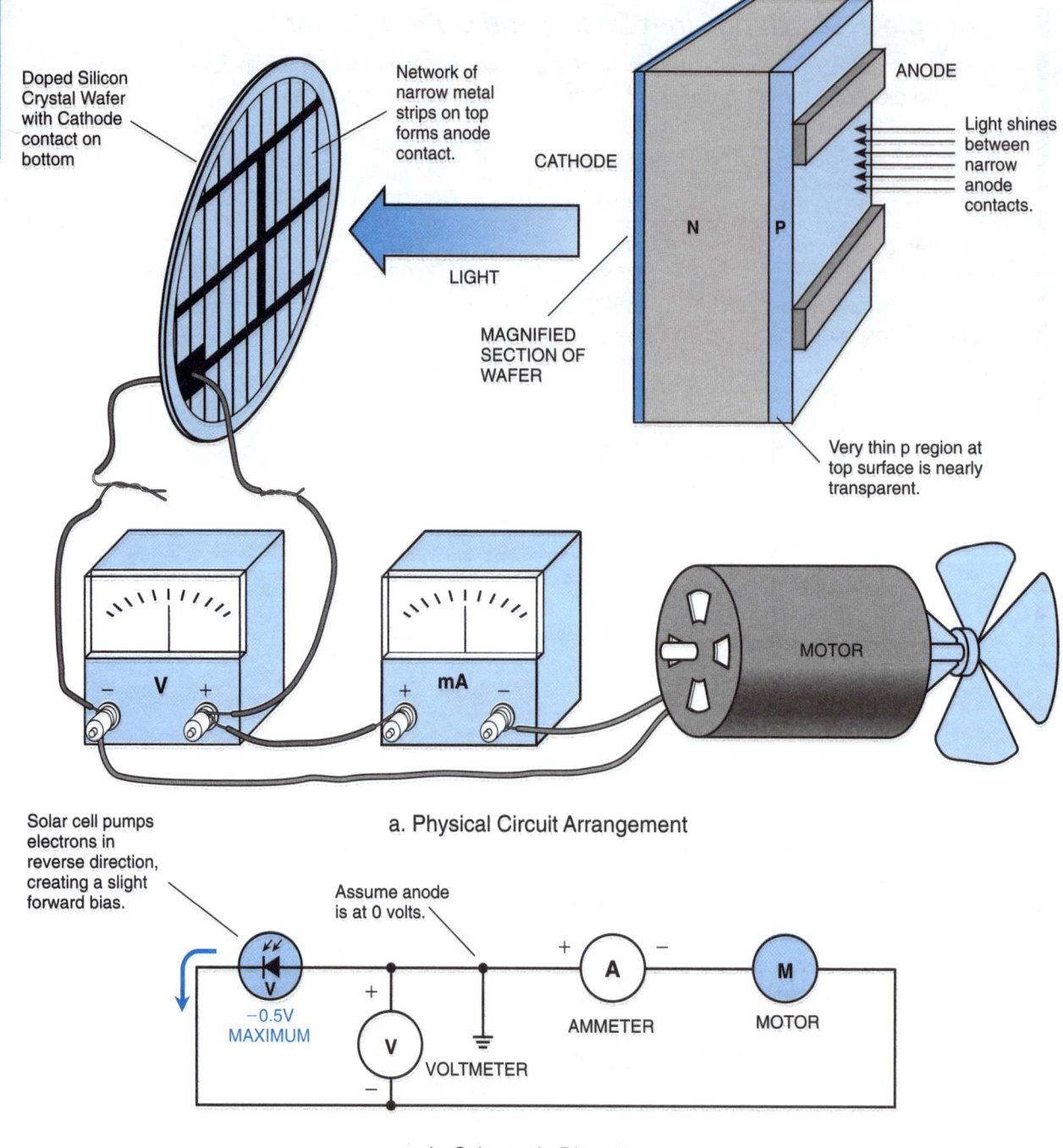

Figure 11-5. How solar cells (photovoltaic diodes) work.

LEDs: Single Devices and Arrays

Finally, let's talk about light-emitting diodes. An LED is simply a p-n junction diode that works as we discussed in Chapter 4, but it is specially made so that *light is created* whenever there is current in the *forward* direction. (We will see how the light is produced in a moment.) The schematic symbol for an LED, as shown in Figure 11-6a, is an ordinary diode symbol with one or two little arrows pointing *outward* at an angle.

It is important to notice that LEDs produce *colored* light instead of white light. The color can be a visible one—usually red, but sometimes green or yellow. Or it can be a color called infrared (pronounced in-fra-red), which is invisible to human eyes.

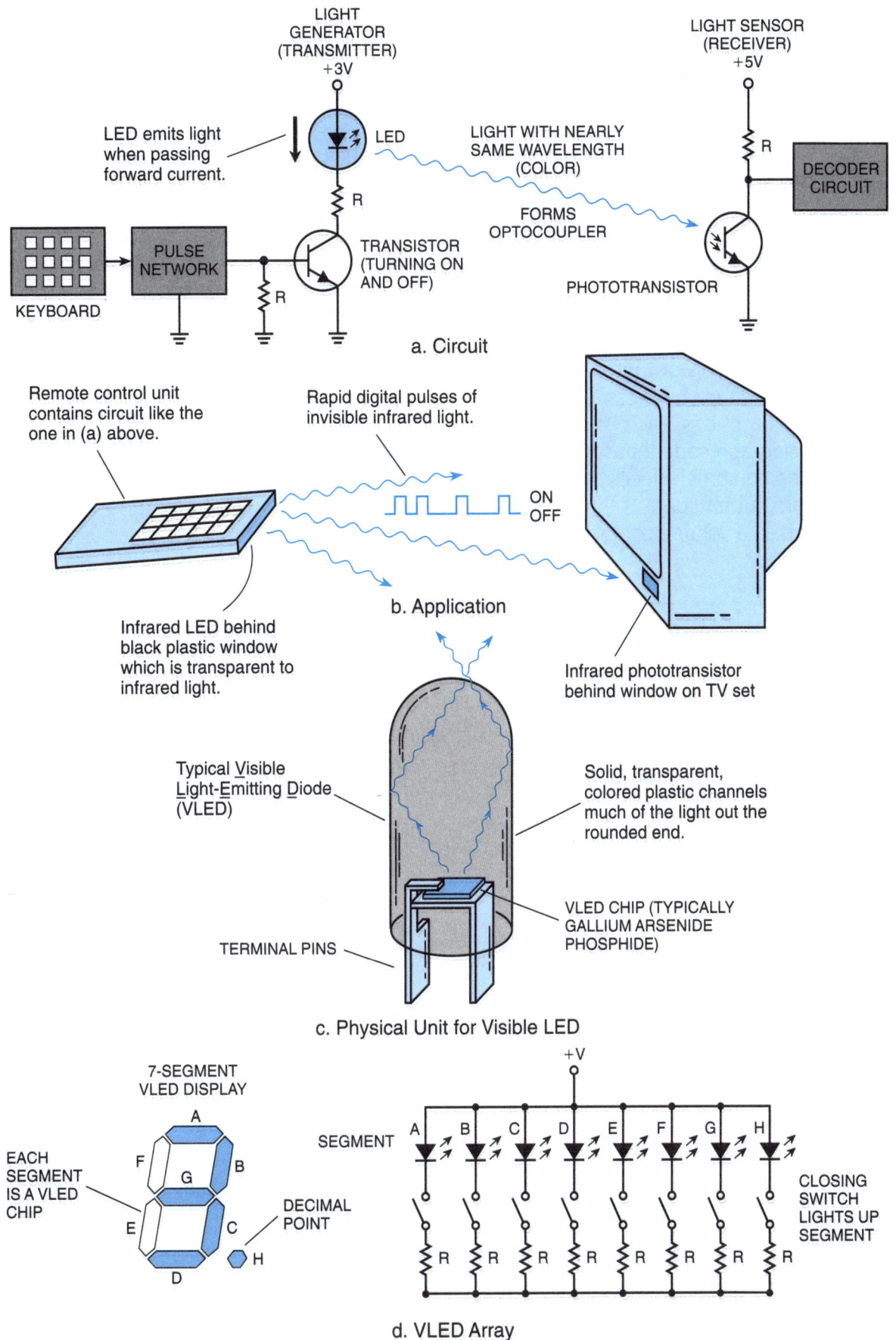

Figure 11-6. Typical Infrared and visible LEDs and their applications.

Infrared color lies below the red end of the rainbow spectrum, as shown in *Figure 11-3*. LEDs that produce *visible* light are called VLEDs (the V for visible). Single VLEDs are commonly used as indicator lights on electronic equipment.

A typical circuit that provides forward current through an LED is shown in *Figure 11-6a*. The light generated by the LED can be sensed with a phototransistor. The combination forms an optocoupler, where the link between the generator, or transmitter and sensor, or receiver, is light. Such couplers are also called optoisolators because there is no wired electrical connection between them; thus, they provide electrical isolation of the two circuits.

For infrared LEDs, one familiar application, shown in *Figure 11-6b,* is in the hand-held remote controller for a television receiver or video cassette recorder. The infrared LED is usually covered with a piece of special plastic that looks black, but to infrared light, that plastic is as transparent as clear glass is to visible light. Behind another little black window somewhere on the TV receiver or VCR is a photodiode or phototransistor that senses the infrared light from the controller.

When you press a key on the controller, the LED transmits infrared light in rapid on-off pulses. Each key has a specific code of pulses. (This is somewhat similar to a telegraph code that specifies a certain letter.) The photodiode or phototransistor on the equipment receives the coded signal and passes the pulsed code to a decoder circuit that controls the equipment's functions. Thus, the TV can be turned on and off, the volume controlled, channels changed, and audio muted by the push of a button.

Example 3. Determine Resistor Values for LEDs

What resistor values need to be assigned for the two LEDs listed when used with the following power supply voltages and turned on to ground?

LED	I_F (mA)	V_F(MAX) (V)	Case 1 V_S (V)	Case 1 R (kΩ)	Case 2 V_S (V)	Case 2 R (kΩ)	Case 3 V_S (V)	Case 3 R (kΩ)
A	20	2	12	0.5	9	0.35	5	0.15
B	10	1.6	12	1.04	9	0.74	5	0.34

$$R = \frac{(V_S - V_F)}{I_F}$$

where: R is in **kΩ**
V_S and V_F are in **volts**
I_F is in **mA**

Individual VLEDs

A typical single VLED looks like a little cylinder of solid, transparent, colored plastic with two wire leads. This is shown in *Figure 11-6c*. The package is usually designed to channel the light produced inside, making most of it come out one end. In some cases, the end where the light emerges is flat rather than spherical.

Inside the plastic package, the light is produced by a tiny chip or bar of a special semiconductor material. The color of the light is determined by the basic crystal material and the dopant elements added to make p and n regions. For infrared LEDs, the undoped crystal may be silicon or a compound (combination of elements) called gallium arsenide. For VLEDs, it is usually some other compound, such as indium antimonide or gallium arsenide phosphide. These other compounds behave much like silicon, but they have somewhat greater differences in energy between the conduction and valence bands that we have discussed. The different visible colors are quite useful as indicators for electronic equipment functions.

VLED Arrays

VLEDs are also used in groups, or arrays, to form a display, such as in a calculator. As shown in *Figure 11-6d,* the displays usually have eight semiconductor elements for each digit position. Seven of them are long and narrow in shape, positioned to formed a squared-off 8. This is called a *seven-segment display*. By lighting various combinations of these LEDs, any numeral from 0 to 9 can be formed. (Some alphabetic characters can also be formed.) For example, to form the numeral 3, segments A,B,C,D,G are lighted. The eighth LED is a little dot which is used as a decimal point. The semiconductor elements in a display are usually made separately, then mounted together on a substrate (flat support). Conductors in the substrate make electrical contact with both terminals of each diode.

Example 4. Determine Resistor Value and Maximum Fanout

In the circuit shown, using AS TTL circuits, what is the value of R and what is the maximum fanout to other AS circuits?

TTL Circuits	V_{OL} (MAX)	I_{OL} (MAX)	I_{IL} (MAX)	V_{CC} (TYP)
Std	0.4V	16mA	−1.6mA	5V
AS	0.35V	20mA	−1.0mA	5V

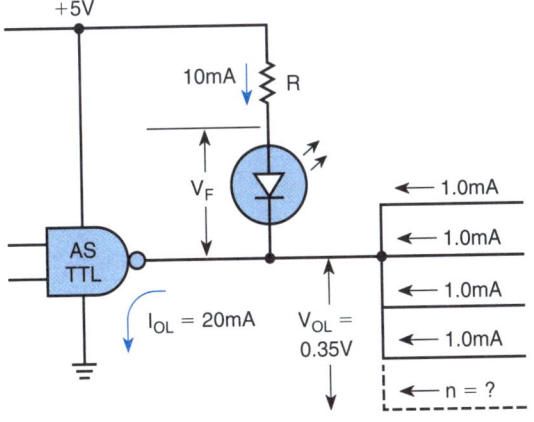

VLED for design intensity
$V_F = 1.6V$
$I_F = 10mA$

Value of R:

$$R(k\Omega) = \frac{(5 - V_F - V_{OL})}{10mA} = 0.305k\Omega$$

Fanout n:

$$n = \frac{I_{OL} - 10mA}{1mA} = \frac{20 - 10}{1} = 10$$

Free Electrons Falling Into Holes Release Light Energy

Now—how can forward current in a p-n junction diode produce light? Do *all* p-n junction diodes light up when they conduct?

The answer to the second question is yes—they all produce light. But ordinary diodes, such as those used as rectifiers, are made of silicon. And the color produced by a typical silicon p-n junction is mainly invisible infrared. Ordinary diodes are not designed to emit the infrared light very efficiently. The light is absorbed by the silicon crystal and the package, which converts the light energy to heat. In an LED, the p and n regions are shaped and positioned so that a lot of the light shines out instead of being absorbed.

As to the first question, *light is produced whenever free electrons fall into holes* in a semiconductor crystal. This process is simply the reverse of the liberation of a free electron and creation of a hole by the absorption of a photon like we discussed in connection with photodiodes and solar cells. The point is that when each free electron drops to the lower energy level, *the potential energy it loses is instantly converted into a photon*, as if by magic. The photon zips away from the junction in some random direction traveling at the speed of light.

The light produced by an LED is colored from infrared to purple. All the free electrons falling into holes give up very nearly the same amount of energy, so all the photons emitted have nearly the same energy. According to what we said earlier in studying photodiodes, that means the photons all have nearly the same wavelength and color. The semiconductor materials used in VLEDs have a wider forbidden gap between the conduction and valence bands than silicon, so the emitted photons have more energy and shorter wavelengths. The wider the gap, the more the color is shifted toward the purple end of the rainbow. Presently, green is about as far as the designers of VLEDs can go without running into some problems.

Completing a Basic Understanding of Electronic Devices and Circuits

This is as far as we will take you in studying basic electronics. Here, at the last, we have looked more deeply into the operation of semiconductor devices by considering energy levels of electrons in terms of voltages. This led us into some practical realities of electronic circuitry, such as leakage currents, and the generation and effects of heat. If you now review the earlier chapters with this in mind, you will be able to understand dc and ac circuits better, and the operation and use of ordinary diodes and transistors will make a bit more sense. With this understanding, you should have a more complete knowledge of the basic theme on which all electronic circuits are based: that electron devices simply act as switches and variable resistors controlled by electrical signals.

Quiz for Chapter 11

1. A photodiode is normally:
   a. emitting light
   b. unbiased
   c. forward biased
   d. reverse biased

2. When the light intensity increases, the reverse minority-carrier current:
   a. increases
   b. decreases
   c. reverses direction
   d. is unaffected

3. What segments of a seven-segment LED are lighted to form the numeral 4?
   a. all seven
   b. a,b,g,c,and d
   c. f,g,b,c,and e
   d. f,g,b, and c

4. A device that acts as a variable resistor controlled by light is the:
   a. photovoltaic diode
   b. solar cell
   c. LED
   d. phototransistor

5. When there is no light applied to a photodiode, the main reason for some reverse leakage current is due to:
   a. noise
   b. heat
   c. sound
   d. holes

6. The frequency or wavelength of a photon determine its:
   a. energy
   b. color
   c. both a and b
   d. neither a nor b

7. In a photodiode, an electron can be knocked from the valence band up to the conduction band by a:
   a. loud sound
   b. photon
   c. proton
   d. neutron

8. Electrons in the conduction band compared to electrons in the valence band are:
   a. higher in potential
   b. lower in energy
   c. longer in wavelength
   d. lower in voltage

9. Of the following colors of light, the color that has the highest frequency is:
   a. red
   b. orange
   c. green
   d. blue

10. Besides the intensity, the energy contained in a light is determined by its:
    a. luminance
    b. color
    c. saturation
    d. distortion

Answers: 1d, 2a, 3d, 4d, 5b, 6c, 7b, 8a, 9d, 10b

Questions for Chapter 11

1. How is current affected if light on a p-n junction is increased?

2. Of the two outermost shells of permitted electron orbits in a silicon atom, the lower band is called the _____ band, and the upper one is called the _____ band.

3. The atoms in a silicon crystal are held together by a _____ bond.

4. Heat energy constantly creates a few free _____ and _____ throughout the chip.

5. For lights of the same brightness, which would produce more photons; a red light or a violet light? Why?

6. A _____ is a photodiode with a built-in current amplifier.

7. How does a phototransistor differ from an ordinary bipolar transistor?

8. A solar cell makes use of the effective _____ between electrons in the conduction band and the valence band of silicon atoms.

9. LEDs produce _____ light.

10. What is the eighth LED in a seven-segment LED used for?

11. What happens when free electrons fall into holes in a semiconductor crystal?

12. What determines the color of light produced by an LED?

13. Light and heat are both forms of _____ .

14. Is forward or reverse voltage used for an LED?

15. The collector-base junction of a phototransistor is designed to work as a _____ .

Appendix

Answers to Chapter Questions and Problems

Chapter 1

1. Sodium (Na)[11]   (2,8,1) = 1 valence electron
   Chlorine (Cl)[17]   (2,8,7) = 7 valence electrons

2. $F = K \left(\dfrac{Q_1 Q_2}{S^2}\right) = (9 \times 10^9)\left(\dfrac{20 \times 30}{(0.02^2)}\right)$

   $F = 1.35 \times 10^{16}$ N (attraction)

3. $W = Fs = (5N)(4M) = 20$ joules

4. $P = \dfrac{W}{t} = \dfrac{Fs}{t} = \dfrac{(8N)(3M)}{(4\text{ sec})} = 6$ watts

5. $P = \dfrac{(5N)(4M)}{(4\text{ sec})} = 5$ watts

   $P = \dfrac{(5N)(4M)}{(2\text{ sec})} = 10$ watts

6. $I = \dfrac{Q}{t} = \dfrac{18C}{6\text{ sec}} = 3$ A

   $I = \dfrac{18C}{2\text{ sec}} = 9$ A

7. $Q = It$ $Q = (7A)(3\text{ sec}) = 21$ coulombs
              $Q = (3A)(7\text{ sec}) = 21$ coulombs

8. $\dfrac{(746\text{ watts})}{Hp} \dfrac{(1\text{ Hp})}{5} = 149.2$ watts or 149.2 joules/sec

9. $w = (6\text{ lb})(4\text{ ft}) = 24$ ft.lbs

10. Unlike charges attract each other; like charges repel each other.

11. Coulomb's Law of Electrostatic Forces: The force (F) of attraction or repulsion exerted between two charged bodies is directly proportional to the product of their charges (Q) and inversely proportional to the square of the distance (s) between them.

    $F_{Q1Q2} = \dfrac{k\, Q_1 Q_2}{s^2}$

12. Proton—positively charged particle in nucleus
    Neutron—neutrally charged particle in nucleus
    Electron—negatively charged particle in orbit

Chapter 2

1. $R = \dfrac{\rho l}{A} = \dfrac{(18.4\text{ ohm-cir mil/ft})(50\text{ ft})}{16^2} = 3.59$ ohms

2. $(+2.5A)+(+6A)-1.4A-3.2A = -3.9A$ away from the terminal

3. $I = \dfrac{E}{R} = \dfrac{24\text{ V}}{8\text{ ohm}} = 3A$

4. $E = IR = (5mA)(4k\Omega) = 20$ Volts

5. $R = \dfrac{E}{I} = \dfrac{8\text{ V}}{60\mu A} = 0.133$ MΩ = 133 kΩ

6. $P = IE = (0.5A)(120\text{ V}) = 60$ Watts

7. $1.2k + 2.4k + 3.7k + 4.1k = 11.4$ kΩ

8. $(2k\Omega)(4) = 8$ kΩ

9. $\dfrac{2}{4}$ kΩ = 0.5 kΩ = 500Ω

10. $R_{eq} = \dfrac{1}{\dfrac{1}{2k\Omega}+\dfrac{1}{4k\Omega}+\dfrac{1}{5k\Omega}+\dfrac{1}{8k\Omega}} = 930$ ohms

11. 27,000 ±5% ohms

12. $5\tau = (5)(0.5\mu F)(1M\Omega) = 0.5$ sec $\times 5 = 2.5$ sec

Chapter 3

1. $f = \dfrac{1}{T} = \dfrac{1}{0.05\text{ sec}} = 20$ Hz

2. $(120\text{ V}_{RMS})(1.414) = 169.7$ volts peak (V_p)
   $= \times 2 = 339.4$ volts peak-to-peak (V_{pp})

3. 50 volts peak (V_p) $\times$ 0.637 = 31.85 volts average (V_{avg})

4. $\dfrac{100}{2}$ V_{pp} = 50 V_p

   50 V_p $\times$ 0.707 = 35.35 V_{RMS}

5. $X_C = \dfrac{1}{2\pi fC} = \dfrac{1}{(6.28 \times 200\text{ Hz} \times 8 \times 10^{-6}F)} = 99.52\Omega$

6. $X_L = 2\pi fL = 6.28 \times 60$ Hz $\times$ 0.5 H $= 188.4\Omega$

Answers to Chapter Questions and Problems (Cont.)

7. $Z = \sqrt{R^2 + X_C^2}$

   $X_C = \dfrac{1}{2\pi f C} = \dfrac{1}{(6.28 \times 10^4 \times 0.4 \times 10^{-6})} = 39.8\,\Omega$

   $Z = \sqrt{25^2 + 39.8^2}$
   $Z = \sqrt{2200} = 47\,\Omega$

8. $Z = \sqrt{R^2 + (X_L - X_C)^2}$
   $Z = \sqrt{30^2 + (80-40)^2} = \sqrt{30^2 + 40^2} = 50\,\Omega$

9. $I = \dfrac{E}{Z} = \dfrac{120\text{ V}}{50\,\Omega} = 2.4\text{ A}$

10. $f = \dfrac{1}{2\pi\sqrt{LC}} = \dfrac{1}{6.28\sqrt{60\times 10^{-3} \times 0.8 \times 10^{-6}}} =$

    $\dfrac{1}{6.28\sqrt{48 \times 10^{-9}}} = 726\text{ Hz}$

11. $\dfrac{N_s}{N_p} = \dfrac{V_s}{V_p} = \dfrac{28 \text{ volts}}{120 \text{ volts}} = \dfrac{1}{4.28}$ 1:4.28

12. $P_s = I_s E_s = (2\text{A})(60\text{V}) = 120$ watts
    $P_p = I_p E_p = 120$ watts

    $I_p = \dfrac{P_p}{E_p} = \dfrac{120 \text{ watts}}{220 \text{ volts}} = 0.545\text{ A}$

Chapter 4

1. switch
2. reverse, forward
3. anode, cathode
4. junction
5. anode, cathode
6. covalent
7. energy
8. valence shell
9. core
10. ions
11. full (very stable)
12. doping
13. N-type
14. hole
15. base

Chapter 5

1. an amplifier
2. bias resistor
3. load resistor
4. quiescent operating point
5. Class A
6. Class B
7. collector feedback, negative
8. negative

9. Advantages: reduced distortion, increased bandwidth
   Disadvantages: reduced gain
10. capacitive
11. to match impedance
12. oscillator
13. $V_{out} = 25 \times 0.040\text{V} = 1.0\text{V}$
14. $I_c = h_{FE} \times I_b = 120 \times 0.05\text{mA} = 6\text{mA}$
15. $V_C = V_{CC} - I_C R_L = 9 - (2\text{mA})(2\text{k}\Omega) = 5\text{V}$

Chapter 6

1. transmission line
2. wavelength
3. speed of light (3.0×10^8 m/sec or 186,000 miles/sec)
4. electric and magnetic waves oscillating 90° crosswise to each other
5. transverse
6. coaxial
7. standing
8. resonate
9. parallel
10. ferrite
11. continuous
12. modulation
13. amplitude
14. frequency
15. diode detector

Chapter 7

1. An *analog* signal varies in a continuous way—a copy of another form. A *digital* signal is switched from one state to another as time passes.
2. analog
3. digital
4. Analog circuits require more accuracy.
5. start bit
6. ASCII
7. baud rate
8. two
9. ones and zeros
10. binary digit (one or zero)
11. serial transmission
12. a group of bits sent at the same time
13. parallel is faster
14. clock signal
15. positive

Chapter 8

1. AND, OR, NOT
2. inverter
3. every possible combination of input bits and the state of the output for each

Answers to Chapter Questions and Problems (Cont.)

4. 1101 = 0; 1000 = 0; 1111 = 1; 0000 = 0
5. 1101 = 1; 1000 = 1; 1111 = 1; 0000 = 0
6. No, they can be hydraulic, pneumatic or mechanical.
7. A truth table shows *logic* states (1 or 0), and a function table shows electrical states (high or low).
8. no effect
9. NOR
10. NAND
11. An inverter is driven by an inverse input.
12. The simplest combination or network
13. It takes more time to perform the addition.
14. N is not used because there is no carry in.
15. S—Sum
    C—Carry

Chapter 9

1. bus
2. controller
3. main memory
4. not according to a definite sequence or pattern
5. (1) when to write a byte
   (2) when to read a stored byte and send out a copy
   (3) which address to use
6. registers
7. The output of one is the input of the other; they are cross-coupled.
8. inverse or opposite
9. sequential
10. gate
11. shift register
12. switching to the opposite state
13. by parallel loading using *preset* and *clear* inputs
14. ripple counter
15. the number of different states through which it counts

Chapter 10

1. a built-in computer that serves a highly specialized function
2. Central Processing Unit (CPU), memory, input/output
3. masking
4. controller
5. instruction cycle
6. program counter
7. instruction register
8. conditional jump
9. RAM is a read-write section of memory and ROM can only be read from, not written to.
10. ROM (or PROM)
11. Multiplication is done by a series of additions and shifts.
    Division is done by a series of subtractions and shifts.
12. Creates a series of machine language instructions to execute the higher-level language instructions.
13. up, down, right, left, fire
14. an analog-to-digital converter (ADC)
15. with a digital-to-analog converter (DAC)

Chapter 11

1. Reverse current is increased.
2. valence, conduction
3. covalent
4. electrons, holes
5. violet light, higher frequency (or shorter wavelength)
6. phototransistor
7. without a base terminal and with a transparent window to let in light
8. voltage difference
9. colored
10. decimal point
11. Light is emitted.
12. the basic crystal material and the dopant elements
13. energy
14. forward
15. photodiode

Determining Resistors Values

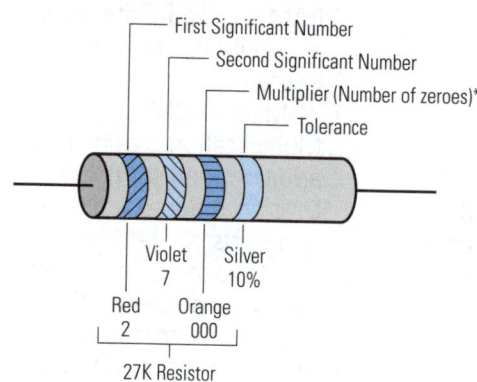

Color Bands

Color	Value
Black	0
Brown	1
Red	2
Orange	3
Yellow	4
Green	5
Blue	6
Violet	7
Gray	8
White	9

Tolerance	
Gold	5%
Silver	10%
No Band	20%

* If multiplier band is gold, the multiplier is 0.1; if it is silver, the multiplier is 0.01.

Color Code

A common way of indicating resistance values for composition resistors is to use color bands on the body of the resistor. A standard color code for the bands has been adopted by resistor manufacturers, and the numerical values they represent are shown. The first band will be nearer one end of the resistor. Read from this band toward the other end. The first band is the first significant number, the second band, the second significant number and the third band the multiplier (number of zeros that follow the first two numbers) to indicate the resistor's value. If the third band is gold or silver, this would indicate a multiplier of 0.1 or 0.01, respectively, rather than additional zeros.

As the resistors are manufactured, there will be some variation in values around the nominal desired value. The fourth band indicates the tolerance of the resistor's value from the indicated (color code) value. A gold fourth band indicates ±5%, a silver band ±10%, and no fourth band indicates a tolerance of ±20%.

The resistor shown is a 27,000 ohm or 27 kilohm resistor. It has a ±10% tolerance, so the actual value of the resistor can be any value from 24,300 ohms to 29,700 ohms. A resistor's physical size usually indicates its power rating. Resistors with less than two watts power rating weigh less than an ounce, but high-power rating resistors have large bulky bodies to dissipate the heat.

0°–90° Trigonometric Functions

ANGLE	SIN	COS	TAN	ANGLE	SIN	COS	TAN	ANGLE	SIN	COS	TAN
0°	0.0000	1.000	0.0000	31°	.5150	.8572	.6009	61°	.8746	.4848	1.8040
1	.0175	.9998	.0175	32	.5299	.8480	.6249	62	.8829	.4695	1.8807
2	.349	.9994	.0349	33	.5446	.8387	.6494	63	.8910	.4540	1.9626
3	.0523	.9986	.0524	34	.5592	.8290	.6745	64	.8988	.4384	2.0503
4	.0698	.9976	.0699	35	.5736	.8192	.7002	65	.9063	.4226	2.1445
5	.0872	.9962	.0875	36	.5878	.8090	.7265	66	.9135	.4067	2.2460
6	.1045	.9945	.1051	37	.6018	.7986	.7536	67	.9205	.3907	2.3559
7	.1219	.9925	.1228	38	.6157	.7880	.7813	68	.9272	.3746	2.4751
8	.1392	.9903	.1405	39	.6293	.7771	.8098	69	.9336	.3584	2.6051
9	.1564	.9877	.1584	40	.6428	.7660	.8391	70	.9397	.3420	2.7475
10	.1736	.9848	.1763	41	.6561	.7547	.8693	71	.9455	.3256	2.9042
11	.1908	.9816	.1944	42	.6691	.7431	.9004	72	.9511	.3090	3.0777
12	.2079	.9781	.2126	43	.6820	.7314	.9325	73	.9563	.2924	3.2709
13	.2250	.9744	.2309	44	.6947	.7193	.9657	74	.9613	.2756	3.4874
14	.2419	.9703	.2493	45	0.7071	0.7071	1.0000	75	.9659	.2588	3.7321
15	.2588	.9659	.2679	46	.7193	.6947	1.0355	76	.9703	.2419	4.0108
16	.2756	.9613	.2867	47	.7314	.6820	1.0724	77	.9744	.2250	4.3315
17	.2924	.9563	.3057	48	.7431	.6691	1.1106	78	.9781	.2079	4.7046
18	.3090	.9511	.3249	49	.7547	.6561	1.1504	79	.9816	.1908	5.1446
19	.3256	.9455	.3443	50	.7660	.6428	1.1918	80	.9848	.1736	5.6713
20	.3420	.9397	.3640	51	.7771	.6293	1.2349	81	.9877	.1564	6.3138
21	.3584	.9336	.3839	52	.7880	.6157	1.2799	82	.9903	.1392	7.1154
22	.3746	.9272	.4040	53	.7986	.6018	1.3270	83	.9925	.1219	8.1443
23	.3907	.9205	.4245	54	.8090	.5878	1.3764	84	.9945	.1045	9.5144
24	.4067	.9135	.4452	55	.8192	.5736	1.4281	85	.9962	.0872	11.43
25	.4226	.9063	.4663	56	.8290	.5592	1.4826	86	.9976	.0698	14.30
26	.4384	.8988	.4877	57	.8387	.5446	1.5399	87	.9986	.0523	19.08
27	.4540	.8910	.5095	58	.8480	.5299	1.6003	88	.9994	.0349	28.64
28	.4695	.8829	.5317	59	.8572	.5150	1.6643	89	.9998	.0175	57.29
29	.4848	.8746	.5543	60	.8660	.5000	1.7321	90	1.0000	.0000	∞
30	.5000	.8660	.5774								

Metric Conversions

International System of Units (SI) – Metric Units

Power

	Btu/h	ft – lb/s	hp	kW	Watt
1 British thermal unit per hour =	1	0.2161	3.929×10^{-4}	2.930×10^{-4}	0.2930
1 foot-pound per second =	4.628	1	1.818×10^{-3}	1.356×10^{-3}	1.356
1 horsepower =	2545	550	1	0.7457	745.7
1 kilowatt =	3413	737.6	1.341	1	1000
1 Watt =	3.413	0.7376	1.341×10^{-3}	0.001	1

Prefix	Symbol		Multiplication Factor
exa	E	$10^{18} =$	1,000,000,000,000,000,000
peta	P	$10^{15} =$	1,000,000,000,000,000
tera	T	$10^{12} =$	1,000,000,000,000
gigi	G	$10^{9} =$	1,000,000,000
mega	M	$10^{6} =$	1,000,000
kilo	k	$10^{3} =$	1,000
hecto	h	$10^{2} =$	100
deca	da	$10^{1} =$	10
(unit)		$10^{0} =$	1

Length

	cm	Meter	km	in.	ft	mi
1 centimeter =	1	10^{-2}	10^{-5}	0.3937	3.281×10^{-2}	6.214×10^{-6}
1 Meter =	100	1	10^{-3}	39.3	3.281	6.214×10^{-4}
1 kilometer =	10^{5}	1000	1	3.937×10^{4}	3281	0.6214
1 inch =	2.540	2.540×10^{-2}	2.540×10^{-5}	1	8.333×10^{-2}	1.578×10^{-5}
1 foot =	30.48	0.3048	3.048×10^{-4}	12	1	1.894×10^{-4}
1 mile =	1.609×10^{5}	1609	1.609	6.336×10^{4}	5280	1

Prefix	Symbol		Multiplication Factor
deci	d	$10^{-1} =$	0.1
centi	c	$10^{-2} =$	0.01
milli	m	$10^{-3} =$	0.001
micro	u	$10^{-6} =$	0.000001
nano	n	$10^{-9} =$	0.000000001
pico	p	$10^{-12} =$	0.000000000001
femto	f	$10^{-15} =$	0.000000000000001
atto	a	$10^{-18} =$	0.000000000000000001

Mass

	gm	kg	oz	lb
1 gram =	1	0.001	3.527×10^{-2}	2.205×10^{-3}
1 Kilogram =	1000	1	35.27	2.205
1 ounce =	28.35	2.835×10^{-2}	1	6.250×10^{-2}
1 pound =	453.6	0.4536	16	1

Force

	dyne	Newton	lb
1 dyne =	1	10^{-5}	2.248×10^{-6}
1 Newton =	10^{5}	1	0.2248
1 pound =	4.448×10^{5}	4.448	1

Schematic Symbols

RESISTORS: FIXED, ADJUSTABLE

CAPACITORS: FIXED, VARIABLE

SWITCHES: SPST, SPDT, TOGGLE, MULTIPOINT, NORMAL OPEN, NORMAL CLOSED, MOMENTARY

BATTERIES: SINGLE CELL, MULTI CELL

WIRING: CONDUCTORS NOT JOINED, CONDUCTORS JOINED

GROUNDS: CHASSIS, EARTH

INDUCTORS

DIODES: LED (DS#), DIODE/RECTIFIER, SCHOTTKY

TRANSFORMERS: AIR CORE, WITH CORE

TRANSISTORS: NPN, PNP, BIPOLAR, P-CHANNEL, N-CHANNEL, SINGLE-GATE DEPLETION MODE MOSFET, SINGLE-GATE ENHANCEMENT MODE MOSFET

RELAYS: SPST, SPDT, DPDT

IC AMPLIFIERS: GENERAL AMPLIFIER, OP AMP

LOGIC (U#): AND, NAND, OR, NOR, XOR, INVERT

Glossary

Access time: Time required in a computer to move information from memory to the computing mechanism.

Active device: A device capable of producing gain or control; for example: transistors, diodes, vacuum tubes, and saturable reactors.

Adder: Switching circuits which combine binary bits to generate the sum and carry of these bits.

Address: Noun: a location, either name or number, where information is stored in a computer. Verb: to select or pick out the location of a stored information set for access.

Alternating current (ac): An electrical current that periodically changes in magnitude and in direction of the current.

Alternation: Either half of a cycle of alternating current. It is the time period during which the current increases from zero to its maximum value (in either direction) and decreases to zero.

Ampere (A): The unit of measurement for electrical current in coulombs (6.25×10^{18} electrons) per second. One ampere results in a circuit that has one ohm resistance when one volt is applied to the circuit.

Amplification: See Gain.

Amplifier: An electrical circuit designed to increase the current, voltage, or power of an applied signal.

Analog-to-Digital Conversion or Converter (ADC or A/D): The process of converting a sampled analog signal to a digital code that represents the amplitude of the original signal sample.

And: A Boolean logic expression used to identify the logic operation wherein given two or more variables, all must be logical "1" for the result to be logical "1". The AND function is graphically represented by the dot (•) symbol, e.g. A • B • C = D

Anode: The positive electrode of a semiconductor diode.

ASCII: A commonly used 7-bit code which represents numbers, letters, and control characters. It is used for transferring information between digital systems, especially those which otherwise would be incompatible.

Asynchronous inputs: Those terminals in a flip-flop which can affect the output state of the flip-flop independent of the clock. Called "Set," "Preset," "Reset" or "DC Set" and "Reset," or "Clear."

Audio and audio frequency (AF): The range of frequencies normally heard by the human ear. Typically, about 20 to 20,000 Hz.

Base: The region between the emitter and collector which received minority carriers injected from the emitter.

Beta (β) (h_{FE}): The current gain of a transistor when connected in a common emitter circuit.

Bias: In an electronic circuit, a voltage or current applied to an active device (transistor, diode, etc.) to set the steady-state operating point of the circuit.

Binary Coded Decimal (BCD): A binary numbering system in which any decimal digit is represented by a group of 4 bits. Each digit in a multi-digit number continues to be identified by its 4-bit group.

Binary digit (Bit): A digit in the binary number system whose value can be either 1 or 0.

Binary logic: Digital logic elements which operate with two distinct states. The two states are variously called true and false, high and low, on and off, or "1" and "0". In computers they are represented by two different voltage levels. The level which is more positive (or less negative) than the other is called the high level, the other, the low level. If the true ("1") level is the most positive voltage, such logic is referred to as "positive true" or "positive logic".

Binary system: A system of mathematical computation based on powers of two.

Bipolar: A semiconductor device having both majority and minority carriers.

Bistable element: Another name for flip-flop. A circuit in which the output has two stable states (output levels "0" or "1") and can be caused to go to either of these states by input signals, but remains in that state permanently after the input signals are removed. The characteristic of two stable states differentiates it from a monostable element, which keeps returning to a specific state, and an astable element, which keeps changing from one state to the other.

Bit: See Binary digit.

Block diagram: A system diagram which shows the relationship between the main functional units of the system represented by blocks.

Boolean algebra: The mathematics of logic which uses alphabetic symbols to represent logical variables and "1" and "0" to represent states. There are three basic logic operations in this algebra: AND, OR and NOT. (See NAND, NOR, Invert, combinations of the three basic operations.)

Breakdown: The condition for a reverse-biased semiconductor junction when its high resistance, under the reverse bias, suddenly decreases, causing excessive current. Not necessarily destructive.

Bridge rectifier: A full-wave rectifier in which the rectifier diodes are connected in a bridge circuit to allow current to the load during both the positive and negative alternation of the supply voltage.

Capacitance (C): The capability to store charge in an electrostatic field. It can be expressed as equal to the charge Q in coulombs that is stored divided by the voltage E in volts that supplied the charge. Capacitance tends to oppose any change in voltage. The unit is farads.

Capacitive reactance (X_c): The opposition that a capacitor offers to a time changing signal or supplied voltage. Its value is $X_c = 1/2\pi fC$

Capacitor (C): A device made up of two metallic plates separated by a dielectric or insulating material. Used to store electrical energy in the electrostatic field between the plates.

Cathode (K): The negative electrode of a semiconductor diode.

Charge (Q): A measurable quantity of electrical energy representing the electrostatic forces between atomic particles. Electrons have a negative charge.

Circuit: A complete path that allows electrical current from one terminal of a voltage source to the other terminal.

Clear: An asynchronous input. Also called "Reset". To restore a memory element or flip-flop to a "standard" state, forcing the Q terminal to logic "0".

Clock input: That terminal on a flip-flop whose condition or change of condition controls the admission of data into a flip-flop through the synchronous inputs and thereby controls the output state of the flip-flop.

Clock or Clock generator: An electronic circuit that generates accurate and precisely controlled, regularly occurring synchronizing or timing signals called clock signals.

Clock rate: The frequency of oscillation of the master clock, or oscillator, in a system.

Coil: The component that is formed when several turns of wire are wound on a cylindrical form or on a metal core.

Collector (C): The element in a transistor that collects the moving electrons or holes, and from which the output usually is obtained. Analagous to the plate of a triode vacuum tube.

Collector cutoff: The operating condition of a transistor when collector current is reduced to the leakage current of the collector-base junction.

Collector junction: The junction between the base and collector regions of a transistor, normally biased in the reverse direction.

Color code: A system in which colors are used to identify the value of electronic components, or other variables, such a component tolerance.

[1]**Common-base amplifier:** An amplifier with the transistor connected into the external circuit so that the base electrode is common to the input and output circuits.

[1]**Common-collector amplifier:** An amplifier with the transistor connected into the external circuit so that the collector electrode is common to the input and output circuits.

[1]**Common-emitter amplifier:** An amplifier with the transistor connected into the external circuit so that the emitter electrode is common to the input and output circuits.

Component: The individual parts that make up a circuit, a function, a subsystem or a total piece of equipment.

Conductivity: The ability of a material to conduct current.

Conductor: A substance through which electrons flow with relative ease.

Continuity: A continuous electrical path.

Controlled rectifier: A four-layer semiconductor device in which conduction is triggered ON by gate current and OFF by reducing the anode voltage below a critical value.

Coulomb (C): The unit of electrical charge, made up of a quantity of 6.25×10^{18} electrons.

Counter, binary: An interconnection of flip-flops having a single input so arranged to enable binary counting. Each time a pulse appears at the input, the counter changes state and tabulates the number of input pulses for readout in binary form. It has a 2^n possible counts where n is the number of flip-flops.

Current (I): The flow of electrons, measured in amperes. One ampere results when one volt is impressed on a circuit that has a resistance of one ohm.

DC operating point: The dc values of collector voltage and current of a transistor with no signal applied.

Decibel (dB): The standard unit for expressing the ratio between powers P_1 and P_2. dB = $10\log_{10}P_1/P_2$, one tenth of a bel.

Dielectric: The non-conducting material used to separate the plates of a capacitor or for insulating electric contacts.

Digital signal: A signal whose level has only discrete values, like on or off, 1 or 0, +5V or +0.2V.

Digital to Analog Conversion (or Converter) DAC or D/A): A circuit that converts a digital input signal to an analog output signal.

Diode: A device which has two terminals and has a high resistance to current in one direction and a low resistance to current in the other direction.

Direct Current (dc): Current in a circuit in one direction only.

Drain: The element in field-effect transistor which is roughly analagous to the collector of a bipolar transistor.

Effective value: The value of ac current that will produce the same heating effect in a load resistor as an equivalent dc current.

Electricity: A form of energy produced by the flow of electrons through materials and devices under the influence of an electromotive force produced electrostatically, mechanically, chemically or thermally.

Electrolytic capacitor: A capacitor whose electrodes are immersed in a wet electrolyte or dry paste.

Electromotive force (E): The force which causes an electrical current in a circuit when there is a difference in potential. Synonym for voltage.

Electron: The basic atomic particle having a negative charge that rotates around a positively charged nucleus of an atom.

Electrostatic field: The electrical field or force surrounding objects that have an electrical charge.

Emitter (E): The semiconductor material in a transistor that emits carriers into the base region when the emitter-base junction is forward biased.

Emitter junction: The junction between the emitter and base regions of a transistor, normally biased in the forward direction to induct carriers in the base region.

Enable: To permit an action or the acceptance or recognition of data by applying appropriate signals (generally a logic "1" in a positive logic) to the appropriate input.

Exclusive OR: A logical function whose output is "1" if either of the two variables is "1", but whose output is "0" if both inputs are "1" or both are "0".

Fan-in: The number of inputs available to a specified logic stage or function.

Fan-out: The number of input stages that can be driven by a circuit output.

Farad (F): The basic unit for capacitance. A capacitor has a value of one farad when it has stored one coulomb of charge with one volt across it.

Field coil: An electromagnet formed from a coil of insulated wire wound around a soft iron core. Commonly used in motors and generators.

Field-Effect Transistor (FET): A 3-terminal semiconductor device where current is from source to drain due to a conducting channel formed by a voltage field between the gate and the source.

Filament: The heated element in an incandescent lamp or vacuum tube.

Filter: A circuit element or group of components which passes signals of certain frequencies while blocking signals of other frequencies.

Flip-flops (storage elements): A circuit having two stable states and the capability of changing from one state to another with the application of a control signal and remaining in that state after removal of signals. (See Bistable Element.)

Fluorescent: The ability to emit light when struck by electrons or other radiation.

Forward current: The net current that flows across the forward-biased PN junction.

Forward voltage (or bias): A voltage applied across a semiconductor junction in order to permit forward current through the junction and the device.

Free electron: An electron in the conduction band of a solid. It is not bound to any atom and is free to move through the crystal.

Frequency (F or f): The number of completed cycles of a periodic waveform during one second.

Gain (G): 1. Any increase in the current, voltage or power level of a signal. 2. The ratio of output to input signal level of an amplifier.

Gate, AND: All inputs must have "1" level signals at the input to produce a "1" level output.

Gates (decision elements): A circuit having two or more inputs and one output. The output depends upon the combination of logic signals at the input.

Gate electrode: The control electrode of a FET, MOS transistor, controlled rectifier, or triac.

Gate, NAND: All inputs must have "1" level signals at the input to produce a "0" level output.

Gate, NOR: Any one input or more than one input having a "1" level signal will produce a "0" level output.

Ground (or Grounded): 1. The common return path for electric current in electronic equipment. Called electrical ground. 2. A reference point connected to, or assumed to be at zero potential with respect to, the earth.

Half-adder: A switching circuit which combines binary bits to generate the sum and carry. (See Adder.)

Half-shift register: Another name for certain types of flip-flops when used in a shift register. Two of these make one stage in a shift register.

Heat sink: A material (usually a metal) placed in contact with a hot body to increase the flow of heat away from the body.

Henry (H or h): The unit of inductance. The inductance of a coil of wire in henries is a function of the coil's size, the number of turns of wire and the type of core material.

Hertz (Hz): One cycle per second.

Hole: A mobile vacancy in the valence structure of a semiconductor created when a valence electron breaks its valence bond and is raised to the conduction band.

Impedance (Z): In a circuit, the opposition that circuit elements present to alternating current. The impedance includes both resistance and reactance.

Inductance (L): The capability of a coil to store energy in a magnetic field surrounding it which results in a property that tends to oppose any change in the existing current in the coil.

Inductive reactance (X_L): The opposition that an inductance offers when there is an ac or pulsating dc in a circuit. $X_L = 2\pi fL$.

Input impedance: The impedance seen by a source when a device or circuit is connected across the source.

[1]Common means "connected to" or "grounded"

Integrated circuit (IC): A complex semiconductor structure that contains all the circuit components for a high-functional-density analog or digital circuit interconnected together on a single chip of silicon.

Inverter: A circuit whose output is always in the opposite state from the input. This is also called a NOT circuit.

Luminescence: Light emission without high heat.

Junction: The region separating two layers in a semiconductor material; e.g., a p-n junction.

Junction diode: A two-terminal, small-area PN junction used for electronic-circuit functions (as opposed to a junction rectifier).

Junction transistor: A PNP or NPN transistor formed from three alternate regions of p and n type material. The alternate materials are formed by diffusion or ion implantation.

Leakage (or Leakage current): The undesired flow of electricity around or through a device or circuit. In the case of semiconductors, it is the current across a reverse-biased semiconductor junction.

Light-emitting diode: A PN junction that emits light when biased in the forward direction.

Linear amplifier: A class A amplifier whose output signal is directly proportional to the input signal. The output is an exact reproduction of the input except for the increased gain.

Load: Any component, circuit, subsystem or system than consumes power delivered to it by a source of power.

Load line: A line drawn on the collector characteristic curves of a transistor on which the operating point of the transistor moves as collector current changes. It is called a load line because the slope of the line depends on the value of the collector load resistance.

Logic diagram: A picture representation for the logical functions of AND, OR, NAND, NOR, NOT present in a circuit.

Loop: A closed path around which there is a current or signal.

Magnetic Field: The force field surrounding a magnet.

Magnetic lines of force: The imaginary lines called flux lines used to indicate the directions of the magnetic forces in a magnetic field.

Megohm (MΩ): A million ohms. Sometimes abbreviated meg.

Metal-oxide semiconductor (MOS) transistor: An active semiconductor device in which conduction is controlled in a region between two electrodes by a voltage applied to an insulated electrode over the region.

Microampere (mA): One millionth of an ampere.

Microelectronics: A broad term covering the entire field of silicon integrated circuits, thick-and thin-film circuits, and any special component of an extremely small nature.

Microfarad (mF): One millionth of a farad.

Milliampere (mA): One thousandth of an ampere.

Millihenry (mH): One thousandth of a henry.

Milliwatt (mW): One thousandth of a watt.

Monolithic: Formed within a single body of material.

NAND: A Boolean logic operation which yields a logic "0" output when all logic input signals are logic "1".

Negative logic: Logic in which the more negative voltage represents the "1" state and the less negative voltage represents the "0" state. (See Binary Logic.)

NOR: A Boolean logic operation which yields a logic "0" output with one or more true "1" input signals.

NOT: A Boolean logic operation indicating negation, not "1". Actually an inverter. If input is "1", output is NOT "1" but "0". If the input is "0", output is NOT "0" but "1". It is graphically represented by a bar over a Boolean symbol such as A. $\overline{A}$ means "when A is not 1 . . ."

NPN Transistor: A bipolar transistor with a p-type base sandwiched between an n-type emitter, and an n-type collector.

N-type semiconductor material (N): A semiconductor material in which the majority carriers are electrons, and there is an excess of electrons over holes.

Ohm (Ω): The unit of electrical resistance. A circuit component has a resistance of one ohm when one volt applied to the component produces a current of one ampere.

Ohms-per-volt: The sensitivity rating for a voltmeter. Also expresses the impedance (resistance) presented to a circuit by the meter when a voltage measurement is made.

One ("1"): See Binary Logic.

Open circuit: An incomplete path for current.

Operating point: The steady-state or no-signal operating point of a circuit or active device.

Operational amplifier (OP AMP): A high-gain analog amplifier with two inputs and one output.

OR: A Boolean logic operation used to identify the logic operation wherein when two or more true "1" inputs only add to one true "1" output. Only one input needs to be "true" to produce a "true" output. The graphical symbol for "OR" is a plus sign (+).

Oscillation: A sustained condition of continuous operation where the circuit outputs a constant signal at a frequency determined by circuit constants and as a result of positive or regenerative feedback.

Parallel: This refers to the technique for handling a binary data word which has more than one bit. All bits are acted upon simultaneously.

Parallel circuit: A circuit connected so that there is current in two or more parallel branches. The branches are said to be in parallel when they have common connections at each end and the same voltage across the branches.

Parallel operation: The organization of data manipulation within computer circuitry where all the digits of a word are transmitted simultaneously on separate lines in order to speed up operation, as opposed to serial operation.

Passive device: A device not capable of producing gain; for example, resistors and capacitors.

Phase reversal: A 180-degree change in phase.

Photodiode: A junction diode that utilizes the photosensitivity of a PN junction.

Phototransistor: A photodetector that incorporates transistor action to give an amplified output.

Pi (π): The mathematical constant which is equal to the ratio of the circumference of a circle to its diameter. Approximately 3.14.

Picofarad (pF): A unit of capacitance that is 1×10^{-12} farads or one millionth of a millionth of a farad.

PN junction: The region of transition between P-type and N-type material in a single semiconductor crystal.

PNP Transistor: A bipolar transistor with an n-type base sandwiched between a p-type emitter and a p-type collector.

Polarity: The description of whether a voltage is positive or negative with respect to some reference point.

Positive logic: Logic in which the more positive voltage represents the "1" state.

Potential difference: The voltage difference between two points, calculated algebraically.

Power (P): The time rate of doing work.

Power (reactive): The product of the voltage and current in a reactive circuit measured in volt-amperes (apparent power).

Power (real): The power dissipated in the purely resistive components of a circuit measured in watts.

Power dissipation: The dispersion or giving up of power in the form of heat.

Power supply: A defined unit that is the source of electrical power for a device, circuit, subsystem or system.

P-type semiconductor material (P): A semiconductor material in which holes are the majority carriers and there is a deficiency of electrons.

Q output: The reference output of a flip-flop. When this output is "1", the flip-flop is said to be in the "1" state; when it is "0", the output is said to be in the "0" state. (See State and Set.)

$\overline{Q}$ output: The second output of a flip-flop. It is always opposite in logic level to the Q output.

Quiescence: The state of a transistor amplifier with no signal applied.

Reactance (X): The opposition that a pure inductance or a pure capacitance provides to current in an ac circuit.

Recombination: The process whereby a free electron fills a hole. The action eliminates two charge carriers; the electron is no longer free and the hole no longer exists.

Rectification: The process of converting alternating current into pulsating direct current.

Register: An interconnection of computer circuitry consisting of a number of storage devices (usually flip-flops) to store a certain number of digits, usually one computer word. For example, a 4-bit register requires 4 flip-flops.

Relay: A device in which a set of contacts is opened or closed by a mechanical force supplied by turning on current in an electromagnet. The contacts are isolated from the electromagnet.

Reset: Also called "clear". Similar to Set except it is the input through which the Q output can be made to go to "0".

Resistance (R): A characteristic of a material that opposes the flow of electrons. It results in loss of energy in a circuit dissipated as heat.

Resistivity: The ability of a material to resist the flow of current.

Resistor (R): A circuit component that provides resistance to current in the circuit.

Reverse bias: An external voltage applied in the nonconducting direction of a PN junction. The connections are opposite to those for forward bias.

Reverse-breakdown voltage: The voltage that produces a sharp increase in reverse current without a significant increase in voltage.

Reverse current: The current when a semiconductor junction is reverse biased.

Ripple counter: A binary counting system in which flip-flops are connected in series. When the first flip-flop changes, it affects the second, which affects the third, etc. If there are ten in a row, the signal must go sequentially from the first flip-flop to the tenth.

Root-Mean-Square (RMS): See effective value. The RMS value of an ac sinusoidal waveform is 0.707 of the peak amplitude of the sine wave.

Saturation: The operating condition of a transistor when an increase in base current produces no further increase in collector current.

Semiconductor: One of the materials falling between metals as good conductors and insulators as poor conductors in the periodic chart of the elements.

Semiconductor laser: A light-emitting diode that uses stimulated emission to produce a coherent-light output.

Serial: This refers to the technique for handling a binary data word which has more than one bit. The bits are acted upon one at a time. It is like a parade going by a review point.

Serial operation: The organization of data manipulation within computer circuitry where the digits of a word are transmitted one at a time along a single line. The Serial mode of operation is slower than parallel operation, but utilizes less complex circuitry.

Series Circuit: In electrical and electronic circuits, arranging the circuit components so that the same current flows through all components in the circuit.

Set: An input on a flip-flop not controlled by the clock (see Asynchronous Inputs) and used to effect the Q output. It is this input through which signals can be entered to get the Q output to go to "1". Note it cannot get Q to go to "0".

Shift: The process of moving data from one place to another. Shifting is done synchronously and by command of the clock. An 8-bit word can be shifted sequentially (serially) in the manner of a bucket brigade. Generally referred to as shifting serially. It takes 8 clock pulses to shift an eight-bit word, or all bits of a word can be shifted simultaneously. This is called "parallel load" or "parallel shift".

Shift register: An arrangement of circuits, specifically flip-flops, which is used to shift serially or in parallel. Binary words are generally parallel loaded and then held temporarily or serially shifted out.

Signal: In electronics, the information contained in electrical quantities of voltage or current that forms the input, timing, or output of a device, circuit, or system.

Silicon Controlled Rectifier (SCR): A semiconductor diode in which current through a third element, called the gate, controls turn-on, and the anode-to-cathode voltage controls turn-off.

Sine (sinusoidal) wave: A waveform whose amplitude at any time through a rotation of an angle from 0° to 360° is a function of the sine of an angle.

Single crystal: A crystal with all its element cubes having the same orientation.

State: This refers to the condition of an input or output of a circuit as to whether it is a logic "1" or a logic "0". The state of a circuit (gate or flip-flop) refers to its output. The flip-flop is said to be in the "1" state when its Q output is "1". A gate is in the "1" state when its output is "1".

Step-down transformer: A transformer in which the secondary winding has fewer turns than the primary.

Step-up transformer: A transformer in which the secondary winding has more turns than the primary.

Synchronous: Operation of a switching network by a clock pulse generator. All circuits in the network switch simultaneously. All actions take place synchronously with the clock.

Thermal generation: The creation of a hole and a free electron by freeing a bound electron through the addition of heat energy.

Thermistor: A resistor whose resistance changes with temperature.

Threshold voltage: The minimum voltage applied to the gate electrode of a MOS transistor that initiates conduction between the source and drain.

Toggle: To switch between two states as in a flip-flop.

Transformer: A set of coils wound on an iron core in which a magnetic field couples energy between two or more coils or windings.

Transistor: A three-terminal semiconductor device used in circuits to amplify electrical signals or to perform as a switch to provide digital functions.

Triac: A device, similar to the controlled rectifier, in which both the forward and reverse characteristics can be triggered from blocking to conducting.

Trigger: A timing pulse used to initiate the transmission of logic signals through the appropriate circuit signal paths.

Truth table: A chart which tabulates and summarizes all the combinations of possible states of the inputs and outputs of a circuit. It tabulates what will happen at the output for a given input combination.

TTL, T²L (Transistor-Transistor-Logic): A logic system which evolved from DTL wherein the multiple diode cluster is replaced by a multiple-emitter transistor commonly applied to a circuit which has an active pullup network.

Turn-off time: Same as Turn-on Time except the output stops sinking current, goes off and/or to a high voltage level (logic "1").

Turn-on time: The time required for an output to turn on (sink current, to ground output, to go to 0 volts). It is the propagation time of an appropriate input signal to cause the output to go to 0 volts.

Turns ratio: The ratio of secondary winding turns to primary winding turns of a transformer.

Valence band: The range of energy levels of a solid crystal in which lie the energies of the valence electrons which bind the crystal together.

Valence electrons: The outer ring of electrons of an atom which form the bond between atoms.

Vector: A line representing the magnitude and time phase of some quantity, plotted on rectangular or polar coordinates.

Voltage (or Volt): The unit of electromotive force that caused current when included in a closed circuit. One volt causes a current of one ampere through a resistance of one ohm.

Voltage drop: The difference in potential between two points caused by a current through an impedance or resistance.

Watt (W): The unit of electrical power in joules per second, equal to the voltage drop (in volts) times the current (in amperes) in a resistive circuit.

Word: A group of bits treated as an entity in a computer.

Zener diode: A junction diode with a sharp reverse breakdown at low voltages, between 3 and 15V.

Index

A
adder circuit: 144, 147-149
address: 179, 182-184
alternating current (ac) defined: 24, 37
American Standard Code for Information Interchange: 129
amplifier, basic: 87, 88
amplifier, using transistor as: 78
analog input to computer 187, 188
analog information 121, 122
analog-to-digital converter (ADC): 187-189
angle, phase: 43, 44, 47, 99
antenna: 107-114
arithmetic and logic unit (ALU): 156, 178-182
ASCII: 129
atom: 3-5, 62-65, 195-198
atomic number: 3, 4
atomic structure defined: 2-4
atoms, charged: 6

B
barrier, semiconductor: 72
baud: 126
biasing methods: 88, 90, 92, 93
binary numbers: 127
binary to decimal: 128
bipolar transistor: 82
bit: 127
Bohr atomic model: 5
bus: 178, 179, 183
byte: 130, 153

C
canonical combinational network: 146
capacitor charging and discharging: 31
capacitor defined: 30
central processing unit (CPU): 178
charged atoms: 6
circuit defined: 18
class, amplifier: 90, 91
clock: 132
coaxial cable: 109
color code: 30

combinational logic: 146
common-emitter configuration: 89
complementary transistors: 81
compound: 3, 4
computer system: 178
computer program: 182
conditional jump: 184
conduction band: 195, 196
conductor defined: 8, 9
controller: 153, 178, 182-184
coulomb defined: 7
counter: 166-169
coupling methods for amplifiers: 93-96
covalent bond: 65, 66, 195-198
current defined: 12, 13, 14
current gain: 78
current, semiconductor: 67
cut-off defined: 124
cycle defined: 40-42

D
decoder: 183
demodulation: 115
depletion mode: 82
depletion zone: 71, 195-198
detector, radio: 117, 118
diffusion: 75, 76
digital information: 122, 123
digital system: 136
digital-to-analog converter (DAC): 189, 190
diode: 59
dipole: 112, 113
direct current (dc) defined: 24, 37
dopant material: 66
doping: 66
dynamic RAM (DRAM): 171-173
dynamic memory elements: 169-171

E
electric charge: 5
electric waves: 103-105
electric field: 105-108
electricity defined: 2
electromotive force (EMF): 10

Index

electron: 3-5, 62-65
electron devices: 1
electron tubes: 1, 2
electronics defined: 1
electrostatic field: 7-9
electrostatic force, Coulomb's law of: 7
electrostatics, law of: 6
element: 3, 4
energy: 10
enhancement mode: 82
external load: 88

F
feedback, positive: 98
field-effect transistor (FET): 82-84
flip-flop: 156-163
forward bias pn junction: 71, 73
forward bias base-emitter junction: 77
free electron: 67
frequency defined: 41, 42
frequency, oscillation: 100
frequency to wavelength: 109
full adder: 147, 149
function table: 140

H
half adder: 149
hertz defined: 41
hole, semiconductor: 70, 195-198

I
impedance: 47-50
impedance matching: 96
inductor current increasing and decreasing: 32
inductor defined: 31, 32
input/output (I/O): 179
instruction cycle: 182
instruction register: 182, 183
insulator defined: 8, 9
integrated circuit: 143
inverter: 137
inverting input: 98
ion: 8, 65
ionic bond: 65

J
joystick input to computer: 187

K
kinetic energy: 10, 11
Kirchhoff's laws: 21, 25, 27

L
law, conservation of energy: 11
law, Kirchhoff's: 21, 25, 27
law, Ohm's: 21
law of electrostatic force: 7
law of electrostatics: 6
light-emitting diode (LED): 193, 202-204
LED color: 205, 206
logic families: 143
logic gates, basic: 136, 137
logic gates using relays: 138-140
logic gates using transistors: 142

M
magnetic field: 105-108
main memory: 153, 154, 178
masking bits: 180, 182
mass memory: 155
master-slave flip-flops: 162, 163
matter: 3, 4
meters (VOM): 24-26
modulation: 115-117
modulus, counter: 168
Morse code: 123
MOSFET: 82

N
n-type silicon: 66-68
negative logic: 141
neutron: 3-5
non-inverting input: 98
NPN transistor: 75, 76
nucleus: 3-5, 62-65, 195-198

O
ohm defined: 17
Ohm's law: 21
Ohm's law circle: 22
on-off input to computer: 187, 188
orbit: 3-5, 62-65, 195-198
oscillator: 98
oscilloscope: 38, 39

P
p-type silicon: 70
parallel circuit: 19, 20, 27-29
parallel transmission: 130, 131
period of waveform defined: 41, 42
phase: 43, 44, 47, 99
photodiode: 193-198
photon: 197, 198
phototransistor: 193, 199, 200

photovoltaic diode: 193, 200-202
piezoelectric crystal: 83, 100
PN junction: 59, 60, 71, 72, 195-198
PN junction plot of current: 74
PNP transistor: 79, 80
polarity: 6
polarization, antenna: 114
positive logic: 128, 140, 141
potential difference: 12
potential energy: 10, 11
power: 10, 33, 34
power circle: 34
power in ac circuit: 52, 53, 54
program counter: 182
program languages: 182
programmable ROM (PROM): 186
propagation delay: 99
proton: 3-5

Q
quiescent operating point: 90

R
R-S latch: 158-160
radio transmitter/receiver: 103
radio waves: 107
random-access memory (RAM): 155, 171-174
reactance, capacitive: 44, 45
reactance, inductive: 45, 46
reactance, R&C: 49, 50
reactance, R&L: 48, 49
reactance, R&L&C: 50-52
read-only memory (ROM): 185
recombination: 71
register: 156, 160, 161, 178
resistance defined: 17
resistor defined: 30
resonance: 54
reverse leakage current: 193, 194
reverse bias pn junction: 73, 74
ripple carry: 149
rms value: 42, 43
routing circuitry: 153, 178

S
$\overline{S}$-$\overline{R}$ latch: 159
saturated defined: 124
semiconductor devices: 1, 2
semiconductor defined: 8, 9
sequential circuits: 156
serial transmission: 126
series circuit: 18, 20, 25, 26
shell: 3-5, 62-65, 195-198
shift register: 164-167
silicon: 60, 61
sine wave: 39, 40
solar cell: 193, 200-202
standing waves: 110, 111
static electricity: 6
static RAM (SRAM): 173, 174
superconductive: 9
switch, using transistor as: 78, 79, 124
switching circuits, coupling: 125

T
time constant: 32, 33
transformers: 54-56
transmission line: 103-110
triangle symbol for amplifier: 97
truth table: 138
turns ratio: 56
twin lead: 109

U
unit prefixes and symbols: 23

V
valence electron: 5, 62-65, 195-198
valence shell: 5, 62-65, 195-198
vector: 44
vectorial addition: 48
visible light-emitting diode (VLED): 204, 205
volt defined: 12
voltage, ac: 40, 42, 43
voltage defined: 12
voltage gain: 89

W
watt defined: 34
waveforms: 38, 98
wavelength to frequency: 109
wavelength defined: 105
work defined: 10